21st Century Nanostructured Materials - Physics, Chemistry, Classification, and Emerging Applications in Industry, Biomedicine, and Agriculture

Edited by Phuong V. Pham

Published in London, United Kingdom

IntechOpen

Supporting open minds since 2005

21st Century Nanostructured Materials - Physics, Chemistry, Classification, and Emerging Applications in Industry, Biomedicine, and Agriculture
http://dx.doi.org/10.5772/intechopen.94802
Edited by Phuong V. Pham

Contributors
Kalyan Singh Kushwah, Deepak Kumar Verma, Muhammad Akhsin Muflikhun, Rahmad Kuncoro Adi, Gil Nonato C. Santos, Sweta Chander, Sanjeet Kumar Sinha, Tarek Fawzi, Ammar A.M. Al-Talib, Esther Hontañón, Stella Vallejos, John W. W. Zeller, Ashok K. K Sood, Latika S. Chaudhary, Nibir K. Dhar, Randy N. Jacobs, Parminder Ghuman, Sachidananda Babu, Harry Efstathiadis, Samiran Ganguly, Avik W. Ghosh, Sheikh Ziauddin Ahmed, Farjana Ferdous Tonni, Dr Gangu Naidu Challa, Srinivasa Rao Y, Vasudha D, Vara Prasada Rao K, Hemali Rathnayake, Gayani Pathiraja, Ece Bayrak, Maha Khayyat, Camila Pia Canales, Mercedes G. Montalbán, Guzmán Carissimi, Marta G. Fuster, Gloria Villora, Parthiban Pazhamalai, Karthikeyan Krishnamoorthy, Sang-Jae Kim, Srini Krishnamurthy, Wan Nura'in Nabilah Noranuar, Ahmad Qushairi Mohamad, Sharidan Shafie, Ilyas Khan, Mohd Rijal Ilias, Lim Yeou Jiann, Phuong Viet Pham, Muhammad Aamir Iqbal, Maria Malik, Wajeehah Shahid, Waqas Ahmad, Kossi A. A. Min-Dianey

Notice
Statements and opinions expressed in the chapters are these of the individual contributors and not necessarily those of the editors or publisher. No responsibility is accepted for the accuracy of information contained in the published chapters. The publisher assumes no responsibility for any damage or injury to persons or property arising out of the use of any materials, instructions, methods or ideas contained in the book.

First published in London, United Kingdom, 2022 by IntechOpen
IntechOpen is the global imprint of INTECHOPEN LIMITED, registered in England and Wales, registration number: 11086078, 5 Princes Gate Court, London, SW7 2QJ, United Kingdom
Printed in Croatia

British Library Cataloguing-in-Publication Data
A catalogue record for this book is available from the British Library

Additional hard and PDF copies can be obtained from orders@intechopen.com

21st Century Nanostructured Materials - Physics, Chemistry, Classification, and Emerging Applications in Industry, Biomedicine, and Agriculture
Edited by Phuong V. Pham
p. cm.
Print ISBN 978-1-80355-084-8
Online ISBN 978-1-80355-085-5
eBook (PDF) ISBN 978-1-80355-086-2

We are IntechOpen,
the world's leading publisher of
Open Access books
Built by scientists, for scientists

5,700+
Open access books available

141,000+
International authors and editors

180M+
Downloads

156
Countries delivered to

Our authors are among the

Top 1%
most cited scientists

12.2%
Contributors from top 500 universities

CLARIVATE ANALYTICS

BOOK
CITATION
INDEX

INDEXED

WEB OF SCIENCE™

Selection of our books indexed in the Book Citation Index (BKCI)
in Web of Science Core Collection™

Interested in publishing with us?
Contact book.department@intechopen.com

Numbers displayed above are based on latest data collected.
For more information visit www.intechopen.com

Meet the editor

 Phuong V. Pham is a pioneering scientist in materials science and electronic devices. He is currently a senior scientist at the School of Micro-Nano Electronics and Hangzhou Global Scientific and Technological Innovation Center (HIC), Zhejiang University, China. He earned a Ph.D. from SKKU Advanced Institute of Nanotechnology (SAINT), Sungkyunkwan University (SKKU), South Korea. Then, he spent a few years as a postdoctoral researcher and research fellow at the School of Advanced Materials Science and Engineering, SKKU and the Center for Multidimensional Carbon Materials (CMCM), Institute for Basic Science (IBS), South Korea, respectively. He is a recipient of the NSF Career Award and the National Postdoctoral Award for Excellent Young Scientists, China. His research interests include low-dimensional materials, 2D material synthesis, twistronics, straintronics, 2D heterostructures, doping technique development, nanocomposites, block copolymers, plasma engineering for flexible display, sensors, photodetectors, transistors, organic light-emitting diodes, and wearable electronics.

Contents

Preface

Nanostructured materials (NMs) are at the heart of 21st-century nanotechnology. They are artificial materials formed as microstructures with length on the nanometer scale. NMs may be defined as those materials whose structural elements, including cluster, crystal, and molecule, have dimensions in the range of 1–100 nanometers. The explosion in both academic and industrial interest in NMs arises from their remarkable variations in fundamental electrical, optical, and magnetic properties, which range from macroscopic solids to accountable atomic particles. This book provides a comprehensive overview of the current state of the art in the synthesis, device design, and investigation of functional NMs as well as their potential applications, focusing on their outstanding physical and chemical aspects at the micro/nanoscale. Carbon materials (CVD graphene, graphene oxides, and carbon nanotubes (CNTs)), plasmonic 2D materials, and other nanostructures (metal nanowires, metal oxides, nanoparticles (NPs), nanofibers, metamaterials, nanocomposites, and nanofertilizers) play an important role in nanoscience and nanotechnology and thus this book examines them in detail. The book also addresses emerging achievements of NM-based applications for industry, biomedicine, and sustainable agriculture. This book is divided into five sections. The first section describes the physic aspects of NMs in photodetectors and transistors, and the superconductivity for quantum technology, while the second section introduces representative plasmonic 2D materials (e.g., CVD graphene, graphene oxide, hBN, MXenes, pnictogens, metal oxides) and related applications in electronics and optoelectronics. The third section discusses emerging NMs. Since many NMs have been developed recently, this section introduces readers to only the topmost emerging NMs. It provides a fundamental demonstration of NMs for applications in industry and biomedicine. The fourth section briefly discusses nanofertilizers and their applications in sustainable agriculture and the environment. The fifth and final section highlights the behaviors of electrochemistry at the interfaces of electrode/solution under in-depth theoretical analysis and features of electrochemical impedance spectroscopy.

Chapter 1 introduces various approaches to using nanostructured metasurface designs to increase the photon capture on a thin detector architecture, which helps to minimize thermally generated dark current and achieve higher absorption efficiency.

Chapter 2 examines the physic impact of noises such as random telegraph noise, thermal noise, flicker noise, and shot noise on Tunnel FET devices and circuits integrated with Si/Ge and III–V groups. The reliability issues of the device have a profound impact on the circuit level design for practical perspectives. Noise is one of the important parameters in terms of reliability and very few research papers address this issue in comparison to other parameter studies. This chapter will help increase understanding of noise issues.

Chapter 3 presents recent advances in plasmonic 2D materials (graphene, graphene oxides, hexagonal boron nitride, pnictogens, MXenes, metal oxides, and nonmetals) and discusses their potential for emerging applications. The chapter is divided into several sections to elaborate on recent theoretical and experimental developments along with potential in the photonics and energy storage industries.

Chapter 4 is a significant contribution to mid-wave infrared (MWIR) detection and imaging for NASA Earth Science applications using high-performance bilayer graphene-HgCdTe detector technology. This chapter discusses the principles, structures, fabrication process, and theoretical modeling for graphene/HgCdTe interface, as well as doping and transfer techniques on bilayer graphene.

Chapter 5 analyzes the heat transfer of water-based CNTs in non-coaxial rotation flow affected by magnetohydrodynamics (MHD) and porosity. Here, single-walled carbon nanotubes (SWCNTs) provide high values of Nusselt number compared to multi-walled carbon nanotubes (MWCNTs). For verification, the chapter presents a comparison between the present solutions and a past study that shows excellent agreement.

Chapter 6 reviews the experimental work and progress of nanowire technology over the last several decades, with more focus on recent work. The final section of this chapter discusses future trends in nanowire research, including nanowire implementation in daily electronic tools to satisfy the demand for electronics of low weight and small size.

Chapter 7 provides a comprehensive summary of fundamental principles of synthesis strategies to control the dimensionality of anisotropic nanowires, their crystal growth, and optical and electrical properties in the fabrication of ultrathin metal hydroxide/oxide nanowires for optoelectronic applications. The chapter highlights the governing theories of crystal growth processes and kinetics that control the anisotropy and dimensions of nanowires.

Chapter 8 begins with a survey of metal-oxide semiconductors (MOS) and their 1D nanostructures with the greatest potential for use in the next generation of chemical sensors, which will be of very small size, low-power consumption, low price, and superior sensing performance compared to chemical sensors on the market currently. The chapter also describes 1D MOS nanostructures, including composite and hybrid structures, and their synthesis techniques. Then, the chapter presents the architectures of current resistive and FET sensors and discusses the methods for integrating 1D MOS nanostructures into these sensors on a large scale and in a cost-effective manner. The chapter concludes with an outlook of the challenges facing chemical sensors based on 1D MOS nanostructures if their massive use in sensor networks is to become a reality.

Chapter 9 deals with the topochemical synthesis of blue titanium oxide (b-TiO2) and its energy storage application as an electrode material for supercapacitor devices in aqueous and organic electrolytes. The mechanism of formation of b-TiO2 via topochemical synthesis and its characterization using X-ray diffraction, UV–visible, photoluminescence, electron spin resonance spectroscopy, laser Raman spectrum, X-ray photoelectron spectroscopy, and morphological studies (FE-SEM and HR-TEM) are discussed in detail. Collectively, this chapter highlights the use of b-TiO2 sheets as advanced electrodes for 3.0 V supercapacitors.

Chapter 10 examines the use of NPs in biomedical fields. It presents the features of biopolymer silk fibroin and its applications in nanomedicine. Silk fibroin, obtained from the Bombyx mori silkworm, is a natural polymeric biomaterial whose main features are its amphiphilic chemistry, biocompatibility, biodegradability, excellent mechanical properties in various material formats, and processing flexibility. All these properties make silk fibroin a useful candidate to act as a nanocarrier. As such, this chapter reviews the structure of silk fibroin, its biocompatibility, and its degradability. In addition, it reviews silk fibroin NP synthesis methods. Finally, the chapter discusses the application of silk fibroin NPs for drug delivery systems.

Chapter 11 reviews the design of NPs proposed as drug delivery systems in biomedicine. It begins with a historical review of nanotechnology including the most common types of NPs (metal NPs, liposomes, nanocrystals, and polymeric NPs) and their advantages as drug delivery systems. These advantages include the mechanism of increased penetration and retention, transport of insoluble drugs, and controlled release. Next, the chapter discusses NP design principles and routes of administration of NPs (parental, oral, pulmonary, and transdermal) as well as routes of elimination of NPs (renal and hepatic).

Chapter 12 studies the green synthesis of different NPs from metal. The inspiring applications of metal oxide NPs have attracted the interest of scientists. Various physical, chemical, and biological methods in materials science are being adapted to synthesize different types of NPs. Green synthesis has gained widespread attention as a sustainable, reliable, and eco-friendly protocol for biologically synthesizing a wide range of metallic NPs. Green synthesis has been proposed to reduce the use of hazardous compounds and as a state of a harsh reaction in the production of metallic NPs. Plant extracts are used for the biosynthesis of NPs such as silver (Ag), cerium dioxide (C2O2), copper oxide (CuO), gold (Au), titanium dioxide (TiO2), and zinc oxide (ZnO). This chapter gives an overview of the plant-mediated biosynthesis of NPs that are eco-friendly and have less hazardous chemical effects.

Chapter 13 explores the development of metamaterials (MMs) and metasurfaces and clarifies their exotic behavior. MMs are artificial materials that obtain their properties from their accurately engineered meta-atoms rather than the characteristics of their constituents. The size of the meta-atom is small compared to the light wavelength. It also addresses the promising applications of MMs in medicine, aerospace, sensors, solar-power management, crowd control, antennas, army equipment, and reaching earthquake shielding and seismic materials.

Chapter 14 illustrates the synthesis of nanocomposite materials through horizontal vapor phase growth (HVPG) and future trends. Since the application of nanomaterials can be found in very wide aspects, the synthesis process of nanomaterials using HVPG can be an alternative method. The future trend as shown in the chapter ensures the sustainability of the synthesis of nanomaterials without compromising the environment and human health.

Chapter 15 provides brief information about production methods (e.g., electrospinning, wet spinning, drawing), characterization methods (e.g., SEM, TEM, AFM), and tissue engineering applications (e.g., core-shell fibers, antibacterial fibers, NPs incorporated fibers, drug-loaded fibers) of nanofibers.

Chapter 16 summarizes recent achievements in nanofertilizers that use nanocomponents such as nanozeolite, nano-hydroxyapatite, macronutrient, NPs, nano-bio fertilizers, and others for innovative applications in horticulture and sustainable agriculture and environment.

Chapter 17 reveals one of the most useful tools to understand the electrode/solution interface: electrochemical impedance spectroscopy (EIS). This tool allows us to describe electrode behavior in presence of a certain electrolyte in terms of electrical parameters such as resistances, capacitances, and so on. With this information, we can infer the electrochemical behavior towards a specific reaction and the capacity of the electrode to carry on the electron transfer depending on its resistance (impedance) values. This chapter presents information on theory, such as Ohm´s Law and its derivations, as well as actual applications.

I acknowledge all the contributing authors for their excellent chapters. I would also like to thank the staff at IntechOpen for their assistance throughout the preparation and publication of this book.

Phuong V. Pham
School of Micro-Nano Electronic,
Hangzhou Global Scientific and Technological Innovation Center,
Zhejiang University,
Hangzhou, China

Physics of Nanostructured Materials in Electronics and Optoelectronics

Chapter 1

Physics of Nanostructure Design for Infrared Detectors

Nibir Kumar Dhar, Samiran Ganguly
and Srini Krishnamurthy

Abstract

Infrared detectors and focal plane array technologies are becoming ubiquitous in military, but are limited in the commercial sectors. The widespread commercial use of this technology is lacking because of the high cost and large size, weight and power. Most of these detectors require cryogenic cooling to minimize thermally generated dark currents, causing the size, weight, power and cost to increase significantly. Approaches using very thin detector design can minimize thermally generated dark current, but at a cost of lower absorption efficiency. There are emerging technologies in nanostructured material designs such as metasurfaces that can allow for increased photon absorption in a thin detector architecture. Ultra-thin and low-dimensional absorber materials may also provide unique engineering opportunities in detector design. This chapter discusses the physics and opportunities to increase the operating temperature using such techniques.

Keywords: infrared detectors, nanostructure, metasurfaces, plasmonics, detector noise, dark current, absorption, thin absorber, photon trap

1. Introduction

Infrared (IR) detectors and focal plane array technologies (FPA) have proven to be at the heart of many defense and commercial applications. Most of these detectors require cryogenic cooling to minimize thermally generated dark current, causing the size, weight, power and cost (SWaP-C) to increase significantly. The objective of this chapter is to discuss physics and development of new approaches in nanostructure engineering of light matter interaction to enhance photon field absorption in very thin infrared absorbing layers, thereby significantly reduce thermally generated noise, thus paving a path to higher operating temperature and eliminating bulky and costly cryogenic coolers. Breakthroughs are necessary in materials design to considerably reduce the detector noise caused by the surface leakage and dark current to enable high operating temperature. The device architecture must be commensurate with improved quantum efficiency and increased lifetime, ideally being only background shot noise limited. This however cannot be accomplished by merely improving the material quality alone. Achieving extremely low dark currents at higher temperature using nanostructured thin absorption layers will lead to practical infrared detector technology that can have wide spread applications.

IntechOpen

Since the early 50s, there has been considerable progress towards the materials development and device design innovations. In particular, significant advances have been made during the past couple of decades in the bandgap engineering of various compound semiconductors that has led to new and emerging detector architectures. Advances in optoelectronics related materials science, such as metamaterials, nanostructures and 2D material designs have opened doors for new approaches to apply device design methodologies, which are expected to offer enhanced performance at higher operating temperature and lower cost products in a wide range of applications.

This chapter discusses advancements in the detector technologies and presents physics of emerging device architectures. The chapter introduces the basics of infrared detection physics and various detector figure of merit (FOM). Advances in pixel scaling, junction formation, materials growth, bandstructure design and processing technologies have matured substantially to make an impact towards higher operating temperature detectors [1]. The concepts presented here are informational in nature for students interested in infrared detector technology. The central ideas discussed are the opportunities to use thin detector designs to reduce thermally generated dark current and increasing radiation absorption techniques. Thus achieving lower SWaP-C.

2. Detector characterization parameters

IR detectors can be categorized as being either a quantum or thermal device. In a quantum detector, electromagnetic radiation absorbed in a semiconductor material generates electron-hole pairs, which are sensed by an electronic readout circuit (ROIC). In a thermal detector, on the other hand, the incident IR photons are absorbed by a thermally isolated detector element, resulting in an increase in the temperature of the element. The temperature is sensed by monitoring an electrical parameter such as resistivity or capacitance. In this chapter we are primarily concerned with quantum detectors.

Due to the various mechanisms used by detectors to convert optical to electrical signals, several FOM are used to characterize their performance. The output of the detector consists of its response signal to the incident radiation and random noise fluctuations. One such FOM is the Responsivity (R) of the detector, defined as the ratio of the root mean squared (RMS) value of the signal voltage (V_s) or output current to the RMS input radiation power (ϕ) incident on the detector. The responsivity is given as:

$$R_v(\lambda, f) = \frac{V_s}{\phi(\lambda)}; R_i(\lambda, f) = \frac{I_s}{\phi(\lambda)} \qquad (1)$$

The spectral current responsivity can also be written in terms of quantum efficiency as:

$$R_i(\lambda, f) = \frac{\lambda \eta}{hc} qg \qquad (2)$$

Where, λ is the photon wavelength, η is the quantum efficiency (QE) (more on QE in Section 4), h is the Plank's constant, c is the speed of light, q is the electronic charge and g is the gain. For photovoltaic (p/n junction) detectors $g = 1$.

The responsivity is usually a function of the bias voltage, the operating electrical frequency, and the wavelength. An important parameter typically used in infrared detectors is the Noise Equivalent Power (NEP). It is the minimum incident radiant power on the detector required to produce a SNR of unity. Therefore, it can be expressed as:

$$NEP = \frac{V_n}{R_v} = \frac{I_n}{R_i} \tag{3}$$

NEP is usually measured for a fixed reference bandwidth. The NEP is typically proportional to the square root of the detector signal, which is proportional to the detector area of optical collection Ad. An important factor in detector operation is the detectivity D, which is the reciprocal of NEP. It is the signal-to-noise ratio. There are multiple approaches to characterize this FOM, a useful one is the specific or normalized detectivity, D^* given by:

$$D^* = D\left(\sqrt{\delta f Ad}\right) = \frac{\sqrt{\delta f Ad}}{NEP} \tag{4}$$

Or

$$D^* = \frac{I_{signal}}{I_{noise}} \frac{\sqrt{\delta f Ad}}{P_{opt}} \tag{5}$$

Where, $I_{signal} = I_{lit} - I_{dark}$ and I_{lit} is the photocurrent, δf is the bandwidth of the measurement, Ad is the detector optical area, P_{opt} is the optical power (ϕ_λ) incident on the detector, and I_{noise} is the RMS noise current. This noise current depends on the dark current I_{dark} and its exact formulation depends on the type of noise dominant in the detector. These noises are primarily of the following four kinds:

a. Thermal or Johnson noise which occurs due to thermal agitation of the electrons in the detector material and can be controlled by cryogenic approaches and is given by $I_{noise}^2 = 4kT/R$ where R is the dark resistance.

b. Flicker or 1/f noise which occurs due to stochasticity of the time of arrival of electrons through the detector film to the contact terminals and is given by $I_{noise}^2 = \alpha_H/nf$ where α_H is Hooge's constant, n is the number of transporting carriers, and f is the measurement frequency. It can be seen that this noise is lower at higher measurement frequencies, and is controlled by high frequency measurements using modulated read signal in kHz ranges, or optical choppers that provide an additional means of reducing the flicker noise.

c. Shot noise occurs due to stochasticity in counting statistics of the number of carriers in the detector material and is given by $I_{noise}^2 = 2q|I|$.

d. Recombination-Generation noise that arises due to stochasticity of recombination and generation processes of electrons and holes and is given as $I_{noise}^2 \propto 4I^2/n(1 + 4\pi^2 f^2\tau^2)$. This can be improved by material processing and detector design to reduce recombination rate.

It is clear from the above discussion that the performance of the detector is heavily impacted by the dark current. Lower is the dark current, higher is the relative signal and lower is the noise. *Therefore, dark current reduction without sacrificing photogeneration is the central goal of detector design and engineering.*

3. Dark current vs. temperature

The ultimate in detector performance is achieved when the noise generated by the background (scene) flux is greater than any thermally generated noise within the detector (for more information see Ref. [2]). Such condition is termed as background Limited Performance or BLIP. It can be seen that the bigger component of noise inherent in the detector is the dark current. It is the output current when no background flux is incident on the detector and depends strongly on the detector temperature. The dark current increases with temperature since the carrier thermal energy KT increases with temperature. When the carrier thermal energy is equal to or greater than the bandgap (E_g) energy, more thermally generated carriers are promoted to the conduction band giving rise to dark current. It is this reason; quantum detectors need to be cooled to a temperature that minimizes the dark current. However, this requires cooling to cryogenic temperature where large coolers are used. This is an inherent problem because it increases the size, weight, power and cost.

The dark current with respect to temperature can be defined as an Arrhenius equation [3]:

$$I_{dark}(T) = A_e J_0 \text{Texp}\left(\frac{-E_a}{K_B T}\right) \tag{6}$$

Where A_e is the electronic area of the detector, J_0 is a constant that depends on the detector absorber materials properties, K_B is the Boltzmann constant, T is the detector temperature and E_a is the activation energy to promote a carrier into the conduction band. This equation clearly illustrates how dark current increases with temperature. The BLIP condition is achieved at a temperature (T_{BLIP}) where photocurrent becomes equal to the dark current.

4. Fundamental photon detector action: absorption, generation, detection

Photodetectors can be built in a variety of device configurations [4]. The major ones include a light dependent conductor (photoconductor), light dependent diode (photodiode, avalanche photo-diode), photovoltaic (such as solar cell), or thermal detector/bolometer (Seebeck effect detectors). At the most fundamental level, all these devices depend on the excitation of matter particles by transferring incident photon energy into the absorber material. In the case of a thermal detector, the excited matter particles are phonons and electrons, in the case of a photovoltaic or photoconductive detector, it is an electron-hole pair also called excitons, and in certain cases it is plasmons acting as an intermediary between the electrons and photons.

It is obvious that the photodetector action depends on the transduction efficiency of the absorbed photons (light) into a detector particle species, primarily an electron (matter) and its subsequent electrical readout. This efficiency is called the internal quantum efficiency (IQE). In most high-quality detectors, this approaches unity, though it is possible to obtain higher than one IQE if the material bandstructure can be engineered to emit more than one electron per absorbed photon, which typically needs special quantum engineered bandstructures, since the electron relaxation post-generation must be "direct" or radiative. External quantum efficiency (EQE) captures the effect of light absorption within this approach, i.e. EQE = Absorption × IQE.

EQE is a function dependent principally on three factors: photon wavelength, detector temperature, and absorption efficiency. We illustrate this using an example as follows. In Ref. [5] we developed a quantum mechanical model of IQE in poly-crystalline PbSe mid-wave IR detectors. This formulation is developed on the lines of the Shockley-Quiesser approach [6]. In this approach the photons of energies larger than the material bandgap first thermalizes to the band edge and then creates a photo-generated carrier. We modeled the absorption efficiency using the classic Moss model [7] which considers non-uniformity of photon absorption in film thickness direction, x, film reflectivity r, and effective absorption coefficient α, and is given by:

$$\kappa(x) = \frac{(1-r)(1-e^{\alpha x})}{1 - re^{\alpha x}} \tag{7}$$

The IQE is given by excitation of matter particle modes. In the same example the carrier generation is given as:

$$\eta_{IQE}(\lambda, T) = \frac{E_g(T)\int_0^\infty D(E,\lambda)N(E,\lambda)\zeta(E,T)dE}{\int_0^\infty ED(E,\lambda)N(E,\lambda)dE} \tag{8}$$

Where E_g is the bandgap (a function of detector temperature, T), D, N are photon density of states and occupancy function, which itself is a function of phonon spectra (and hence a function of wavelength λ), and a band edge disorder function ζ which accounts for Urbach tails [8] into the bandgap (again dependent on T). The overall EQE is then given by:

$$\eta_{EQE}(\lambda, T, x) = \eta_{IQE}(\lambda, T)\kappa(x) \tag{9}$$

It can be seen from Eq. (7), that for thin films the absorption efficiency can be quite small; in fact. as $x \to 0, \kappa \to 0$. This is evident in many thin film "2D materials" where quantum efficiency can be quite poor, and as an illustration, in a single layer graphene, the absorption coefficient is below 3%. In Section 5 below we give a brief description of how thin absorber layers can be used to increase the detector operating temperature.

EQE allows us to calculate the rate of photo-generated carriers. The generation rate can be numerically calculated by:

$$G(T) = \int_0^t \int_0^\infty \eta_{EQE}(\lambda, T, x)n_{opt}(\lambda, T)d\lambda dx \tag{10}$$

Where, $n_{opt}(\lambda, T)$, the photon density can be calculated from Planck's distribution function for a black body source or some other appropriate distribution function for an artificial or structured light source. The generated carrier density is given by $\Delta n = G(T)A_{opt}\tau$ where A_{opt} is the optical area, i.e. the area open to photon collection and generation, and τ is the carrier lifetime. Specific detector design converts this generated carrier density into an equivalent photoresistance, photo-current, or photovoltage which depends on the modality and details of the specific design of the detector. It can be seen that the carrier lifetime plays a critical role; long lived the carrier, better is the signature of photogeneration.

5. Approach to increase the detector operating temperature

As discussed above the photon detector action involves electron-hole pair creation when photons impinge on the absorber layer. Typically, the absorber layer thickness x requires to be in the order of the wavelength of interest. For example, a 10 µm radiation detection would require at least a 10 µm thick absorber layer. The photons interact in the absorber volume of $A_e x$, where A_e is the electronic area of the detector. Typically, the detector's electronic and optical areas (A_{opt}) are same as shown in **Figure 1a**, and usually denoted as A_d. As shown in **Figure 1b**, the detector absorber layer can be thinned substantially to reduce the absorber volume and thus dark current by using a metastructured stack with optical collection area A_{opt}.

The dark current, which strongly depends on temperature is also proportional to this volume. The total noise current generated in this volume for a noise bandwidth of δf and a gain of g is given as [9]:

$$I_n^2 = 2(G+R)A_e x \delta f q^2 g^2 \tag{11}$$

Where, G and R are the generation and recombination rates and other parameters are defined previously. At equilibrium, $G = R$, and denoted as $2G$.

The D^* from Eq. (4) can now be calculated using Eqs. (2) and (3) and yields:

$$D^* = \frac{\eta \lambda}{2hc(xG)^{\frac{1}{2}}} \left(\frac{A_{opt}}{A_e}\right)^{1/2} \tag{12}$$

In Eq. (12) we can see that thickness x is in the denominator. Reducing x will increase D^*, which is what we would want. Another key observation is the ratio of optical collection area to the absorber electronic area. Increasing this ratio also help increase the D^* and thus SNR. However, this can only be true if the quantum efficiency with respect to absorption efficiency $k(x)$ is maintained, see Eq. (9). Thinner x comes at a penalty of lower photon absorption. Therefore, the approach is to reduce the volume by reducing the thickness (x) while maintaining sufficient photon absorption. This can be achieved by the use of a secondary matter particle that is not affected by the thickness and in the contrary prefers a thinned absorber: plasmons. Plasmons [10–12] are collective electromagnetic excitation of the electrons in a material that like to live near the "skin" of the material, characterized by the skin depth δ. The excitation of plasmons by the incident photon then couples with excitonic modes (collectively called polaritons) which can translate into increased electrical response of the material through increased carrier generation that can be detected as increased photocurrent. This approach also yields interesting quantum phenomena such as topologically protected optical behavior, which is not

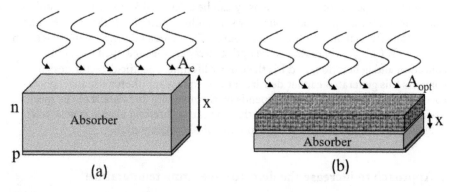

Figure 1.
Illustration of thick (a) and thin (b) design for increasing the detector operating temperature, where A_{opt} is enhanced absorption using nanostructured materials stack such as metamaterial design.

covered here further. Coupling of photon-plasmon-exciton physics is an exciting field that has led to a whole host of new technologies for detection, including metamaterials [13], negative index materials used in metalenses, and metasurfaces [14–16]. These artificially constructed surfaces (illustrated in **Figure 1b** with area A_{opt}) use carefully designed structure of varying dielectric constants to obtain an artificial composite material that shows behavior not normally seen in any "natural" material. One example of such a material is a matrix of anti-dots, which are voids or physical holes [17, 18] in the detector material that show the physics of absorption enhancement through plasmon polaritons. As shown in **Figure 1b**, the detector absorber volume is reduced considerably and by use of metastructure stack design above one can gain back the photon absorption. More discussions on metastructures is given in Section 9.

6. Leveraging bandstructure engineering for better detectors: artificial materials

Advances in fabrication capabilities at atomic levels, using techniques such as molecular beam epitaxy (MBE), atomic layer deposition (ALD), high precision sputtering and chemical vapor deposition (CVD), and colloidal chemical deposition has brought to us the marvels of building novel materials from ground up; materials such that nature does not build on its own. The use of such materials have proliferated many fields of applied sciences and engineering and new applications come up every few years. In the context of Electrical Engineering, such high precision artificially created nanostructured materials include superlattices, which are candidates for the "next transistor" material. One such example are high mobility GaAs| AlGaAs quantum wells which have been candidates for such devices [19] for a long time, and high electron-mobility transistors (HEMT) [20] built from such quantum wells have proven useful in many applications in telecommunications and low power amplifiers. These composite material stacks are of particular interest due to the tunability of their physical behavior through material design, therefore there is growing interest in such stacks for detector designs. A lot of the recent efforts in exploration of such novel materials have been in lower dimensional materials. Of course, physically every piece of a material has all 3 dimensions; the dimensionality in this context means the physical behavior demonstrated by this material can be captured using condensed matter theories of materials at lower dimensionality. We now briefly discuss the physics that is demonstrated by lower dimensional materials, and discuss the possibilities in sensing applications that have been developed or proposed using such materials.

6.1 Energy levels and 0-D material

First pedagogical problem presented in almost all quantum mechanics text is solving the Schrodinger Equation for a particle in an infinite potential well (i.e. particles cannot leak out from the well) and the resulting quantization of allowed "sharp" energy states which a particle may possess [21]. An immediate extension of this problem in the context of an atom led to the accurate prediction of ionized Hydrogen spectra by Niels Bohr and observation of electron diffraction in Stern-Gerlach experiments, which sealed the deal for quantum mechanics, so to say. The story is more complex in condensed matter where interactions of many electrons in close consort lead to a continuum of allowed states with some gaps in between them. In mid 1980s advances in semiconductor processing allowed fabrication of

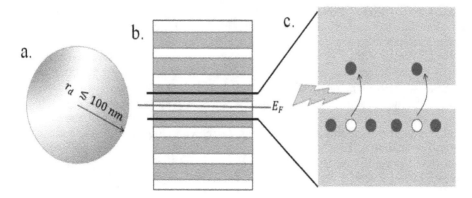

Figure 2.
(a) A quantum dot is made of a semiconducting material with dimensions typically less than 100 nm. (b) The energy spectra of a quantum dot showing a series of thin narrow bands. (c) The two bands of interest: valence and conduction bands participate in photodetection action where incident light promotes electrons from valence to conduction band, from where it can be detected in an external circuitry.

ultra-small dots of particles (**Figure 2a.**) which resemble the pedagogical problem described earlier. The allowed particle states are similarly confined to discrete or very small bands of allowed energies, which are determined primarily by the material type and size of the dots. From the point of view of energy-eigenstates, this is analogous to the particle in a well, only this time it can be physically fabricated with high accuracy and control, hence the name "Quantum Dots" for this kind of a material. It is also immediately obvious why we can call such materials artificial— we can physically select the energy eigenstates rather than depend purely on the intrinsic configuration of the atoms in the material. **Figure 2b** shows an illustrative example of such a material's energy spectra. Ability to tune the eigenstates in quantum dots allows us to build photon detectors of specific wavelength, since the de-Broglie relation links the wavelength to the corresponding energy: $E = \frac{hc}{\lambda}$ where symbols have usual meaning. As an example, the mid-wave infrared (MWIR) corresponds to around 0.25–0.35 eV of energy gaps between conduction and valence bands. Photodetection action then utilizes the light-matter interaction (**Figure 2c**), where incident photon kicks a "bound" electron from valence band to the conduction band which can then be detected -in an external electron counting circuitry.

6.2 From 0-D to 3-D

It is possible to build from bottom-up a picture of evolution of bands and density of states as a function of dimensionality, an illustrative example shown in **Figure 3** (see [21] for a textbook level exposition). Starting from the Gedanken two energy level "simple" model, we obtain a collection of the full 3-D bands. The 0-D can be thought of as a collection of two such bands generally falling on top of each other. However Pauli's Exclusion principle of Fermions do not allow each of these levels to fully overlap, which provides a finite spread of these levels (sub-bands) which are smeared together into a continuum within the bandwidth due to Heisenberg's Uncertainty principle. These two quantum mechanical principles give rise to all bandstructure phenomena in materials.

A 1-D material can be thought of as a collection of multiple 0-D materials, which leads to stacking of multiple thin bands near each other (sub-bands) which merge

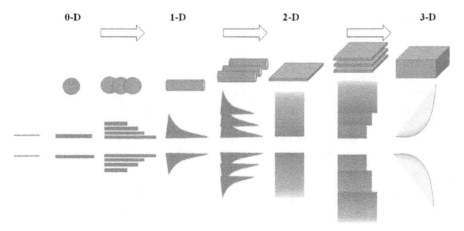

Figure 3.
Evolution of density of states spectra from 0-D sharp bands to 3-D bulk materials.

into a continuous density of states/bands with an energy dependence of $D \propto 1/\sqrt{E}$, this being the density of states of a nanowire. Similarly stacking multiple nanowires together gives us sub-bands that yield into a constant density of states of 2-D materials, prime examples being graphene, MoS_2, black phosphorus etc. Stacking of these 2-D materials then yields the familiar world of 3-D or "bulk" materials with density of states $D \propto \sqrt{E}$.

The advantage of low dimensional materials lie in the density of states and bandstructure that can give rise to unique electrical properties that can be advantageous in building high quality detector. We next discuss some of the engineering opportunities and challenges faced by photodetector materials of low dimensionality.

7. Engineering opportunities and challenges of low dimensionality materials

The current efforts towards building high quality photodetectors are thinned absorbers as discussed before. Lowering the bulk dimensions, particularly the depth, reduce dark current and hence noise. This trend then naturally leads to the exploration of low dimensionality of the material for better detectivity. It should be noted that quantum dots as controlled 0-D materials are popular in photovoltaics and some recent works in IR detectors as well, primarily because of precision in control of detection spectra. Quantum dots are also popular in quantum photonics as coherent photon sources and single spin state storage, and also as detectors of high-frequency shot noise by exploiting transport characteristics in such systems which make them particularly sensitive as detectors. We briefly discuss some of the transport signatures of interest that emerge from low dimensionality.

7.1 Sub-bands and conductance quantum

Confinement in dimensions quantize the corresponding momenta, i.e. the number of "wave modes" of the electrons, similar to confinement of EM waves in waveguides. This can be seen in the intermediate dimension pictures and their corresponding density of states show in **Figure 3**. As discussed before, a 1-D to 2-D transition shows the signatures of multiple "1-D" density of states pieces coming together to form the 2-D density of states. This can have profound effect on the transport signatures. These discrete transverse modes are activated as a Fermi level crosses through them on the application of an external voltage and the famous "staircase" transmission structure was observed [22] by von Klitzing as integer quantum Hall effect, giving rise to the notion of conductance quantum for each ballistic transmission mode. These signatures can determine the overall current voltage characteristics of a detector structure, which should be considered when a detector structure is designed. These include negative differential resistance or NDR effect, as seen in resonant tunneling diodes. It should be noted that there are further signatures of fractional quantum Hall effects which again arise in ultraclean superlattices at cryogenic temperatures due to topological nature of such Hamiltonians. These states, in particular 2/5 fraction are of great interest for topological quantum computing [23]. It remains to be seen how these two fields can intersect technologically in a meaningful way.

7.2 Contacts

Contacts play a critical role in replenishing and diminishing the carriers in the detector material. This action depends on the relative band alignments between the detector material and the contact. For a low resistance "Ohmic" contact, the corresponding carrier band should be close to each other, since any mismatch leads to resistance as the carriers have to jump "up" or "down", resulting in a change in the momentum and energy, which gives rise to resistance from a mesoscopic point of view [24]. A mismatched contact is called a Schottky contact and works through thermionic emission over the barrier or through a tunneling mechanism. Choice of the contact material plays a critical role in building a high-quality detector, see [25] for a detailed exposition. A low dimensionality material may show narrow bandwidths and low density of states, which do not necessarily make good contacts with bulk wiring leads in a circuit. Choice of a well-matched contact material requires considerable effort. As an example, in quantum dot detectors, the contact is not made directly from the wire lead to the dots themselves, rather they live in a matrix of PMMA and graphite complexes [26]. Recent developments include use of graphene sheets and ribbons to form conformal contacts with the dots and then to a metallic lead. In fact, graphene itself required considerable efforts to find an appropriate contact metal [27]. The central engineering challenge therefore is to build contacts of appropriate quality; in a bipolar structure it may mean building good contacts for both electrons and holes, while in a unipolar, the ideal contact might be an Ohmic one with the transporting carrier, say electron and a blocking Schottky contact with the holes.

7.3 Low density of states

Low dimensional materials often end up showing very low density of states, resulting in low currents even in the case of ultra-low scattering or near ballistic transport. An equivalent analogy of such a case is two cities being connected by a narrow rural one lane road as compared to a four-lane divided highway. The carrying capacity of the former is a bottleneck in transport. This can lead to high shot noise in the material, though cutting down dimensionality may reduce flicker noise due to cutting down of number of possible paths for a carrier to take from one terminal to the other, if the sample is clean. Additionally, 2-D materials may show Anderson localization effect which shows up as an extra resistance in the case of disordered materials, which can then increase and even exceed the flicker noise seen in bulk samples. It must be noted that while this low current is a critical issue in logic circuits, this is much less of an issue in detectors, and in fact an advantage of low dimension materials by cutting the dark current significantly. However increased noise in disordered materials dictates fabricating high-quality material samples for detector applications.

7.4 Blockaded transport

0 and 1-D materials may show a behavior called the blockaded transport regime. In this regime, due to ultra-low density of states in the material, presence of one electron in the channel "charges up" the channel so much that it prevents another one to enter due to Coulombic repulsion [28]. This is also called the Coulomb blockade. A similar phenomenon can be seen in spin-polarized transport as well, where certain spins are not allowed to go in and out of the channel [29]. Blockaded transport may raise interesting applications for detection since it allows electrical spectroscopy and is a popular method for doing so in STM experiments and molecular electronics [30], though a controlled mechanism for the process will need to be demonstrated before any such application is realized in sensing domain.

7.5 Quantum capacitance

Another consequence of low density of states is the appearance of quantum capacitance in the transport behavior in the material. Quantum capacitance is given as $C_q = q^2 D$ [31]. It can be seen that as the density of states D goes down, the capacitance goes down. This capacitance then effects the electrostatics and transient behavior of the detector material with consequences for detector design and operation. As per basic circuit laws, series capacitances add up as harmonic sum, which is dominated by the smallest capacitance in the system. For large bulk systems, C_q is substantial and can be ignored in device capacitance considerations, however in nanowires or nanosheets it starts to play a substantial role. Central consequence of this phenomena is diminished control over the detector electrostatics, particularly if the detector is working in photoFET mode. Therefore, it is critical to design a detector keeping in consideration of quantum capacitance in low dimensions.

8. Modeling and simulation approaches for nanostructures: quantum effects, light-matter interaction, carrier transport

Modeling approaches in nanostructures can be roughly divided into three categories of techniques, dealing with the different particles that come into picture in the detector physics. We broadly divide the discussion into bosonic and fermionic particles, as their physics are very different and therefore their modeling approaches have differed in their developments. It is impossible to cover all these myriad approaches in a single volume monolog, much less in a section of a chapter, therefore we provide only a very brief overview of these approaches along with some textbook references that can serve as starting point for deeper education and understanding.

For "bosonic" particles such as photons, plasmons (and various associated polaritons), phonons etc. it is customary to use a wavelike approach i.e. by solving a wave equation in a classical physics sense [32], and it is rarely necessary to approach them with quantum field theoretic approaches, i.e. second quantization methods. Photons are indeed extremely well modeled using a classical Lagrangian written by formalizing the Maxwell's equation in terms of a 4-vector gauge theory [33]. Similar approaches work very well for plasmons, which are the collective oscillations of electrons, or phonons which are mechanical oscillations of the material lattice. Both plasmons and phonons show a dispersion relation, i.e. an energy-momentum relation with similar structure (optical and acoustic branches) and therefore are amenable to a unified approach to modeling and simulations.

This above fact is exploited by the many available computational electromagnetics software that convert the wave equation, which is a vector PDE boundary value problem into a linear algebra problem using the technique of Finite Element Analysis (FEA). This approach converts a physical domain into a collection of packed small triangular (most widely used shape in FEA) regions over which the solution is considered to be constant. In one iteration the solution over the "finite elements" are updated based on the previous step values, and this iteration continues till the error value over the whole grid reaches a minimum threshold, set as per the specifications of the problem. See [34] for a classic text on FEA. Other similar approaches include finite difference method, which instead of creating a mesh of triangles that can cover any possible physical geometry, uses a Cartesian grid on which the derivatives are discretized over "grid points" instead of triangular regions as used in FEA, and the problem is converted into a linear algebra problem. An associated method called the finite volume method (FVM) generalizes FEA and FEM to 3D problems. However these approaches are general enough that they can be extended to any dimensions, with associated approaches to discretization. The relative "mechanistic" approach of these methods that depend on linear algebraic methods have allowed development of extremely power simulators, including Ansys HFSS, COMSOL Multi-Physics, MATLAB etc. which can now utilize GPUs to accelerate the solution of very large problems.

For "fermionic" particles such as electrons and holes, the approaches differ in the sets of equations which are solved, depending on the application. Traditionally Boltzmann transport equation (BTE) has been a popular method in solving electron transport problems. However powerful many-body quantum statistical mechanics approaches have been developed as well, the central one being the Non-Equilibrium Green's Function (NEGF) method that has shown tremendous success in solving the problem of electron transport in complex material stacks such as superlattices and nanostructures. A classic exposition that covers both these approaches in an extremely approachable way is [35]. We next describe the fundamental principles underlying these methods and their application spaces.

The central entity of the Boltzman transport equation is the carrier distribution function $f(t, p, q)$ which is a function of time t, generalized position p, associated momentum q. The equation can be written as:

$$\dot{f} + \dot{p}\nabla_p f + \dot{q}\nabla_q f = Sf \tag{13}$$

Where ∇f denotes the gradient of f with respect to the subscripted variable, S is a scattering operator that mixes the distribution function in various p and q states, depending on the particular problem being addressed. In general solving this problem for even one particle involves keeping track of the distribution function in a 7D phase-space (6D in a steady state problem), which can quickly become intractable for even moderate sized problems, especially interacting systems. Therefore, the most commonly used approach to solve this PDE is to use Monte Carlo methods which allows us to construct the full solution by evaluating only certain trajectories through the phase-space, instead of calculating the full phase-space. Careful application of importance sampling can lead to fast tractable solutions to f from which it is possible to then calculate current-voltage relations (first moment), noise currents (second moment), and even higher order moments as necessary. It is possible to also calculate heat flows by energy weighted first moment of f.

NEGF method can be seen as a quantum analog of Boltzman transport equation. The central quantity of the interest in this formalism are the retarded Green's function G^r, which the "causal" propagator of the carrier in the system, and the correlation function $G^<$ which is the quantum density matrix. The NEGF method evolves the $G^<$ under the influence of various scattering phenomena (Büttiker probes are a good example) under the action of the G^r. The two equations are written as:

$$G^r = [E - H - \Sigma]^{-1} \tag{14}$$

$$-iG^< = G^r \Sigma^{in} G^a \tag{15}$$

Where E is the energy, H is the system Hamiltonian, Σ is the total "self energy" for all the scattering mechanisms applicable to the system, including physical contacts, momentum scattering, dephasing, inelastic scattering etc. Σ^{in} is the corresponding sum of the "inflows" from these mechanisms into the system. We leave the details of how these may be written to the reference mentioned above and other follow on works from the author, a recognized authority on the subject.

The NEGF approach is well suited to handle any arbitrary combinations of material stacks, i.e. an arbitrary H, not necessarily confined to materials with parabolic bands. This method also forgoes any equilibrium or near equilibrium assumption that is commonly applied in BTE. Transport phenomena of lower dimensional, and indeed any quantum system is automatically incorporated in the NEGF approach. This makes NEGF a very powerful tool for modern detector structures which are increasingly based on superlattices. However, incorporating appropriate self-energy and inflow terms in the NEGF equation requires a non-trivial intellectual effort. As the field develops further we expect phenomena such as Auger recombination, Impact ionization, SRH recombination, multi-electron generation etc. will be included systematically in the toolset of NEGF methods, which will then yield a truly bottom-up quantum mechanical simulation approach for photonic devices.

9. Metasurfaces for LWIR devices

In this section, we now explore the possibility of exploiting the fast-developing area of metasurfaces—more generally known as flat optics—to add functionality to the existing IR sensors without affecting the performance or increasing SWaP. In addition, metasurfaces offer a possibility of increasing the temperature of operation. In reference to **Figure 1b**, the metasurface-coupled detector is an emerging technology to enhance optical absorption in a thin detector architecture. This becomes even more important where absorber layer is naturally thin—such as quantum dots, 2D materials— or require to be thin bulk absorber to minimize dark current and overall detector materials noise.

As the electromagnetic radiation passes through a resonant element, both its amplitude and phase undergo a large change. Interestingly the phase can be varied from 0 to π either for the frequencies in the vicinity of the resonant frequency as shown in **Figure 4** or equivalently for element size in the vicinity of size that is resonant at a given frequency. Particularly the latter observation along with little or no variation in amplitude can be exploited to introduce phase variation in the XY-plane (for the light propagating in Z direction) of the optoelectronic device. With

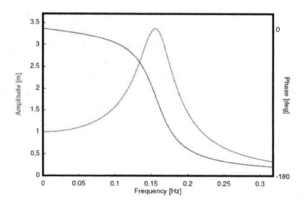

Figure 4.
Schematic variation in amplitude (red) and phase (blue) near the resonant frequency of the metaelement.

appropriate phase gradient on a dielectric surface—called metasurface—and the generalized Fresnel's equations [36, 37], we can change the propagation wave front to arbitrary shape and direction, leading to a number of interesting applications [36, 37].

Interestingly, the functions such as focusing, defocusing, total reflection, total transmission, total absorption, frequency filtering, non-specular reflection and/or transmission, polarization filtering, which often requires bulky optical elements, can be achieved with sub-wavelength-thick metasurface [38–44]. The metaelements include sub-wavelength and appropriately shaped metal—known as plasmonics—and all-dielectric, size-dependent Mie resonant scatterer. The advantage of metal plasmonics is that the size can be exceedingly small (often 1/10th of the wavelength) and field concentration/redirection can be huge. However, ohmic loss in metals often affects the optical absorption required in the optical devices. On the other hand, Mie resonant scatterer with a transparent dielectric can provide required phase change without any absorption but requires high index contrast for efficient light modulation. Since the Mie element size is $\sim \lambda/n$ (where λ is the design wavelength and n is refractive index of the dielectric) for resonance, the metasurface unit cell can contain only a few elements and hence a broadband design is often difficult. For illustration, we consider three applications—near-perfect reflection [42, 43] polarization filter [44] and broadband absorption [45, 46] for possible high temperature operation.

Our work in metasurface area is motivated by adding functionalities and operability to the existing IR devices. For example, a metasurface for near-perfect narrow band reflection. Highly reflective surfaces are of great interest for several applications, including sensor and eye protection under laser illumination, hardware hardening against laser irradiation, and optical elements in high-energy laser systems. In these applications, the surfaces require high reflectivity under intense illumination with little or no absorption or transmission. To protect the IR device from the high intensity radiation at a wavelength in the absorbing band, an ultrathin metasurface can be fabricated on the optical devices and the elements of the metasurface can be chosen to fully reflect the high intensity wavelength. We have used shape-dependent Mie scattering to design ultra-thin metasurfaces and demonstrated [43] near-perfect reflection (>99.5%) in a narrow band (around 1550 nm) as shown in **Figure 5**. Note that ~500 nm-tall Si nanopillars (with index 3.5) on low index SiO_2 (index 1.45) are sufficient to achieve the high reflection without any metal component. The design principle is applicable to all reflection windows with appropriate changes to the element size and materials set.

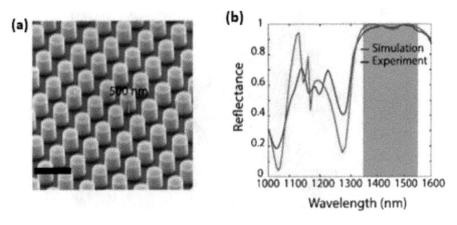

Figure 5.
(a) Fabricated Si nanopillars on SiO₂ and (b) the simulated (green) and measured (black) reflectance [43].

The constituent materials for short-wavelength infrared to the long-wavelength infrared, can be, for example Si, InP, or GaAs for the high index and ZnS, BaF₂, CaF₂, or MgF₂ for the low index materials that have high transparency with little or no absorption to 12 μm.

The polarization of the radiation can be exploited to improve imaging quality, particularly when the thermal contrast between a target and its background is insufficient for polarization-insensitive sensors [47]. Two kinds of polarizers—absorptive polarizers and polarization beam splitters—are currently used. Absorptive polarizers such as wire grids, dichroic materials, or nanoparticle composites provide high degrees of polarization, but both polarizations cannot be analyzed simultaneously as needed in applications such as imaging and quantum information. The polarization beam splitters such as birefringent cubes or Brewster angle reflection in multilayer dielectric films preserve the rejected polarization by reflection or diffraction. However, they are bulky for chip-scale photonic and optoelectronic devices. We have slightly modified the design of **Figure 5** to introduce lattice asymmetry as shown in **Figure 6a** so that resonance frequency for S- and P-polarizations are separated. Consequently, one polarization is reflected near-perfectly while the other is transmitted near totally as demonstrated (**Figure 6b**). The small difference in the transmitted component between simulations (dashed) and measured (solid) are because of over-etching of SiO₂ around the pillars which can be corrected. Notice that mere 500 nm-thick metasurface, which can be monolithically integrated with the sensor can provide near perfect polarization splitting to add polarimetric functionality to the existing IR devices. The concept can be extended to other wavelength bands with appropriate choice of transparent materials. As can be noted, the reflectivity changes in a narrow band and the structures are useful in applications where the large bandwidth is not a requirement. However, several options including the optimization in height to diameter ratio, periodicity of the pillars, multiple element with differing size in the unit cell are available to increase the bandwidth.

Metasurfaces for broadband absorption and transmission will be more useful in IR sensors in increasing the operating temperature. We discuss the issues limiting the temperature of operations and then some metasurface designs that offer the possibility of addressing these issues.

Sensing technology in the infrared (IR) spectrum enables substantial advances in several applications including security, surveillance, industrial monitoring, and

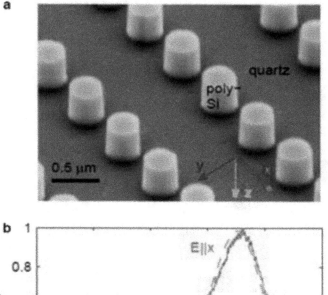

Figure 6.
(a) Fabricated Si nanopillars on SiO$_2$ and (b) the simulated (dashed) and measured (solid) polarization-dependent reflectance [44].

autonomous vehicle navigation. The reduction in SWaP of these sensors without sacrificing performance is particularly important. Among various IR technologies, long wavelength infrared (LWIR) imaging systems are crucial for target-acquisition tasks because of their advantages in adverse environments. Current detection and imaging technologies that cover the LWIR spectral regions typically operate at cryogenic temperatures of 77 K or lower. This requires expensive and bulky cooling systems that increase the overall SWaP of imagers. Uncooled LWIR detector would greatly alleviate the SWaP constraint and enhance the imaging system's portability, enabling broad applications in surveillance and reconnaissance under adverse conditions.

The increase in operating temperatures often requires unfavorable tradeoff between performance, cost, and power consumption. In photon detectors, the dark current increases exponentially with temperature as depicted in Eq. (6), resulting in more power consumption, high noise, deeper cooling requirements and thus poorer performance. Uncooled microbolometers for imaging have a considerably lower performance as compared to cooled photodetectors. Novel and nanocrystalline materials such as colloidal quantum dots (CQDs) show promise, but the synthesis of large sized particles for absorption in the LWIR with fewer surface states needs to be mastered. There is a large number of publications [48, 49] demonstrating LWIR sensing at room temperature with 2D materials—often exploiting high absorption in thin layers and the Schottky barrier between them and Si—and also by measuring the changing properties resulting from high absorption in structured graphene [50], for example. While these results are encouraging, considerable improvements are needed to increase the QE—because only carriers within a few nm from the interface participate in absorption—and the path forward for integrating with ROICs for imaging is not clear.

For many years, there has been a continuing effort to increase the operating temperatures (T_d) of IR detectors and arrays with conventional materials such as HgCdTe and type II superlattices (T2SL). The efforts include high-quality materials to reduce defect-mediated dark currents, low n-doping to achieve high depletion and suppress Auger and radiative-mediated dark currents, strained layer superlattices to reduce the Auger mediated dark currents, novel device architectures such as nBn and CBIRD, and others to reduce the diffusion currents, as well as optical immersion lenses, photon trapping structures, and plasmonic structures to reduce the collection volume and thus dark currents. However, the epitaxial layers of LWIR photodetectors are still very thick, requiring long growth times and creating sensitivity to chamber and source maintenance, inevitably promoting the numbers of SRH centers. Beyond some critical photodetector thickness, dark current begins to dominate and the signal-to-noise ratio of a photodetector decreases.

Current LWIR sensors employing p-n junctions require absorbers with a thickness equal to or greater than the targeted cut off wavelength (~12 μm-thick for LWIR) for ~80% absorption of the incident radiation [51]. However, as mentioned before, the thick absorption region results in a large dark current, arising from generation mediated by the Auger, radiative and Shockley-Read-Hall (SRH) mechanisms [52]. Efforts to reduce the dark current via higher bias (for depletion) increases the band-to-band and/or trap-assisted tunneling and thus do not offer a robust solution [53]. For example, higher depletion by very low doping of LWIR detectors did not increase the operating temperatures beyond 100 K [54]. Various prior efforts to pattern the entry surface—known as photon trapping or PT structures— succeeded in minimizing Fresnel reflection but achieved only a moderate decrease in absorption region thickness and a slight increase in operating temperature of larger band gap mid-wavelength infrared (MWIR) devices to near 200 K [55–57]. Since the absorption in the structured surface did not increase the photocurrent, the overall QE actually becomes poorer.

Recently, LWIR and MWIR photodetectors based on T2SLs, especially Ga-free ones, have shown promising performance [58]. However, the LWIR operation temperature is still limited to 77 K [59]. GaSb-based T2SLs have a higher operation temperature, but the temperature increase (up to 110 K) is still modest for a 10-μm cut-off [60]. In addition, epitaxially grown quantum dots (QDs), which are sometimes coupled with quantum wells, have been used to increase the operation temperature up to 200 K [61]. The challenge of epitaxial QDs lies in their low QE. CQDs have shown promise in the MWIR with a high QE. Still, the dark current and noise are high owing to their poor transport properties [62]. The requirements for high temperature of operation are illustrated [63] in **Figure 7**. The assumptions in this illustrative calculation include—8–10 μm spectral band, $f/1$ optics, 10 μm detector cut off, 70% in-band QE and 19 Me-integrated charge (30 Me-well). We note that when the absorber thickness t is 10 μm, the lowest noise equivalent differential temperature (NEdT) of 30 mK (red) is above the preferred value of 20 mK. When the collection volume is reduced *without loss* of QE (blue), we can increase T_d by about 20–40 K. However, when this reduction in volume is combined with Auger suppression by a factor of 50, for example expected in superlattices because of flat

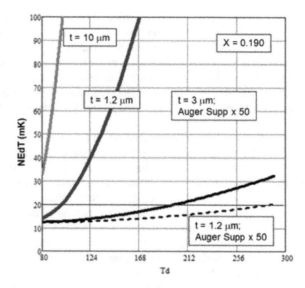

Figure 7.
Effect of collection volume and auger on detector performance [63].

bands or in a lightly depleted absorber, the T_d can be increased (dotted) to ~270 K. More specifically, the designs for high temperature operation should meet the following criteria:

1. Large reduction in collection volume *without loss* of QE.

2. Full depletion at lower bias.

3. Reduction in minority carrier leakage current.

Having examined the aforementioned efforts of improving the operation temperature of LWIR photodetectors, we conclude that any approach to increase the operating temperature will require a substantial reduction in absorption thickness as we know thinner absorber reduces the collection volume, is easier to fully deplete with small voltages, and result in few defect centers.

The resonant pixel approach [64] developed by Choi et al. attempts to reduce the collection volume by allowing the light to enter through thin metallic layer and force it to go through multiple reflection by sandwiching the absorber between two metal layers. This effort has been successful, for example, in reducing the absorber thickness to 0.7–1.2 µm and still achieve quantum efficiency of 50–70% with about 2 µm bandwidth in LWIR region [63, 65]. The use of metal, required for resonant cavity, causes ohmic loss and further increase in bandwidth and quantum efficiency have not yet been demonstrated.

Several metasurfaces are proposed or demonstrated to achieve broadband absorption. However most of them use metallic elements, limiting the absorption in the dielectric absorber layer which cannot increase the photocurrent, and are not considered here. We will discuss two promising all-dielectric designs [45, 46]. To enable high absorption for solar application with thin MoS_2 layers, the authors [45] proposed a structure shown in **Figure 8a** with corresponding absorption measurements shown in **Figure 8b**. Although this structure resembles that of Fabry-Perot cavity structure, they designed a photonic band gap (PBG) in standoff oxide layer to increase the absorption in MoS_2 through increasing the interaction with guided

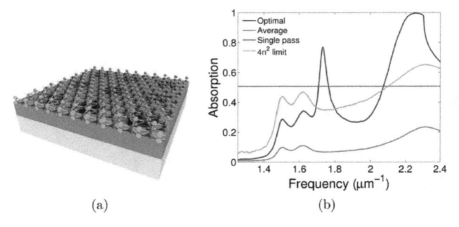

(a)   (b)

Figure 8.
(a) Modified Fabry-Perot structure to include a photonic band gap design (periodic holes) in the oxide layer and (b) the predicted absorption spectrum (blue) in the monolayer MoS_2 [45].

resonance of the PBG. With a single layer MoS_2 (of thickness ~0.6 nm) and an absorption coefficient of ~2×10^5 cm^{-1}, a broadband average of 50% absorption in the visible is predicted. The scalable design can be extended to mid- and long-wave infrared region and even with 100× smaller absorption coefficient, the absorber thickness can be estimated to be ~65 nm for this level of absorption. However, a detailed modeling absorption dispersion in the LWIR band is required to ensure that guided mode traverses through the entire absorber thickness required for even higher QE.

Interestingly, broadband all-dielectric transmitting metasurface—in the visible band—has been demonstrated [46]. Appropriately designed Si nanopillars on Si absorber (**Figure 9a**) has been fabricated (**Figure 9b**) and reflectance has been measured to be very low and broad band (**Figure 9c**, black). All the rest are transmitted and absorbed in Si. Notice that absorption is on the average over 95% broadband covering entire visible spectrum. The nanopillars behave like an antenna funneling the radiation over larger area, covering the nanopillar and the surrounding metal in the unit, through the pillars into the absorbing substrate. The funneling effect increases the electric field right below the nanopillars, resulting in absorption

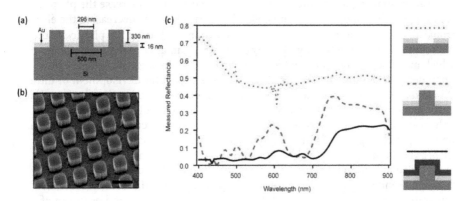

Figure 9.
(a) Cross section of the absorber (si) with metasurface (b) fabricated metasurface, and (c) measured reflectance without absorbing nanopillar (red), with absorbing nano pillar (blue) and with antireflection coated (black) [46].

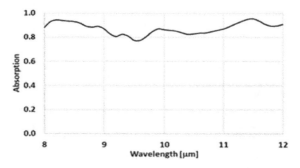

Figure 10.
Predicted absorption spectrum in 2 μm-thick absorber with a metasurface.

enhancement in this layers. Note another important consequence of the funneling effect is that radiation does not impinge on the metal, allowing the metasurface to be transparent and conduction. This feature is particularly very important for all electrically operated IR devices.

Noting that above design is scalable, we have designed a similar structure with appropriate choice of materials for LWIR and the calculated absorption spectrum with 2 μm-thick absorber is shown in **Figure 10**. We see that about 80% broadband absorption is possible. Absorber without the metasurface will require to be at least 12 μm thick to achieve this level of absorption. In essence, the absorber volume can be reduced by a minimum of 6-fold reducing the dark current proportionately. Further, 2 μm low doped region is far easier to fully deplete, resulting in considerable reduction in Auger scattering, which further reduces the dark currents. Also, smaller bias is sufficient to achieve depletion, reducing the power consumption. The device development is in progress for possible 300 K operation.

10. Summary

In summary, the fast-developing field of metasurfaces show considerable promise for easier integration with the detector absorber layer to both, increase the SWAP performance and the operating temperature. This chapter discussed various approaches to using nanostructured metasurface designs to increase the photon capture by enhancing the apparent collection area A_{opt} while decreasing the electronic absorption volume, $A_e x$. We discussed from ground-up the physics of the photodetection as an interplay of energy transfer from light to matter particles, which may take many routes, which all contribute to the large phase space of photodetector design. We discussed why there is a need for development of ultra-thin and even lower dimensional materials for photodetection, to cut down on dark current and the corresponding decrease in the noise. This allows us to trade D^* enhancements with higher operating temperatures. We discussed the implications on the light absorption of ultra-thin and low dimensional materials, compared to their bulk counterparts. We also discussed some unique transport signatures that are enabled at low dimensions, including quantized conductance, reduction of flicker noise in ultra-clean 2D materials. We also mentioned issues of material integration and electrical controllability due to contact physics and quantum of capacitance. We briefly discussed the numerical approaches taken for simulation of photonic and electronic components. We provided a detailed exposition of the physics of metasurfaces to bring back the quantum efficiency in ultra-thin and low dimensional detectors. We further discussed the increasing trend of using bandstructure engineering (T2SL) in building detectors, in particular nBn and complementary barrier detectors which engineer the electron transport properties of the detector, and how even such "novel" materials benefit from thinness of the detector material in achieving high operating temperature.

We hope that in this chapter we have provided a broad overview of the rich physics of photodetectors to the reader and provided further points to ponder over and references to follow for more in-depth education. The future prospects of infrared detectors are clearly in the direction of nanostructure designs to develop, not only systems that eliminates bulky cryogenic coolers, but also looks into thin sensor architectures that can be conformal. In addition, there will be a significant advancement in the area of focal plane array and readout electronic designs with subsequent system designs that will push intelligence near the sensor. The emergence of Artificial Intelligence, Machine Learning and neuromorphic approaches will also drive more and more functionalities in the sensor that will reduce the data pipeline, post processing and lag associated with current technologies.

Author details

Nibir Kumar Dhar[1]*, Samiran Ganguly[2] and Srini Krishnamurthy[3]

1 Virginia Commonwealth University, Richmond, VA, USA

2 University of Virginia, Charlottesville, VA, USA

3 Sivananthan Laboratories, Bolingbrook, IL, USA

*Address all correspondence to: dharnk@vcu.edu

IntechOpen

References

[1] Dhar NK et al. Optoelectronics— Advanced Materials and Devices. Rijeka: IntechOpen; 2013 (Chapter 7)

[2] Kinch MA. Tutorial Texts in Optical Engineering. Vol. TT76. Bellingham, WA: SPIE Press; 2007

[3] Palaferri D et al. Ultra-subwavelength resonators for high temperature high performance quantum detectors. New Journal of Physics. 2016; **18**:113016

[4] Rogalski A. Infrared Detectors. CRC Press; 2010

[5] Ganguly S et al. A multiscale materials-to-systems modeling of polycrystalline PbSe photodetectors. Journal of Applied Physics. 2019; **126**(14):143103

[6] Shockley W, Queisser HJ. Detailed balance limit of efficiency of p-n junction solar cells. Journal of Applied Physics. 1961;**32**:510

[7] Moss TS. Photoconductivity. Reports on Progress in Physics. 1965; **28**:15

[8] Urbach F. The long-wavelength edge of photographic sensitivity and of the electronic absorption of solids. Physics Review. 1953;**92**:1324

[9] Martyniuk P et al. HOT infrared photodetectors. Opto-Electronics Review. 2013;**21**(2):239-257

[10] Zayats AV, Smolyaninov II, Maradudin AA. Nano-optics of surface plasmon polaritons. Physics Reports. 2005;**408**(3-4):131-314

[11] Pitarke JM, Silkin VM, Chulkov EV, Echenique PM. Theory of surface plasmons and surface-plasmon polaritons. Reports on Progress in Physics. 2006;**70**(1):1

[12] Ditlbacher H, Krenn JR, Schider G, Leitner A, Aussenegg FR. Two-dimensional optics with surface plasmon polaritons. Applied Physics Letters. 2002;**81**(10):1762-1764

[13] Cai W, Shalaev VM. Optical Metamaterials. Vol. 10(6011). New York: Springer; 2010

[14] Choi K-K, Dutta A, Dhar N. Novel IR photodetectors and energy harvesting devices. IEEE Journal of Quantum Electronics. 2021;**1172303**

[15] Grayer J, Ganguly S, Yoo S-S. Plasmonics: Design, Materials, Fabrication, Characterization, and Applications XVII. Vol. 11082. San Diego, CA: International Society for Optics and Photonics; 2019

[16] Holloway CL, Kuester EF, Gordon JA, O'Hara J, Booth J, Smith DR. An overview of the theory and applications of metasurfaces. IEEE Antennas and Propagation Magazine. 2012;**54**(2): 10-35

[17] Rabiee-Golgir H et al. Infrared Technology and Applications XLV. Vol. 11002. Baltimore, MD: International Society for Optics and Photonics; 2019

[18] Safaei A et al. Wide angle dynamically tunable enhanced infrared absorption on large-area nanopatterned graphene. ACS Nano. 2018;**13**(1):421-428

[19] Shur MS. GaAs Devices and Circuits. Berlin/Heidelberg, Germany: Springer Science & Business Media; 2013

[20] Smith, P. M., et al. 1987 IEEE MTT-S International Microwave Symposium Digest. Vol. 2. Las Vegas, NV: IEEE; 1987.

[21] Ghosh A. Nanoelectronics: A Molecular View. Singapore: World Scientific; 2016

[22] Klitzing KV, Dorda G, Pepper M. New method for high-accuracy determination of the fine-structure constant based on quantized hall resistance. Physical Review Letters. 1980;**45**(6):494-497

[23] An S, Jiang P, Choi H, Kang W, Simon SH, Pfeiffer LN. arXiv:1112.3400

[24] Datta S. Lessons from Nanoelectronics: A New Perspective on Transport. Vol. 1. World Scientific Publishing Co Inc; 2012

[25] Ganguly S, Yoo S-S. On the choice of metallic contacts with polycrystalline PbSe films and its effect on carrier sweepout and performance in MWIR detectors. Journal of Electronic Materials. 2019;**48**(10):6169-6175

[26] Zheng W, Wong S-C. Electrical conductivity and dielectric properties of PMMA/expanded graphite composites. Composites Science and Technology. 2003;**63**(2):225-235

[27] Xia F et al. The origins and limits of metal-graphene junction resistance. Nature Nanotechnology. 2011;**6**(3): 179-184

[28] Muralidharan B et al. Theory of high bias coulomb blockade in ultrashort molecules. IEEE Transactions on Nanotechnology. 2007; **6**(5):536-544

[29] Muralidharan B, Datta S. Generic model for current collapse in spin-blockaded transport. Physical Review B. 2007;**76**(3):035432

[30] Dorogi M et al. Room-temperature Coulomb blockade from a self-assembled molecular nanostructure. Physical Review B. 1995;**52**(12):9071

[31] Guo J et al. Electrostatics of nanowire transistors. IEEE Transactions on Nanotechnology. 2003; **2**(4):329-334

[32] John David Jackson. Classic Electrodynamics. 3rd ed. New York: Wiley; 1999

[33] Hehl F, Obukhov Y. Foundations of Classical Electrodynamics. Basel, Switzerland: Birkhäuser; 2003

[34] Thomas JR. Hughes: The Finite Element Method: Linear Static and Dynamic Finite Element Analysis. Hoboken, NJ: Prentice-Hall; 1987

[35] Datta S. Electronic Transport in Mesoscopic Systems. Cambridge, UK: Cambridge University Press; 1997

[36] Yu N et al. Light propagation with phase discontinuities: Generalized laws of reflection and refraction. Science. 2011;**334**:333-337

[37] Aieta F et al. Out-of-plane reflection and refraction of light by anisotropic optical antenna metasurfaces with phase discontinuities. Nano Letters. 2012;**12**: 1702-1706

[38] Glybovski SB et al. Metasurfaces: From microwaves to visible. Physics Reports. 2016;**634**:1

[39] Jung J et al. Broadband metamaterials and metasurfaces: A review from the perspectives of materials and devices. Nanophotonics. 2020;**9**:3165-3196

[40] Arbabi A et al. Dielectric metasurfaces for complete control of phase and polarization with subwavelength spatial resolution and high transmission. Nature Nanotechnology. 2015;**10**:937

[41] Khorasaninejad M et al. Efficient polarization beam splitter pixels based on a dielectric metasurface. Optica. 2015;**2**:376

[42] Slovick B, Yu ZG, Berding M, Krishnamurthy S. Perfect dielectric-metamaterial reflector. Physical Review B. 2013;**88**:165116

[43] Moitra P et al. Experimental demonstration of a broadband all-dielectric metamaterial perfect reflector. Applied Physics Letters. 2014;**104**:171102

[44] Slovick B, Zhou Y, Yu ZG, Kravchenko I, Briggs D, Moitra P, et al. Metasurface polarization splitter. Philosophical Transactions of the Royal Society A. 2017;**375**:20160072

[45] Piper JR, Fan S. Broadband absorption enhancement in solar cells with an atomically thin active layer. ACS Photonics. 2016;**3**:571

[46] Narasimhan et al. Hybrid metalsemiconductor nanostructure for ultrahigh optical absorption and low electrical resistance at optoelectronic interfaces. ACS Nano. 2015;**9**:10590

[47] Tyo JS, Goldstein DL, Chenault DB, Shaw JA. Review of passive imaging polarimetry for remote sensing applications. Applied Optics. 2006;**45**(22):5453-5469

[48] Lan C et al. 2D materials beyond graphene towards Si integrated infrared optoelectronics devices. Nanoscale. 2020;**12**:11784 and the references cited therein

[49] Long M et al. PdSe$_2$ long wavelength infrared photodetector with high sensitivity and stability. ACS Nano. 2019;**13**:2511

[50] Safaei S, Chanda D. Dynamically tunable extraordinary light absorption in monolayer graphene. Physical Review B. 2017;**96**:165431

[51] Lee D et al. High operating temperature HgCdTe: a vision for the near future. Journal of Electronic Materials. 2016;**45**:4587

[52] For the latest review see, Rogalski A. HgCdTe Photodetectors, Mid-infrared Optoelectronics. Amsterdam, Netherlands: Elsevier; 2020 (Ch. 7)

[53] Krishnamurthy et al. Tunneling in LWIR Photodiodes. Journal of Electronic Materials. 2006;**35**:1399

[54] Strong RL et al. Performance of 12-μm- to 15-μm-pitch MWIR and LWIR HgCdTe FPAs at elevated temperatures. Journal of Electronic Materials. 2013;**42**:3103

[55] Martynuik P et al. New concepts in infrared photodetector designs. Applied Physics Reviews. 2014;**1**:041102

[56] Choi KK et al. Resonant structures for IR photodetection. Applied Optics B. 2017;**56**:1559

[57] Choi KK. Metastructures for VLWIR SLS detectors. In: Proc. SPIE 11407, IR Tech. and Applications XLVI, 114070K. Anaheim, CA: SPIE; 2020

[58] Haddadi A et al. Bias–selectable nBn dual–band long–/very long–wavelength infrared photodetectors based on InAs/InAs1- xSbx/AlAs1- xSb x type–II superlattices. Scientific Reports. 2017;**7**:3379

[59] Nguyen BM, Hoffman D, Huang EKW, Delaunay PY, Razeghi M. Background limited long wavelength infrared type-II InAs/GaSb superlattice photodiodes operating at 110 K. Applied Physics Letters. 2008;**93**:123502

[60] Pour SA et al. High operating temperature midwave infrared photodiodes and focal plane arrays based on type-II InAs/GaSb superlattices. Applied Physics Letters. 2011;**98**:143501

[61] Yamanaka T et al. Gain-length scaling in quantum dot/quantum well infrared photodetectors. Applied Physics Letters. 2009;**95**:093502

[62] Lhuillier E, Guyot-Sionnest P. Recent progresses in mid infrared nanocrystal optoelectronics. IEEE

Physics of Nanostructure Design for Infrared Detectors
DOI: http://dx.doi.org/10.5772/intechopen.101196

Journal of Selected Topics in Quantum
Electronics. 2017;**23**:1-8

[63] Velichu S. Private communication

[64] Choi KK et al. Resonator-quantum
well infrared photodetectors. Applied
Physics Letters. 2013;**103**:201113

[65] Choi KK et al. Small pitch resonator-
QWIP detectors and arrays. Infrared
Physics and Technology. 2018;**94**:118

Chapter 2

Noise Analysis in Nanostructured Tunnel Field Devices

Sweta Chander and Sanjeet Kumar Sinha

Abstract

Tunnel Field Effect Transistors (TFETs) have appeared as an alternative candidate of "beyond CMOS" due to their advantages like very low leakage current and steep sub-threshold slope i.e. <60 mV/dec., etc. From past decades, researchers explored TFETs in terms of high ON current and steep subthreshold slope at low supply voltage i.e. < V_{DD} = 0.5 V. The reliability issues of the device have profound impact on the circuit level design for practical perspectives. Noise is one of the important parameters in terms of reliability and very few research papers addressed this problem in comparison to other parameter study. Therefore, in this chapter, we discussed the impact of noise on Tunnel FET devices and circuits. The detail discussion has been done for the random telegraph noise, thermal noise, flicker noise, and shot noise for Si/Ge TFET and III-V TFETs. Recent research work for both low frequencies as well high frequency noise for different TFET device design has been discussed in details.

Keywords: Band-to-Band Tunneling, Flicker noise, Shot noise, Thermal noise, Random Telegraph Noise

1. Introduction

The semiconductor industry has been on a continuous run to search for miniaturized devices without compromising the electrical parameters [1]. Scaling down the device dimensions of *metal–oxide–semiconductor field-effect transistor* (MOSFET) has been very challenging due to the short channel effects like high subthreshold slope, high leakage current, and threshold voltage roll-off, etc. [2]. Tunneling-field-effect-transistor (TFET) has appeared as a potential substitute to replace already existing MOSFET because of small leakage current and steep Subthreshold Swing (SS) [3, 4]. The basic difference between the MOSFET and TFET is the working principle. In MOSFET the current through the channel is set by an energy barrier and the height of the barrier determines the amount carriers injected into the channel thermionically. But in TFET, the charge carriers tunnel through the barrier to reach the channel [5]. TFET is gated reverse-biased pin diodes that operate on Band to Band (BTBT) tunneling principle [6, 7]. In the miniaturized devices operating on low voltages, the electrical noise has a serious effect on the device performance [8, 9]. The non-white noises like flicker noise and burst noise become more vigorous as we scale down the device dimensions, which degrades the performance of semiconductor memories [10] and analog circuits [11]. Also, high-frequency white noise sources like shot noise and thermal noise are desirable in analog/Radio-frequency (RF) applications [12]. In the literature, the effect of electrical noise on the performance of TFETs has not been reported much and still under exploration. In the presence of noise, it is quite difficult to comprehend the behavior of TFETs. In the literature, very few works related to the modeling of the impact of high-frequency noise and low-frequency noise on the performance of TFETs have been reported [13, 14]. This paper provides a detail study of types of electrical noise in TFET structures and the modeling of electrical noise in various TFET structures. This paper also elaborates on the basic working principle of TFET along with the device operation.

2. Tunnel field effect transistors

TFET is gated reverse-biased pin diodes that operate on BTBT) tunneling principle [15]. **Figure 1** shows the cross-sectional view of double gate TFET and energy band diagram of TFET for ON and OFF state, in which p-type silicon substrate of length L_{ch} is taken. The source and drain regions have lengths of L_s and L_d with heavily doped p-type and n-type materials. Also, gate oxide of thickness T_{ox} is deposited over a p-type silicon substrate.

When no voltage is applied at the gate terminal, no current flows through the channel. In TFETs, the off-current/leakage current is very small that makes it an energy-efficient device. When a positive bias is applied at the gate terminal, it pushes the conduction band down in the channel region. Thus, a tunneling path is formed between the source band and channel conduction bands.

In heterojunction TFET (HTFET), high ON current and low leakage current can be achieved. The source and channel may have large band gaps in the case of HTFET but at the source-channel interface, the tunneling barrier reduces significantly [16]. The energy band diagram of homojunction and heterojunction TFET is shown in **Figure 2**. In homojunction TFET, the same material is used for the source,

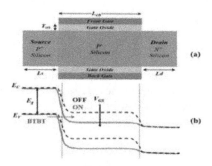

Figure 1.
(a) Cross-sectional view of double gate TFET (b) energy band diagram of TFET for ON and OFF state.

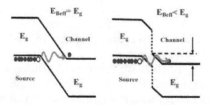

Figure 2.
Energy band diagram of homojunction and heterojunction TFETs [16].

channel, and drain. The barrier height and the bandgap of material are the same. In heterojunction TFET, at the source-channel junction, the barrier height reduces significantly.

3. Noise in MOS devices

It is very significant to study the sources of noise as noise degrades the quality of the desired signal. In MOSFET, the various noise sources are: (a) thermal noise introduced by the channel/polysilicon gate resistance/source-drain resistance/ distributed substrate resistance, (b) flicker noise from the channel. In MOSFET device thermal noise dominates at high frequency and in the channel, it further gives rise to both drain channel noise and induced gate noise. In the nMOS structure, excess thermal noise exhibits in the channel. As drain to source voltage is increased, the noise in the channel increases [17].

D.P. Triantis et al. presented a systematic formulation of the high frequency noise in short-channel MOSFET. The reported MOSFET structure was operating in the saturation region [18]. The small-signal behavior and noise analysis of nanoscale MOSFET at RF was reported by M.A. Chalkiadaki and C.C. Enz [19]. A.G. Mahmutoglu and A. Demir presented an idealized trap model to encounter the behavior of traps present in the gate oxide [20]. H. Tian and A. EL. Gamal proposed a nonstationary extension noise model to analyze flicker noise in MOSFET circuits more accurately [21]. C. Hu et al. studied the low-frequency noise characteristics of MOSFET to investigate the effect of noise by changing metal interconnect perimeter length, device W/L ratio, and gate-biasing voltage [22].

Renuka Jindal developed noise mechanisms in MOSFET for both intrinsic and extrinsic noise. The study of intrinsic noise mechanisms is essential to study the effect of channel thermal noise, induced gate noise, and induced substrate noise on device operation [23].

4. Noises in tunnel FETs

In Tunnel FETs, different types of noise from low-frequency to high-frequency must be considered. Low-frequency noise can be further classified into random telegraph noise (RTN)/burst noise and flicker noise, while high-frequency noise can be further classified into shot noise and thermal noise.

4.1 Low frequency noise sources

4.1.1 Random telegraph noise

Random telegraph noise (RTN) is a non-white noise that mainly occurs due to the presence of impurities in semiconductors and thin gate-oxide films. The trapping and de-trapping of carriers in the channel are the source of RTN [24]. If the trap is located on the source-channel interface, RTN is more pronounced because the trapped charge can change the junction electric field which in turn affects BTBT [25]. It is a function of temperature, radiation, and induced mechanical stress. In audio amplifiers, the burst noise sounds as random shots, which are similar to the sound associated with making popcorn. The noise spectral density is given by RTN is given by Eq. (1) [26]:

$$S_{RTN}(f) = C \frac{4(\Delta I)^2}{1 + \left(\frac{2nf}{f_{RTN}}\right)^2} \tag{1}$$

f_{RTN} is RTN noise corner frequency. The intensity of RTN noise depends upon the site at which the trap center is located concerning the Fermi level and the center area of the Fermi level is responsible for the generation of RTN noise.

4.1.2 Flicker noise

Flicker noise is a low-frequency noise that arises from the trapping and de-trapping of charge carriers in the trap states in the gate oxide around the quasi-fermi level. It is mostly generated at the interface of Si-substrate and gate oxide. At the interface of Si substrate, there exist dangling bonds. These dangling bonds give rise to extra energy states. The charge carriers that move across these energy states, get trapped in these sites. The noise spectral density is given by surface model is given by Eq. (2) [26]:

$$S_{1_f} \alpha (\overline{\Delta N})^2 \int_{\tau_1}^{\tau_1} \frac{1}{\tau} \frac{4\tau}{1 + \omega \tau^2} d\tau = (\overline{\Delta N})^2 \frac{1}{f} \tag{2}$$

where, $\frac{1}{\tau_2} \ll \omega \ll \frac{1}{\tau_1}$ and the noise spectral density for surface model remains constant up to $f_2 = \frac{1}{2\pi\tau_2}$.

4.2 High frequency noise sources

Thermal noise and shot noise are the two types of noise sources that degrade the device's performance operating at high frequency. For the analysis of thermal noise effect on TFET structure, the thermal noise model of MOSFET in ON-state is used, when the applied drain-source voltage is zero [27]. Though, shot noise is the dominant form of high-frequency noise/white noise in heterojunction TFET. The modeling of shot noise can be done in the same way as it is done in tunnel diodes [28].

4.2.1 Shot noise

In electronic devices, the noise that arises due to the unavoidable random fluctuations of electric current when the charge carriers travel a gap is known as shot noise [29]. Shot noise exhibits since the current is not a continuous flow but it is the sum of discrete pulses in time where each pulse corresponds to the transfer of electron through a conductor. Shot noise is caused by the thermal motion of electrons and occurs in any conductor having resistance R. The most dominant white noise is shot noise and in TFET, it can be modeled similarly as it is modeled in tunnel diode.

$$i_{shot}^2 = 2qI_D\Gamma \tag{3}$$

where, Γ is Fano factor that indicates the deviation in the magnitude of shot noise from the nominal value I_D. Fano factor value depends upon applied voltage and it is frequency independent [30].

4.2.2 Thermal noise

In electronic devices, the noise generated inside a conductor at equilibrium due to the thermal agitation of charge carriers is known as thermal noise. In 1928, Johnson experimentally verified the theory of the fluctuating movement of charges in thermal equilibrium that was initially proposed by Einstein in 1905 [31]. The magnitude of random motion of free electrons and resistance of elements increases with an increase in temperature. Due to this thermal motion of charge carriers, a fluctuating voltage is created at the terminals of the conductor [12].

5. Recent study of noises in tunnel FETs

The electrical noise analysis especially the low-frequency noise analysis of TFET structure has been done by many researchers from universities and electronic companies. But the high-frequency analysis of TFET is yet to be explored a lot. In the past a few years, the research of various TFET structures has been accelerated. The brief review of impact of noise on various TFET structure is presented in this section.

The impact of single acceptor-type and donor-type interface trap induced RTN on TFET was reported in [32]. Using a charge-based approach, an analytical model for thermal noise and induced gate noise was proposed in junctionless FETs, whereas the power spectral density of noise was the same for the same drain voltage [33]. Neves et al. reported the effect of low-frequency noise on the behavior of vertical tunnel FETs (TFETs) experimentally and compared with MOSFET [34]. Chen et al. reported the amplitude of RTN on the characteristics of TFET experimentally and reveals that nonuniform distribution of BTBT generation rate along device width direction is responsible for high-amplitude of RTN. It also shows the high source doping concentration fluctuates the BTBT generation rate and reduces the RTN amplitudes [35]. High-frequency noise shows different behavior for the devices having different gate lengths. The scaling of gate length of the High-electron-mobility-transistor (HEMT) was performed in [36]. **Figure 3** shows the change in the value of the noise suppression factor by changing length.

Ghosh et al. presented analysis of flicker noise of TFET structure. A selective buried oxide (SELBOX) SiGe layer has been used at the source channel junction in the presence of trap concentrations as shown in **Figure 4a**. In **Figure 4b**, the presence of trap degrades both ON–OFF current ratio and SS has been studied. Thus, it degrades the overall performance of the device [37].

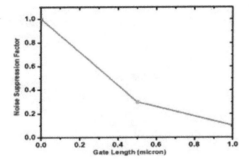

Figure 3.
Gate length vs. noise suppression factor [36].

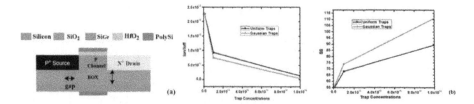

Figure 4.
(a) Proposed SELBOX structure, (b) effect of change in trap concentrations on ON–OFF current ratio and SS [37].

S.Y. Kim et al. reported the dependency of low frequency noise on the applied gate voltage of TFET device. At low gate voltage, the noise in TFET is high and for high gate voltage, electrons directly tunnel from source to drain and noise in TFET is low [38].

Another noise analysis of Ferroelectric Dopant Segregated Schottky Barrier TFET has been done for different types of materials in the dopant segregated layer and concluded that flicker noise affects the low frequency and mid-frequency range [39].

6. Noise in Si/Ge tunnel FETs

The conventional Si-based TFET exhibits SS of less than 60 mV/decade and a very small leakage current but practically implementation is still questionable because of low ON current. Thus, to improve the ON current, a thin layer of SiGe can be used on top of the Si source. It is found from, study that the proposed device is free from short channel effects. Using the strained SiGe layer improves on current while leakage current was still very low (fA) [40]. In **Figure 5**, the proposed TFET structure with a strained SiGe layer is shown.

The features like VLSI compatibility, mature synthesis techniques, and tunable bandgap makes SiGe compatible with the TFET structure. Li et al. experimentally demonstrated a TFET structure in which a thin layer of SiGe is inserted between source and channel. By using the SiGe layer, it has been found that the ON current increases and SS reduces. The barrier height of tunneling reduces and the lateral electric field increases that further increases the transport of carriers across tunneling junction. The performance of the proposed TFET device has been improved by using SiGe [41].

Vandooren et al. presented the electrical performance of vertical Si homo-junction and SiGe source hetero-junction TFETs. The analysis of trap-assisted tunneling through simulations and characterization has been performed. The trap-assisted tunneling worsens the onset characteristics and SS of the TFET. The simulation results of Ge TFET are in agreement with experimental data up to some

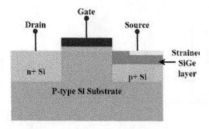

Figure 5.
TFET structure with strained SiGe layer [40].

extent. The results demonstrated that SiGe TFET exhibits improved performance than Si TFET because of the lower bandgap [42].

For HTFET and FinFET, an electrical noise model for thermal noise, flicker noise, and shot noise was realized in a differential amplifier with a capacitive load circuit. The circuit shown in **Figure 6** was studied for subthreshold HTFET design to obtain the same range of gain. From the analysis, it evident that the HTFET structure offers to gain twice the gain of FinFET structure. Also, the cut-off frequency is higher in TFET. Also, small-signal amplitudes can be detected by HTFET design. Based on the overall design analysis, it has been found that the proposed device design is more suitable for the applications operating in low-voltage/low-power [26].

Apart from noise, ambipolar behavior also affects the performance of the device. The comparison of different structures of TFET has been reported by Sathish et al. In TFET structures, heterojunctions have been used to form a source/body junction. By doing so, the tunneling distance can be lowered and the ON current can be increased. It can also be used to control the ambipolar behavior of TFET. Heterojunction TFET characteristics can be determined by using different materials structures [43].

For the fabricated devices at the nanoscale, flicker noise becomes very problematic because flicker noise increases the reducing the device dimensions. The measure of device quality and reliability is done by checking the level of flicker noise. Das et al. presented the analysis of low-frequency flicker noise on dual dielectric pocket HTFET as shown in **Figure 7**. Three different gate-source underlap lengths and different thicknesses were considered. This proposed device exhibits a super steep slope and high current ratio. In the analysis, two different trap distributions (uniform and gaussian) were considered. The study specifies that the ON current of

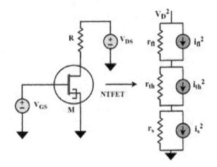

Figure 6.
Representation of noise model implemented at transistor level [26].

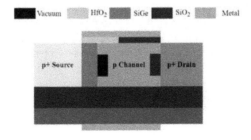

Figure 7.
2-D structure of dual dielectric pocket HTEFT [44].

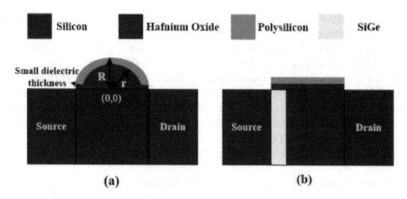

Figure 8.
(a) 2-D structure of novel circular gate TFET (b) and hetero-junction TFET [44].

the reported HTFET structure is not affected by presence of interface traps as compared to the OFF current [44].

Goswami et al. proposed a new architecture of TFET with a circular gate. The electrical noise analysis was performed for both structures; (a) proposed structure (circular gate), (b) HTFET as shown in **Figure 8** [45]. A comparative analysis for both the structures was done also. The uniform and gaussian trap distributions were simulated and based on the simulation results it can be concluded that structure 'a' shows the lesser impact of noise than structure 'b'. Flicker noise shows an intense effect at low frequency and mid-frequency range. While at high-frequency, diffusion noise effects. Nonetheless, the drain current in structure 'a' is more prone to traps than the drain current in structure 'b'. The cut-off frequency of structure can be improved by using gate-drain underlap and it is well-suited for digital applications.

7. Noise in III-V tunnel FETs

In the TFET device, during ON state, very low drive current/ON current flows from source to drain. The drive current can be improved by using low bandgap materials like SiGe, InGaAs, InAs. The use of SiGe has been already discussed in the previous section. **Figure 9** shows the schematic diagram of homojunction and heterojunction TFET using III-V materials.

Pandey et al. reported RTN study analysis of HTFET using III-V materials with 20 nm long and 40 nm wide channels. RTN due to the presence of a single charge trap has been modeled by placing the trap charges to pre-defined coordinated mesh. When the electron got captured in a trap, it causes a reduction in the drain current. Also, the analysis of the relative amplitude dependence of RTN on trap location has been performed. Different locations of the trap were considered for the analysis. Firstly, when the trap is present in the channel, from the source towards drain region. Secondly, three different cases can be considered by changing the depth of

Figure 9.
Homojunction and heterojunction TFET using III-V materials.

the trap like when the trap is (a) present in oxide (b) present in the oxide-channel interface (c) present inside the channel. From the analysis, it has been observed that the relative amplitude of RTNnoise is maximum for case c followed by case b. But the relative RTN amplitude is minimum when the trap is present in gate oxide [13].

Bijesh et al. presented GaAsSb/InGaAs heterojunction TFET using III-V materials. In this work, a high ON current has been achieved at the low drain to source voltage. Also, in the case of InGaAs homojunction, the ON current increases to double the value in heterojunction. Because in heterojunction TFET, reduction in the effective tunneling barrier has been observed. Flicker noise analysis for both homo and heterojunction TFET structure has been performed. The effect of flicker noise on drive current is lower in the case of heterojunction TFET than in homojunction TFET structure. An analytical model to analyze the flicker noise characteristics for both the TFET structure has been developed [46].

Bu et al. developed an analytical model to determine the variation of the electrostatic potential because of the presence of charged trap in the gate oxide of the TFET device. A noise model based on the flow of current through tunneling of carriers in the channel has been proposed. The power spectral density of the TFET device is presented that shows dependency on the frequency and applied gate voltage. It is evident from the analysis that the noise power spectral density of because of the tunneling is more affected by voltage applied at gate terminal than the movement of traps through the channel. The noise observed in the channel is due to the variation in the mobility of traps [47].

8. Impact of noise in TFET based circuit and memory design

In the processors, SRAM memory cell have been broadly used as data caches and these memory cells are the most significant digital building blocks. Fan et al. widely reviewed the effect of single-trap-induced RTN on TFET, and FinFET. The effect of noise on BTBT dominated current conduction in TFET and thermionic based current conduction has been presented in both device and circuit level. The trap location has an intensive effect and that effect varies with applied bias variation and type of trap. The worst-case analysis of different parameters for RTN noise has been investigated for TFET based 6 T/8 T SRAM cells [48]. For different trap locations A, B, C, A', B' the analysis of TFET device has been performed. **Figure 10** demonstrates the diagram of reverse-biased TFET with a single charge trap and having asymmetric source and drain dopant concentrations.

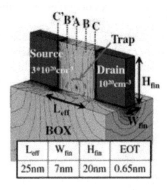

Figure 10.
Potential contour by charge trap [48].

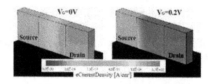

Figure 11.
TFET electron density profile for two different voltages [50].

S.H. Fani et al. presented a new low-power TFET 8 T-SRAM cell with an improved noise margin. The stability of the 8 T-SRAM cell was improved by using supply feedback. The proposed structure exhibits 33% in reading noise margin and 26% in write margin as compared to conventional 6 T SRAM cell for the supply voltage of 0.3 V. The area of the proposed SRAM is larger than the existing one but the features like stability and high performance at very low voltage supply make it useful. The use of the TFET device has limited the working of SRAM cells as it is a unidirectional device but this issue has been resolved by using transistors (n-type and p-type) placed parallelly and the use of one bit-line [49].

The investigation of the effect of RTN noise present in TFET-based 8 T SRAM cell was done by Fan et al. To account the effect of negatively charged trap, the atomistic 3D TCAD simulations were performed for the analysis of TFET-based SRAM. From the analysis, it has been observed that if the trap is present near the tunneling junction, fluctuation in drain current has been observed. The RTN causes 16% additional variation for 8 T SRAM circuit configuration [50]. The electron current density profile for gate voltage values of 0 V and 0.2 V has been shown in **Figure 11**. An increase in gate voltage increases the electron current density.

Luong et al. fabricated half SRAM (HSRAM) cells to examine the capability of TFETs for 6 T-SRAM for the first time. This reported structure has been strained with Si nanowire. The proposed TFET structure does not work up to the mark even when the ambipolar behavior has not been included. Also, analysis of the proposed structure has restricted the static figure of merit [51].

Pandey et al. investigated the effect of a single charge trap RTN in HTFET-based SRAM. This study focused on the analysis of Schmitt trigger mechanism-based variation tolerant 10 T SRAM. A comparison of Si-FinFET and HTFET in terms of iso-area SRAM cell configurations has been done. It has been clear from the analysis that HTFET based SRAM cells show very good performance even in the presence of RTN. For the applied voltage of 0.2 V, the proposed structure offers 15% more improvement as compared to Si-FinFET. Also, the HTFET ST SRAM structure shows less delay in a read operation and consumes less power [52].

The performance of MOSFET devices degrades on scaling down the dimensions. All the issues can be solved using TFET structures as TFET devices offer low leakage current and steeper SS. Nonetheless, TFETs are ambipolar and produce low ON current. However, this behavior can be overcome by using increasing the doping concentration. The main challenges in TFETs are to achieve high ON current, low OFF current, and low average SS. By choosing an accurate predictive model, proper choice of materials, and dimensions of the device, these challenges can be overcome.

9. Conclusion

Noise is one of the important parameters in terms of reliability. This review reported the impact of noise in Tunnel FET devices to understand the reliability issues. The detail discussion has been done for the random telegraph noise, thermal noise, flicker noise, and shot noise for Si/Ge TFET and III-V TFETs. Recent research work for both low frequency as well high frequency noise for different TFET device design has been discussed in details. The effect of noise for memory and circuit based on Tunnel FET devices as also been discussed. The effect of noise on BTBT dominated current conduction in TFET and thermionic based current conduction has been presented in both device and circuit level. The analytical models for noise analysis of different TFETs structures has also been reported which is required for circuit implementation and memory design.

Acknowledgements

This work is supported by Science and Engineering Research Board (SERB), Department of Science & Technology, Government of India, CRG/2020/006229, dated: 05/04/2021.

Author details

Sweta Chander and Sanjeet Kumar Sinha*
School of Electronics and Electrical Engineering, Lovely Professional University, Punjab, India

*Address all correspondence to: sanjeetksinha@gmail.com

IntechOpen

References

[1] R. Dreslinski, M. Wieckowski, D. Blaauw, D. Sylvester and T. Mudge, "Near-threshold computing: reclaiming Moore's law through energy efficient integrated circuits", Proc. IEEE, vol. 98, no. 2, pp. 253–266, 2010.

[2] S. Chander, P. Singh and S. Baishya, "Optimization of direct tunneling gate leakage current in ultrathin gate oxide FET with High-K dielectrics", International Journal of Recent Development in Engineering and Technology, vol. 1, no. 1, pp. 24-30, 2013.

[3] S. O. Koswatta, M. S. Lundstrom and D. E. Nikonov, "Performance comparison between p–i–n tunneling transistors and conventional MOSFETs", IEEE Trans. Electron Devices, vol. 56, pp. 456–465, 2009.

[4] W. Y. Choi, B. G. Park, J. D. Lee and T. J. K. Liu, "Tunneling field-effect transistors (TFETs) with subthreshold swing (SS) less than 60 mV/dec", IEEE Electron Device Lett, vol. 28, pp. 743–745, 2007.

[5] S. Banerjee, W. Richardson, J. Coleman and A. Chatterjee, "A new three-terminal tunnel device", *IEEE Electron Device Lett*er, vol. 8, pp. 347–349, 1987.

[6] K. K. Bhuwalka, J. Schulze and I. Eisele, "Scaling the vertical tunnel FET with tunnel bandgap modulation and gate workfunction engineering", IEEE Trans. Electron Devices, vol. 52, pp. 909-917, 2005.

[7] S. Chander, B. Bhowmick and S. Baishya, "Heterojunction fully depleted SOI-TFET with oxide/source overlap", Superlattices and Microstructures, vol. 86, pp. 43-50, 2015.

[8] G. Ghibaudo and T. Boutchacha, "Electrical noise and RTS fluctuations in advanced CMOS devices", Microelectron. Rel., vol. 42, no. 4, pp. 573–582, 2002.

[9] M. J. Knitel, P. H. Woerlee, A. J. Scholten and A. Zegers-Van Duijnhoven, "Impact of process scaling on 1/f noise in advanced CMOS technologies", IEDM Tech. Dig., pp. 463–466, 2000.

[10] M. Agostinelli, J. Hicks, J. Xu, B. Woolery, K. Mistry, K. Zhang, *et al.*, "Erratic fluctuations of SRAM cache Vmin at the 90 nm process technology node", IEDM Tech. Dig., pp. 655–658, 2005.

[11] K. K. Hung, P. K. Ko, C. Hu and Y. C. Cheng, "A physics-based MOSFET noise model for circuit simulators", IEEE Trans. Electron Devices, vol. 37, no. 5, pp. 1323–1333, 1990.

[12] H. F. Teng, S. L. Jang and M. H. Juang, "A unified model for highfrequency current noise of MOSFETs", Solid-State Electron., vol. 47, no. 11, pp. 2043–2048, 2003.

[13] R. Pandey, B. Rajamohanan, H. Liu, V. Narayanan and S. Datta, "Electrical noise in heterojunction interband tunnel FETs", IEEE Trans. Electron Devices, vol. 61, pp. 52–560, 2014.

[14] R. Bijesh, D. K. Mohata, H. Liu and S. Datta, "Flicker noise characterization and analytical modeling of homo and hetero-junction III–V tunnel FETs", Device Res. Conf. Dig., pp. 203–204, 2012.

[15] L. Zhang and M. Chan, "SPICE modeling of double-gate tunnel-FETs including channel transports", IEEE Transactions on Electron Devices, vol. 61, no. 2, pp. 300-307, 2014.

[16] M. R. Tripathy, A. K. Singh, A. Samad, S. Chander, K. Baral, P. K. Singh and S. Jit, "Device and circuit-level

assessment of GaSb/Si Heterojunction vertical tunnel-FET for low-power applications", IEEE Transactions on Electron Devices, vol. 67, no. 3, pp. 1285-1292, 2020.

[17] R. P. Jindal, "Hot-electron effects on channel thermal noise in fine-line NMOS field-effect transistors. IEEE Transactions on Electron Devices, vol. 33, no. 9, pp. 1395-1397, 1986.

[18] D. P. Triantis, A. N. Birbas and D. Kondis, "Thermal noise modeling for short-channel MOSFETs", IEEE Transactions on Electron Devices, vol. 43, no. 11, pp. 1950-1955, 1996.

[19] M. A. Chalkiadaki and C. C. Enz, "RF small-signal and noise modeling including parameter extraction of nanoscale MOSFET from weak to strong inversion", IEEE Transactions on Microwave Theory and Techniques, vol. 63, no. 7, pp. 2173-2184, 2015.

[20] A. G. Mahmutoglu and A. Demir, "Analysis of low-frequency noise in switched MOSFET circuits: Revisited and clarified", IEEE Transactions on Circuits and Systems I: Regular Papers, vol. 62, no. 4, pp. 929-937, 2015.

[21] H. Tian and E. L. Gamal, "Analysis of 1/f noise in switched MOSFET circuits", IEEE Transactions on Circuits and Systems II: Analog and Digital Signal Processing, vol. 48, no. 2, pp. 151-157, 2001

[22] C. Hu, G. P. Li, E. Worley, and J. White, "Consideration of low-frequency noise in MOSFETs for analog performance", IEEE Electron Device Letters, vol. 17, no. 12, pp. 552-554, 1996.

[23] R. P. Jindal, "Compact noise models for MOSFETs", IEEE Transactions on Electron Devices, vol. 53, no. 9, pp. 2051-2061, 2006.

[24] K. K. Hung, P. K. Ko, C. Hu, and Y. C. Cheng, "Random telegraph noise of

deep-submicrometer MOSFETs," IEEE Electron Device Lett., vol. 11, no. 2, pp. 90–92, 1990.

[25] T. G. M. Kleinpenning, "1/f noise and random telegraph noise in very small electronic devices", Physica B, vol. 164, pp. 331–334, 1990.

[26] A. Konczakowska and B. M. Wilamowski, "Noise in semiconductor devices", Fundamentals of Industrial ElectronicsCRC Press, pp. 11-11, 2018.

[27] R. P. Jindal, "Gigahertz-band high-gain low-noise AGC amplifiers in fine-line NMOS", IEEE Journal of Solid-State Circuits, vol. 22, no. 4, pp. 512-521, 1987.

[28] J. Tiemann, "Shot noise in tunnel diode amplifiers," Proc. IRE, vol. 48, no. 8, pp. 1418–1423, 1960.

[29] B. E. Turner, "Noise in the tunnel diode," Ph.D. dissertation, Dept. Electr. Current Rectifiers, Diodes, Electron., Univ. British Columbia, Vancouver, BC, Canada, 1962.

[30] F. Wu, T. Tsuneta, R. Tarkiainen, D. Gunnarsson, T. H. Wang, P. J. Hakonen, "Shot noise of a multiwalled carbon nanotube field effect transistor", *Physical Review B*, vol. 75, no. 12, pp. 125419, 2007.

[31] N. E. Flowers-Jacobs, A. Pollarolo, K. J. Coakley, A. E. Fox, H. Rogalla, W. L. Tew and S. P. Bez, "A Boltzmann constant determination based on Johnson noise thermometry," *Metrologia*, vol. 54, no. 5, pp. 730, 2017.

[32] M. L. Fan, V. P. H. Hu, Y. N. Chen, P. Su and C. T. Chuang, "Analysis of single-trap-induced random telegraph noise and its interaction with work function variation for tunnel FET", IEEE transactions on electron devices, vol. 60, no. 6, pp. 2038-2044, 2013.

[33] F. Jazaeri and J. M. Sallese, "Modeling channel thermal noise and

induced gate noise in junctionless FETs", IEEE Transactions on Electron Devices, vol. 62, no. 8, pp. 2593-2597, 2015.

[34] F. S. Neves, P. G. Agopian, J. A. Martino, B. Cretu, R. Rooyackers, A. Vandooren, E. Simoen, A.V.Y. Thean and C. Claeys, "Low-frequency noise analysis and modeling in vertical tunnel FETs with Ge source", IEEE Transactions on Electron Devices, vol. 63, no. 4, pp. 1658-1665, 2016.

[35] C. Chen, Q. Huang, J. Zhu, Y. Zhao, L. Guo and R. Huang, "New understanding of random telegraph noise amplitude in tunnel FETs", IEEE Transactions on Electron Devices, vol. 64, no. 8, pp. 3324-3330, 2017.

[36] M. W. Pospieszalski, "On the limits of noise performance of field effect transistors", International Microwave Symposium IEEE, pp. 1953-1956, 2017.

[37] P. Ghosh and B. Bhowmick, "Low-frequency noise analysis of heterojunction SELBOX TFET", Applied Physics A, vol. 124, no. 12, pp. 1-9, 2018.

[38] S. Y. Kim, H. S. Song, S. K. Kwon, D. H. Lim, C. H. Choi, G. W. Lee and H. D. Lee, "Gate Voltage Dependence of Low Frequency Noise in Tunneling Field Effect Transistors", Journal of nanoscience and nanotechnology, vol. 19, no. 10, pp.6083-6086, 2019.

[39] P. Ghosh and B. Bhowmick, "Deep insight into material-dependent DC performance of Fe DS-SBTFET and its noise analysis in the presence of interface traps", AEU-International Journal of Electronics and Communications, vol. 117, pp. 153124, 2020.

[40] N. Patel, A. Ramesha and S. Mahapatra, "Drive current boosting of n-type tunnel FET with strained SiGe layer at source", Microelectronics Journal, vol. 39, no. 12, pp. 1671-1677, 2008.

[41] W. Li and J. C. Woo, "Vertical P-TFET with a P-type SiGe pocket", IEEE Transactions on Electron Devices, vol. 67, no. 4, pp. 1480-1484, 2020.

[42] A. Vandooren, D. Leonelli, R. Rooyackers, A. Hikavyy, K. Devriendt, M. Demand, R. Loo, G. Groeseneken and C. Huyghebaert, "Analysis of trap-assisted tunneling in vertical Si homo-junction and SiGe hetero-junction tunnel-FETs", Solid-State Electronics, vol. 83, pp. 50-55, 2013.

[43] M. Sathishkumar, T. A. Samuel and P. Vimala, "A Detailed Review on Heterojunction Tunnel Field Effect Transistors", International Conference on Emerging Trends in Information Technology and Engineering IEEE, pp. 1-5, 2020.

[44] D. Das and U. Chakraborty, "A Study on Dual Dielectric Pocket Heterojunction SOI Tunnel FET Performance and Flicker Noise Analysis in Presence of Interface Traps", Silicon, vol. 13, no. 3, pp. 1-12, 2021.

[45] R. Goswami, B. Bhowmick and S. Baishya, "Electrical noise in circular gate tunnel FET in presence of interface traps", Superlattices and Microstructures, vol. 86, pp. 342-354, 2015.

[46] R. Bijesh, D. K. Mohata, H. Liu and S. Datta, "Flicker noise characterization and analytical modeling of homo and hetero-junction III–V tunnel FETs", Device Research Conference IEEE, pp. 203-204, 2012.

[47] S. T. Bu, D. M. Huang, G. F. Jiao, H. Y. Yu and M. F. Li, "Low frequency noise in tunneling field effect transistors". Solid-State Electronics, vol. 137, pp. 95-101, 2017.

[48] M. L. Fan, S. Y. Yang, V. P. H. Hu, Y. N. Chen, P. Su and C. T. Chuang, "Single-trap-induced random telegraph noise for FinFET, Si/Ge Nanowire FET,

Tunnel FET, SRAM and logic circuits",
Microelectronics Reliability, vol. 54, no.
4, pp. 698-711, 2014.

[49] S. H. Fani, A. Peiravi, H. Farkhani
and F. Moradi, "A novel TFET 8T-
SRAM cell with improved noise margin
and stability", *IEEE 21st International
Symposium on Design and Diagnostics of
Electronic Circuits & Systems,* pp. 39-44,
2018.

[50] M. L. Fan, V. P. H. Hu, Y. N. Chen,
P. Su and C. T. Chuang, "Investigation
of single-trap-induced random
telegraph noise for tunnel FET based
devices, 8T SRAM cell, and sense
amplifiers", *IEEE International
Reliability Physics Symposium,* pp. CR-1,
2013.

[51] G.V. Luong, S. Strangio, A. T.
Tiedemann, P. Bernardy, S.
Trellenkamp, P. Palestri, S. Mantl and
Q. T. Zhao, "Experimental
characterization of the static noise
margins of strained silicon
complementary tunnel-FET SRAM",
European Solid-State Device Research
Conference IEEE, pp. 42-45, 2017.

[52] R. Pandey, V. Saripalli, J. P.
Kulkarni, V. Narayanan, S. Datta,
"Impact of single trap random telegraph
noise on heterojunction TFET SRAM
stability", IEEE Electron Device Letters,
vol. 35, no. 3, pp. 393-395, 2014.

Section 2

Plasmonic 2D Materials and Functional Applications

Chapter 3

Plasmonic 2D Materials: Overview, Advancements, Future Prospects and Functional Applications

Muhammad Aamir Iqbal, Maria Malik, Wajeehah Shahid, Waqas Ahmad, Kossi A. A. Min-Dianey and Phuong V. Pham

Abstract

Plasmonics is a technologically advanced term in condensed matter physics that describes surface plasmon resonance where surface plasmons are collective electron oscillations confined at the dielectric-metal interface and these collective excitations exhibit profound plasmonic properties in conjunction with light interaction. Surface plasmons are based on nanomaterials and their structures; therefore, semiconductors, metals, and two-dimensional (2D) nanomaterials exhibit distinct plasmonic effects due to unique confinements. Recent technical breakthroughs in characterization and material manufacturing of two-dimensional ultra-thin materials have piqued the interest of the materials industry because of their extraordinary plasmonic enhanced characteristics. The 2D plasmonic materials have great potential for photonic and optoelectronic device applications owing to their ultra-thin and strong light-emission characteristics, such as; photovoltaics, transparent electrodes, and photodetectors. Also, the light-driven reactions of 2D plasmonic materials are environmentally benign and climate-friendly for future energy generations which makes them extremely appealing for energy applications. This chapter is aimed to cover recent advances in plasmonic 2D materials (graphene, graphene oxides, hexagonal boron nitride, pnictogens, MXenes, metal oxides, and non-metals) as well as their potential for applied applications, and is divided into several sections to elaborate recent theoretical and experimental developments along with potential in photonics and energy storage industries.

Keywords: graphene, metal oxides, pnictogens, hBN, MXenes, non-metal plasmonics, photonics

1. Introduction

Plasmonics is the emerging research field, indicating the ability of materials to control light at nanoscale range to examine them for various properties and functions. The plasmonic materials exploit the surface plasmon resonance effects to achieve astonishing optical properties that originate with light-matter interaction and leads to remarkable results. Surface plasmon can confine electromagnetic fields at very small scales whereas various structures can be employed to control surface plasmons. Previously, Ag, Au, and Al metals were used as plasmonic materials but they did not perform well because of radiative losses, high amount of energy dissipation, and their poor tuneability. To overcome these problems for efficient plasmonic applications, a class of two-dimensional (2D) materials is proposed which presents a significant light-matter interaction phenomenon resulting in efficient quantum confinement effects. A variety of materials including semiconductors, conductive oxides, and dielectric materials have been investigated as plasmonic materials owing to their extra-ordinary plasmonic properties. Considering the advanced properties along with bandgap manipulation and electron transfer, 2D materials got higher attention for plasmonic applications [1, 2].

Graphene was the first 2D material investigated with zero bandgap having exceptional conductivity because of its high electron mobility. Considering graphene's achievements and enormous applications at the laboratory and industry level, researchers have started investigating further 2D materials to explore their potential for plasmonic applications. Currently, almost 150 members of the 2D materials family are serving in elementary and advanced technologies such as light-emitting diodes (LEDs), Field-effect transistors (FETs), environmental applications, sensing applications, and physical catalysis [3–5]. Some important under discussion members of 2D materials, analogous to graphene are; hexagonal boron nitride (hBN), black phosphorene, metal oxides, metal carbides and nitrides (MXenes), metal halides, pnictogens, and non-metals which are being considered as potential plasmonic materials [6–8]. This 2D materials family exhibits a broad electronic and plasmonic characteristic spectrum covering a wide range of properties such as; high surface area, surface state nature, minimum dangling bonds, spin-orbit coupling, and quantum spin Hall effects [9, 10].

On the other hand, stacking of different 2D materials is also the emerging part of the material industry which yields novel heterostructure materials capable of introducing some building blocks in a materials family with enhanced physical and chemical properties. The novel 2D materials such as metal carbides and nitrides, metal oxides and graphene-based materials have mixed properties and can be further tuned by adjusting bandgap that would result in increased light-harvesting efficiency which is the basis to achieve desired optical, electronic, and optoelectronic properties, making them promising materials for plasmonic applications [11, 12]. In addition, the plasmonic efficiency of 2D materials can also be enhanced by injecting plasmonic hot electrons to alter carrier intensity in 2D materials for higher photocatalysis output [13]. The recent extension of plasmonic materials from traditional metals to semiconductors to semi-metal graphene are identified as an ideal materials for surface plasmon resonance in plasmonic structures and their subsequent applications needed to be addressed accordingly. Moreover, the coupling effects between excitons and plasmons for 2D materials are the growing research interests that profound further studies for light-matter interactions to discover novel materials for innovative device applications.

2. Overview of 2D materials

The radiation-matter interaction is more prominent in 2D materials because of their thin sheet structures and significant quantum confinement effects that lead to enhanced electronic and optical properties. Owing to their advanced nature, 2D materials are advantageously evaluated for plasmonic characteristics, and multiple studies have been conducted for hBN to investigate plasmon molecular vibration coupling, plasmon substrate phonon coupling, and graphene plasmon-phonon polaritons coupling [14–16]. The 2D graphene structure exhibits exciting results due to its single-atom thickness and their environmental sensitivity. Other than environmental sensitivity, graphene plasmons can also be tuned with external magnetic and electric fields [17]. The effectiveness of graphene-based plasmonics can be determined by charge carrier density, and heavily doped graphene exhibits high efficiency which is required for plasmonic applications [18]. As a result, graphene is an excellent plasmonic material, and combining graphene with other 2D materials is favorable to obtain optimum efficiency [19].

Hexagonal Boron Nitride (hBN) is one of the most intriguing findings in 2D materials for plasmonics, having the unique ability to be fabricated within the host material, and can be used as a promising substrate for graphene-based plasmonic applications because of its graphene-matched crystal structure [18, 20]. The hBN-graphene mixture is helpful to enhance grapheme-plasmon lifetime when compared with other 2D materials and can maintain its bandgap even in varied thicknesses depicting a wide range of plasmonic properties including electro-optic and quantum-optics [21, 22]. Moreover, the point defects in hBN at room temperature demonstrate single-photon emission properties that can be used to integrate plasmonic nanostructures. Despite its wide bandgap, hBN offers high quantum efficiency, optical nonlinearity, and novel plasmonic properties to make it the best choice as 2D plasmonic material [23]. Its structure is shown in **Figure 1** [24].

The MXenes are a new class of 2D materials that contain carbides and nitrides, and they are the biggest family currently available, as seen in **Figure 2** [25]. This family of materials is substantially more stable than graphene with high metallic conductivity, folding and molding properties, and good electromagnetic properties, possessing the unique property of being combined with other materials to tune

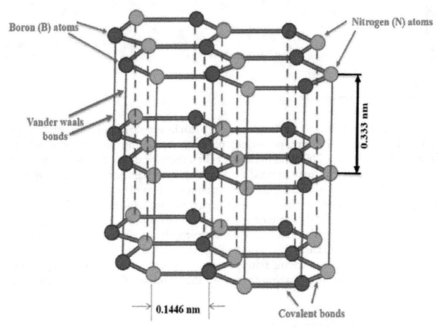

Figure 1.
Schematics of hBN structure [24].

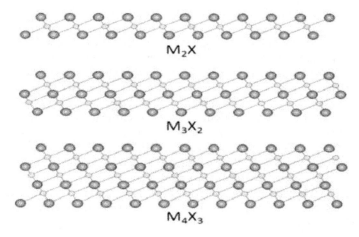

Figure 2.
MXene sheets with multiple layers where larger spheres represent transient metal, while smaller spheres are C, N, or CN [25].

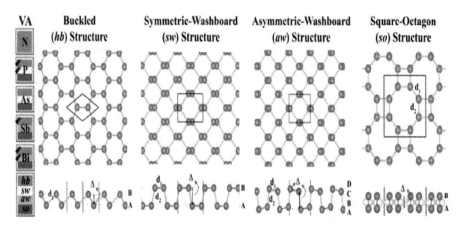

Figure 3.
Pnictogens and their schematic 2D structures [27].

their properties for desired applications. They can be employed in a variety of applications including energy storage devices, photonic-plasmonic structures as well as photocatalytic devices [26], and because of their metallic character along with high conductive nature, they may be used as plasmonic materials equivalent to graphene.

Researchers anticipated the group VA elements such as nitrogen, arsenic, antimony as well as bismuth as single-layer 2D structures with the introduction of 2D materials synthesis, and these elements are referred to as pnictogens and are shown in **Figure 3** [27]. These materials feature a honeycomb, washboard, and square-octagon structure, and they offer outstanding electrical, optical, electro-optical, and plasmonic properties having strong spin-orbit coupling, a narrow bandgap, and band inversion properties, making them ideal for plasmonic device applications [27].

3. Theoretical advancements

Photonics deals with the light-matter interaction which usually results in the formation of a single electron–hole pair by interacting light photons with free charge carriers in a metal, whereas in plasmonics, there is a large number of charge Carriers present that leads to collective oscillations which is the fundamental problem in plasmonics because all charge carriers are not part of the solid and can be influenced by structural defects as well as other materials defects such as dislocations. As a result, multi-scale modeling at various structural complexity levels is required for theoretical exploration of these complex models and plasmonic excitations in bulk materials and localized plasmons in metallic structures. To analyze this complicated issue, several theoretical and numerical models have been presented, although only a few of them are described here.

The Drude-Lorentz model which gives a theoretical insight into a material and can be employed in plasmonic applications is an intuitive way to study the underlying dielectric characteristics of solids [28, 29]. The Drude-Lorentz model, also known as the oscillator model, entails representing an electron as a driven damped harmonic oscillator in which the electron is connected to the nucleus by a hypothetical spring with an oscillating electric field acting as a driving force. It also describes the behavior of electrons in terms of their electro-optical characteristics when light interacts with them [30]. The Drude-Lorentz model's predictions are completely supported by the classical oscillator model as well as quantum mechanical features like electronic dipole moments of materials. To justify the microscopic qualities exhibited by classical and quantum techniques using this model, it is necessary to understand the Ehrenfest theorem which shows that quantum mechanical predicted values fundamentally follow classical mechanical conditions [31]. The Kohn-Sham approach is praised for its ease of use in relating a many-body system to a non-interacting system and thereby solving it using Kohn-Sham density functional theory [32]. **Figure 4** illustrates such a model [32].

With the developments in computing, new algorithms are being devised to accomplish difficult jobs rapidly and accurately. Different approaches and models for electrical and photonic systems are being explored to accurately anticipate their characteristics, which were previously explored using differential equations. This section discusses the frequency domain approach and the time domain method to have a better knowledge of computational model advancements [33].

The decomposition of periodic systems into harmonic time-dependent eigenmodes is a basic approach for understanding the optoelectronic and plasmonic characteristics of materials. The frequency-domain approach is a subset of these decompositions that enlarges electromagnetic fields into Fourier eigenmodes which may be used to comprehend optical material properties in the absence of nonlinear effects [34]. This approach is usually started from fundamental photonic systems

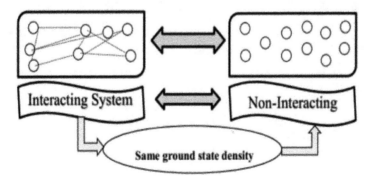

Figure 4.
Kohn-sham mapping of the interacting and non-interacting systems [32].

with translational symmetry which produces electromagnetic states and photonic band structures using Maxwell equations and wave Equations [35]. Although the frequency-domain technique is effective for defining material characteristics, but it is an expensive method that restricts its application in numerical models and as a result, the finite-difference time-domain method was presented as an alternative. The time-domain technique is a grid-based method that is linked to several other finite methods. This approach models electromagnetic wave propagation in dielectric media without needing derivation methods, making it easier to be utilized in complicated geometrical simulations, such as non-linear systems, which were previously difficult to manage using the frequency-domain method. Furthermore, in this approach, Maxwell equations are discretized by differences arising from spatial and time derivatives, and the obtained results are solved in a leapfrog fashion on a staggered grid which is a good method being utilized in fluid dynamics [36]. Using the Phyton modules, these simulations can also be used to determine the plasmonic characteristics of materials [37].

The plasmonic material's behavior can be determined by studying the frequency-dependent dielectric factor linked with the excited state of the material. It is essential to analyze both the ground and excited states of material while calculating optical transitions based on material states. For the analysis of material characteristics based on these facts, *ab initio* methodologies like density functional theory (DFT), Hartree-Fock theory, and Green's function method [38–40] are essential. DFT [32] is founded on the concept that, for a quantum mechanical system, ground state charge density provides a complete comprehension of the system's ground state because the charge density of the state is mapped to the total energy of the system. The degree of freedom and functional complexity would be reduced if energy density is properly approximated. **Figure 5** [41] shows the bandgap difference between graphene and hBN, which seems comparable but differs significantly at K-vector space as predicted within DFT (PBE) approximation computed by Warmbier et al. [41]. Green's function method is a quasi-particle approach for

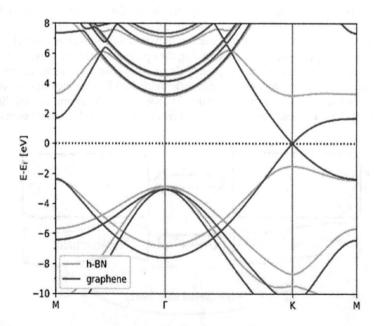

Figure 5.
Band structure of graphene and hBN computed within DFT (PBE) [41].

improving bandgap findings and reproducing band structures, with Hedin's GW as the most frequent implementation. All of these approaches hold promise for studying plasmonic material characteristics and predicting specific plasmonic device applications.

4. Experimental progress in the synthesis of 2D materials

The material characteristics such as its dimension, morphology, physical as well as chemical properties, orientation, and crystallinity mainly depend on the material's electronic properties which are impacted by the synthesis technique and experimental conditions. Mechanical exfoliation techniques have been tried in the past, but they have failed due to insufficient van-der-wall forces between 2D material layers, which limit uniformity and quality control, as well as their inability to scale-up [42]. Physical and chemical synthesis methods with controlled structural fabrication can be used with the top-down approach having the disadvantage of poor product yield and sheet restacking, limiting its application, and the bottom-up approach yielding promising results by assembling materials in a substrate using vapor deposition techniques such as physical vapor deposition (PVD), chemical vapor deposition (CVD), and atomic layer deposition. These are the most often utilized promising ways for fabricating 2D materials with customized thickness, dimensional control, high conductivity, and flexibility for electron transport; all of which are highly sought quantities for plasmonic applications [43].

CVD allows for the controlled synthesis of large areas of 2D materials with the added benefit of step-by-step film synthesis on various substrates and adjustable growth parameters to get the desired output. Metal–organic chemical vapor deposition (MOCVD) is a modified version of CVD that is used to synthesize high-quality, large-area 2D materials for a variety of applications [44]. New research shows that a metal gas-phase precursor might be employed for regulated and uniform thickness instead of a powder precursor, which results in inhomogeneous nucleation and hence uncontrolled synthesis. Furthermore, temperature and pressure have an important impact on deposition uniformity, for example, high temperature and moderate pressure would result in excellent precursor coverage with regulated dimensions while excessively high temperatures might have negative consequences [45]. A comparison of various synthesis techniques is shown in **Figure 6** [46].

Many techniques are available for producing uniform 2D materials and heterostructures with atomic layer deposition (ALD) being a refinement of the vapor-based deposition method in which the self-limiting reaction of the precursor is an essential aspect of ALD and the fact that a self-saturating surface monolayer is created after each precursor exposure distinguishes it from other deposition processes. In addition, ALD allows for the creation of 2D materials with fewer flaws and the synthesis of 2D material heterostructures with a small atom size thickness [47]. As a result of the advancements, ALD has opened up new means of synthesis with decreased interfacial impurities and large area deposition conformity with improved structural properties [48].

Apart from vapor-based synthesis approaches, liquid-phase exfoliation is another good way to get 2D materials in which the surface oxide of liquid metals produces unexpected results from a combination of physical and chemical features of liquid alloys [49]. While interacting with their ambient conditions, liquid metals with electron-rich metallic cores form a natural 2D film; the self-limiting surface oxide film with a thickness of a few atoms [50]. These liquid ingredients serve as host materials for the production of high-quality, one-kind films for innovative applications. The origin to define its development characteristics is the host

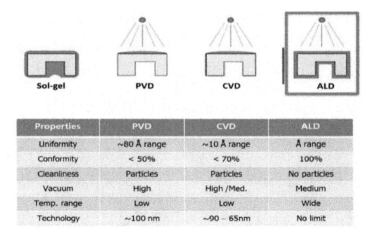

Properties	PVD	CVD	ALD
Uniformity	~80 Å range	~10 Å range	Å range
Conformity	< 50%	< 70%	100%
Cleanliness	Particles	Particles	No particles
Vacuum	High	High /Med.	Medium
Temp. range	Low	Low	Wide
Technology	~100 nm	~90 – 65nm	No limit

Figure 6.
Comparison of various 2D materials synthesis techniques [46].

materials fluidity, chemical composition, and thermodynamic properties that are the building block for examining the resultant 2D materials [51].

5. Plasmonic materials

Graphene, graphene oxides (GO), MXenes, pnictogens, and hBN are just a few of the commonly utilized plasmonic materials described in this section.

5.1 Graphene

Graphene is proving itself a revolutionary material for a wide range of applications since 2004 because of its electronic behavior which is responsible for exceptional features including, high mobility charge carriers, optical transmission, and tunable carrier densities [52–60]. The ability of graphene's structure to strongly confine excited surface plasmons in comparison to other materials as well as its ability to tune surface plasmons by manipulating charge densities is remarkable for prospective applications primarily in optoelectronics and plasmonics [61]. Experimental studies [62] show that graphene surface plasmons may be coupled with electrons and photons, allowing them to be used in more promising applications. This graphene coupling is in the form of quasi-particles that hold an intense interest in optoelectronics and condensed matter physics [63]. Surface plasmons in graphene offer a variety of significant advantages over other plasmonic materials, including high confinement, high tuneability, reduced frequency loss, improved electron relaxation time, and high many-body interactions. **Figure 7** shows the structure and band description of graphene [64].

5.2 Graphene oxides (GO)

Graphene oxide (GO) is an amorphous insulator with carbon network bases on the hexagonal rings with both sp^2 and sp^3 hybridization as well as hydroxyl and epoxide groups on sheet sides and carboxyl and carbonyl groups on sheet edges. This kind of morphological structure is responsible for its wide range of technological applications mainly in nano-electronics, nano-photonics, and

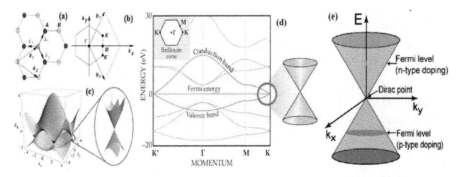

Figure 7.
(a) Graphene lattice structure, (b) BZ of graphene, (c) lattice electronic dispersion, (d) graphene electronic structure Bloch band description, (e) Low EBS approximation [64].

nano-composites. Because of the oxygen functionalities present, GO has a significant benefit when mixed with other polymeric or ceramic materials, resulting in improved electrical and mechanical properties [65]. The GO is in high demand in both industry and academics because of its zero-bandgap feature and excellent flexibility with superior thermal and electrical conductivity. In addition, one of GO's unique properties is the coexistence of its size, shape, and hybridization domains via a reduction mechanism that may eventually control the bandgap and convert GO to a semi-metal form. The main distinction between GO and reduced graphene oxide (r-GO) is that GO has oxygen-containing functional groups, whereas reduced graphene oxide does not [66]. Owing to heterogeneous electrical structure, GO is fluorescent throughout a wide wavelength range, while reduced GO allows quick response in GO-based electronics [67]. The GO is a promising choice for novel photonic materials, solar cells, optical devices, and a range of other applications due to its unique features. **Figure 8** depicts the difference between GO and r-GO [66].

5.3 Hexagonal boron nitride (hBN)

The hBN is a traditional 2D heterostructure material that was previously used as a substrate for thin-layered materials but now has the potential to be employed as an active plasmonic material. It is an excellent encapsulant for graphene because it protects it from the environment and increases its electrical mobility, extending the

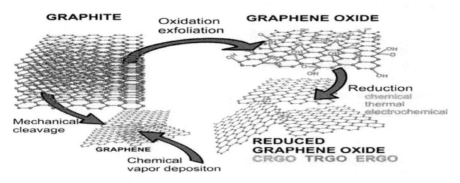

Figure 8.
Schematic flow from graphene-to-graphene oxide (GO) and reduced graphene oxide (r-GO) [66].

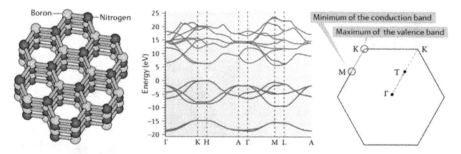

Figure 9.
hBN and its band structure properties [70].

life of surface plasmon polaritons in plasmonic applications [18, 68]. It has a crystal nature and polar bonding, allowing it to perform a wide range of optical, electrical, optoelectronic, and quantum optic functions for device applications. Also, it has a wide bandgap, high internal quantum efficiency, and significant optical nonlinearities that depend on material thickness and is specified by the rotation angle between heterostructure material layers [23, 69]. Its structure and band properties are displayed in **Figure 9** [70].

5.4 MXenes

MXenes are new types of 2D plasmonic materials made up of nitrides and carbides that were discovered in 2011. They are more stable than graphene, can be readily shaped and folded, have superior electromagnetic properties, and can be coupled with other materials to exhibit a wide range of applications in energy storage, supercapacitors, photonics, and plasmonics. The properties of MXenes can be determined by exploring surface termination, composition, doping, or mixing with other materials, resulting in adjustable conductivity that can change the material's properties as a metal or semiconductor [71]. The relative dielectric permittivity of MXenes for plasmonic applications can be studied by inter- and intra-band transitions that define optical parameters such as, absorption coefficient, refractive index, transmittance, and reflectance, and are linked to the material's electrical conductivity, which has already been demonstrated computationally and experimentally to find a place in electronic and optoelectronic applications [72, 73].

5.5 Pnictogens

Pnictogens are monolayer stable structures found from elements of group VA (nitrogen, arsenic, antimony, and bismuth) after the discovery of black phosphorene. These materials are named as nitrogen in hexagonal buckled structure, arsenic in hexagonal buckled as well as symmetric washboard structure, antimony, and bismuth in either hexagonal buckled or asymmetric washboard structure, but later on, these elements occurred to have other exotic structures [74]. The pnictogens stability can be depicted using molecular dynamic simulations performed at high temperatures and materials phonon frequencies [75]. Pnictogens, as contrasted with group IV elements, are significantly more stable semiconductor materials with an appropriate bandgap for numerous device applications. Also, in contrast to black phosphorus (BP), they are thermodynamically stable monolayer structures with rhombohedral structural

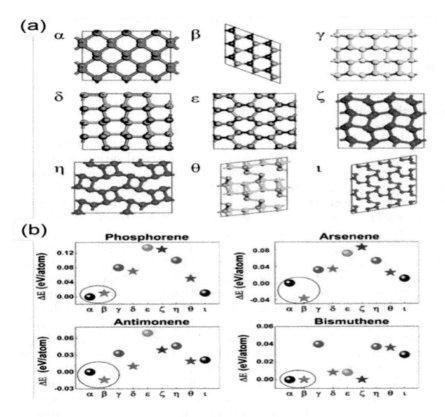

Figure 10.
(a) Honeycomb structures, (b) average binding energies of 2D pnictogens [77].

characteristics and interlayer covalent connections which decrease as anisotropy decreases and metallic character increases from arsenic to antimony to bismuth [76]. The 2D monolayer structures of pnictogens exhibit strong directionality in various physical properties that can be implied on plane lateral heterostructures to produce parallel strips of 2D pnictogens with advanced technological applications while the same effect can be observed by a monoatomic chain of group VA elements attached to their monolayers. **Figure 10** represents structures as well as binding energies of pnictogens [77].

5.6 Metal oxides

Metal oxides show strong metallic behavior owing to stable charge carrier concentration when doped with various significant dopants such as aliovalent, oxygen vacancies, or interstitial dopants that result in localized surface plasmon resonance and by carefully choosing the host material as well as doped material, these surface plasmonic resonances can be tuned in the range of near- and mid-infrared (IR) region spectrum. The optical modeling of metal oxides illustrates the importance of defects and their impact on charge carrier mobility and the electronic structure of the material which reveals the choice of dopant as an important factor for metal oxides as plasmonic materials. Metal oxides are different from ordinary metals in the sense that they may change their localized surface plasmon resonance by changing their elemental composition, regardless of material size or shape, and these plasmon resonances can also be adjusted by altering external stimuli, resulting in the unique features of plasmonic materials as a result of crystal and morphological configurations that are useful for a variety of device applications [78].

6. Functional applications

The 2D materials can generally be categorized on the basis of electrical and optoelectronic properties in device applications such as flash memories, sensors, tunnel junctions, photodetectors, photonic crystals, optical metamaterials, nanophotonics, and quantum optics. The 2D graphene-based photonic and optoelectronic devices dragged much attention because of their versatile applications in broad fields such as sensing, communication, and imaging technologies [79]. In terms of density of state and band structure, graphene has an adjustable light absorption spectrum and carrier density which may be used in waveguide-integrated graphene photonic devices and molecular sensor detection. Graphene with optical adjustments has also been used for light modulation and detection, and its derivatives are proving to be a feasible alternative for a variety of applications. The GO can be utilized in the manufacture of electrical devices such as FETs and GFETs, LEDs, and solar cells, while r-GO dispersed in solvents may be utilized to replace FTO and ITO electrodes in transparent electrode manufacture; moreover, their large surface area and conductivity allow them to function as energy storage devices for longer periods with a greater capacity [80]. Recent investigations have shown that near-field IR optical microscopy and IR microscopy of graphene are responsible for surface plasmon modeling in plasmonic applications [81]. An overview of 2D plasmonic materials-based devices is shown in **Figure 11** [82].

The hBN has exciting technological applications including, photonics, and its nanostructures feature weak polaritons that interact poorly with light and might be utilized to control the optical angular momentum of hyperbolic phonon polaritons, implying maximum optical density of state and/or improving molecular IR vibrational absorption through surface enhancement [83]. Hybridization of hBN phonon-polariton with graphene surface plasmon-polaritons resulted in the active

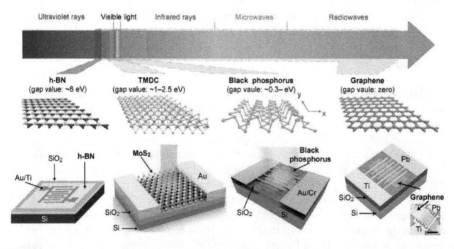

Figure 11.
2D plasmonic material-based devices for optoelectronic applications [82].

tuning of the polaritons which is a potential characteristic for chip-based nano-photonics [84], photonic devices, modulators, and hyper-lensing [85]. The hBN defects that reduce phonon lifespan have uses in single-photon emitters (SPEs), which have certain appealing properties such as high quantum efficiency, optical stability, linear polarization, and high brightness. In addition, owing to its high efficiency and extended life in device applications, hBN has been utilized to replace AlGaN in deep-UV applications. Another method for incorporation of hBN is to link emission with plasmonic resonator-based structures in which localized surface plasmons cause broad field confinements throughout a wide range of emission, resulting in substantial Purcell amplification for dipole emission coupled to these resonators. When compared to uncoupled devices, the hBN quantum emitter coupling with plasmonic arrays has previously been demonstrated, with studies revealing PL enhancement and lifespan reduction with a quantum efficiency of around 40% and enhanced saturation count rates [86].

MXenes plasmonics is a relatively new field with a wide range of possible applications, including surface-enhanced Raman spectroscopy, conductive substrates, and plasmonic sensing [72]. Nonlinear optical applications based on the nonlinear absorption process by plasmonic illumination near plasma frequency have been suggested for MXenes, and these nonlinear applications include ultrafast lasers, optical switching, and optical rectification devices like optical diodes. At near-infrared frequencies, arrays of two-dimensional titanium carbide ($Ti_3C_2T_x$) MXene nanodisks exhibit strongly localized surface plasmon resonances, which have been exploited to produce broadband plasmonic metamaterial absorber [87]. MXenes are also used as super-absorbers in broadband plasmonic metamaterials, and these super-absorbers may be used for photodetection and energy harvesting. **Figure 12** shows an MXenes super-absorber [88] with configurable nano-aperture width for broadband applications.

Pnictogens with a 2D structure are in great demand for high-performance device applications, since they have a midrange tunable bandgap and unparalleled mobility, allowing them to be employed in FETs for more efficient response than other materials. Because 2D materials lack a suitable bandgap, photodetectors are a major challenge, but pnictogens direct and tunable bandgap has solved this problem acting as a bridge between narrow and wider bandgap materials, attracting a lot of interest in photodetectors with improved photo-responsivity

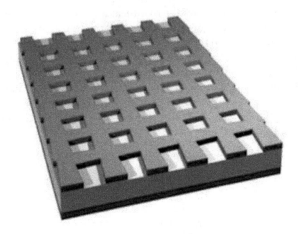

Figure 12.
MXene broadband super-absorber [88].

for telecommunication applications. The BP has an intriguing direct bandgap from visible to IR, making it a potential material for optoelectronic applications [89], while ultra-thin BP FETs have been described as effective NO_2 gas sensors with remarkable stability of pnictogens-based FETs during sensing [90, 91]. The 2D metals and metal oxide semiconductors, whether conducting or insulating, are useful for thin-film transistors and numerous device applications, where they can be employed in any component such as the source, drain, gate, electrodes, or gate dielectrics. Metal oxides are also useful in p-n junction device fabrications for diode rectifiers, solar cells, and organic photovoltaic applications, where they operate as a charge extracting interfacial layer to improve power conversion efficiency.

7. Future challenges and prospects

Plasmonics has advanced to the forefront of science due to technical advances in the experimental and computational fields as well as contributions to scientific applications. These contributions also confront some challenges that must be addressed in the future for effective plasmonic applications. To begin with, plasmonic nanostructures of controlled size and features cannot simply embed in their surroundings because they change the dielectric function of the surrounding medium, affecting plasmonic switching and hence plasmonic applications. Another problem is optical pumping, which has the potential to deliver ultrafast plasmonic switching but has the drawbacks of destructive heat accumulation and high-power consumption. A major shortcoming of plasmonic materials is that self-tuned plasmonic structures lack effective plasmon coupling control abilities. Also, it is difficult to fabricate colloidal metal nano-crystals in controlled symmetry for plasmonic device applications on a large scale, even though lithography techniques performed well but had some drawbacks such as high cost, long-time consumption, and difficulty with damped plasmonic properties on a large scale [92].

Plasmonics must control light at the nanoscale with minimal losses, and to do so, light localization must be pushed to new heights without jeopardizing its propagation nature. Similarly, advances in topological plasmons must be incorporated in nanophotonic circuits by maintaining plasmon propagation stability and improving manufacturing techniques. For the experimental process to be effective, theoretical models must be improved to acquire the nonlinear and nonlocal physics of plasmonic devices. In short, both light and matter quantization are required to make a fine path toward a better understanding of light-matter interactions for advanced large-scale applications. The numerical approaches outlined are strong tools in terms of computing but they have conceptual limitations and their validity range becomes inefficient when a heterogeneous system is studied. To tackle plasmonic multi-scale challenges, the validity of numerical models must be improved by combining them with other numerical tools which is not well understood at this time and requires future considerations.

8. Conclusions

We have briefly addressed 2D plasmonic materials and their active properties in this chapter that are responsible for their wide range of applications in the electrical, photonic, and optoelectronic fields such as, FETs and GFETs, LEDs, and solar cells, modulators, hyper lensing, metamaterial absorbers, super-absorbers as well as nonlinear applications including ultrafast lasers, optical switching, and optical rectification devices like optical diodes. The synthesis techniques employed for 2D plasmonic materials have also been reviewed, with pulsed laser deposition (PLD) and CVD being the most extensively used and promising approaches for more controlled and conformational film growth. Also, these techniques have the advantage to provide desirable results by tuning their functional parameters such as temperature, pressure, substrate angles, and deposition time. Computational models have to be examined to carry out a successful experiment, and there is a need to update simulation approaches to address problems in achieving desired plasmonic device features. Finally, we have outlined new prospective applications of 2D plasmonic materials and their significance in the industry as well as the drawbacks of materials that prohibit them from performing properly while providing the possible directions for future research.

Conflict of interest

There are no conflicts of interest to declare.

Note

US spelling with serial comma.

Author details

Muhammad Aamir Iqbal[1], Maria Malik[2], Wajeehah Shahid[3], Waqas Ahmad[1,4], Kossi A. A. Min-Dianey[5,6] and Phuong V. Pham[6*]

1 School of Materials Science and Engineering, Zhejiang University, Hangzhou, China

2 Centre of Excellence in Solid State Physics, University of the Punjab, Lahore, Pakistan

3 Department of Physics, The University of Lahore, Lahore, Pakistan

4 Institute of Advanced Materials, Bahauddin Zakariya University, Multan, Pakistan

5 Département de Physique, Faculté Des Sciences (FDS), Université de Lomé, Lomé, Togo

6 Hangzhou Global Science and Technology Innovation Center, School of Micro-Nano Electronics, College of Information Science and Electronic Engineering, and Zhejiang University-University of Illinois at Urbana-Champaign Joint Institute (ZJU-UIUC), Zhejiang University, Hangzhou, China

*Address all correspondence to: phuongpham@zju.edu.cn; maamir@zju.edu.cn

IntechOpen

References

[1] Geim AK, Novoselov KS. The rise of graphene. In: Nanoscience and Technology: A Collection of Reviews from Nature Journals. Singapore: World Scientific; 2010. pp. 11-19

[2] Novoselov KS, Morozov SV, Mohinddin TMG, Ponomarenko LA, Elias DC, Yang R, et al. Electronic properties of graphene. Physica Status Solidi. 2007;**244**(11):4106-4111

[3] Late DJ, Huang YK, Liu B, Acharya J, Shirodkar SN, Luo J, et al. Sensing behavior of atomically thin-layered MoS$_2$ transistors. ACS Nano. 2013;**7**(6):4879-4891

[4] Zhang Y, Chang TR, Zhou B, Cui YT, Yan H, Liu Z, et al. Direct observation of the transition from indirect to direct bandgap in atomically thin epitaxial MoSe$_2$. Nature Nanotechnology. 2014;**9**(2):111-115

[5] Pospischil A, Furchi MM, Mueller T. Solar-energy conversion and light emission in an atomic monolayer p–n diode. Nature Nanotechnology. 2014;**9**(4):257-261

[6] Nair RR, Blake P, Grigorenko AN, Novoselov KS, Booth TJ, Stauber T, et al. Fine structure constant defines visual transparency of graphene. Science. 2008;**320**(5881):1308-1308

[7] Wrachtrup J. Single photons at room temperature. Nature Nanotechnology. 2016;**11**(1):7-8

[8] Liu Y, Weiss NO, Duan X, Cheng HC, Huang Y, Duan X. Van der Waals heterostructures and devices. Nature Reviews Materials. 2016;**1**(9):1-17

[9] Tang Q, Zhou Z. Graphene-analogous low-dimensional materials. Progress in Materials Science. 2013;**58**(8):1244-1315

[10] Butler SZ, Hollen SM, Cao L, Cui Y, Gupta JA, Gutiérrez HR, et al. Progress, challenges, and opportunities in two-dimensional materials beyond graphene. ACS Nano. 2013;**7**(4): 2898-2926

[11] Zu S, Bao Y, Fang Z. Planar plasmonic chiral nanostructures. Nanoscale. 2016;**8**(7):3900-3905

[12] Zhu H, Yi F, Cubukcu E. Plasmonic metamaterial absorber for broadband manipulation of mechanical resonances. Nature Photonics. 2016;**10**(11):709-714

[13] Cheng F, Johnson AD, Tsai Y, Su PH, Hu S, Ekerdt JG, et al. Enhanced photoluminescence of monolayer WS2 on Ag films and nanowire–WS$_2$–film composites. ACS Photonics. 2017;**4**(6):1421-1430

[14] Yang X, Kong XT, Bai B, Li Z, Hu H, Qiu X, et al. Substrate phonon-mediated plasmon hybridization in coplanar graphene nanostructures for broadband plasmonic circuits. Small. 2015;**11**(5):591-596

[15] Rodrigo D, Limaj O, Janner D, Etezadi D, De Abajo FJG, Pruneri V, et al. Mid-infrared plasmonic biosensing with graphene. Science. 2015;**349**(6244):165-168

[16] Brar VW, Jang MS, Sherrott M, Kim S, Lopez JJ, Kim LB, et al. Hybrid surface-phonon-plasmon polariton modes in graphene/monolayer h-BN heterostructures. Nano Letters. 2014;**14**(7):3876-3880

[17] Yan H, Li Z, Li X, Zhu W, Avouris P, Xia F. Infrared spectroscopy of tunable Dirac terahertz magneto-plasmons in graphene. Nano Letters. 2012;**12**(7):3766-3771

[18] Woessner A, Lundeberg MB, Gao Y, Principi A, Alonso-González P, Carrega M, et al. Highly confined low-loss plasmons in graphene–boron

nitride heterostructures. Nature Materials. 2015;**14**(4):421-425

[19] Yeung KY, Chee J, Yoon H, Song Y, Kong J, Ham D. Far-infrared graphene plasmonic crystals for plasmonic band engineering. Nano Letters. 2014;**14**(5):2479-2484

[20] Caldwell JD, Novoselov KS. Mid-infrared nanophotonics. Nature Materials. 2015;**14**(4):364-366

[21] Dai S, Ma Q, Liu MK, Andersen T, Fei Z, Goldflam MD, et al. Graphene on hexagonal boron nitride as a tunable hyperbolic metamaterial. Nature Nanotechnology. 2015;**10**(8):682-686

[22] Tran TT, Bray K, Ford MJ, Toth M, Aharonovich I. Quantum emission from hexagonal boron nitride monolayers. Nature Nanotechnology. 2016;**11**(1):37-41

[23] Ni GX, Wang H, Wu JS, Fei Z, Goldflam MD, Keilmann F, et al. Plasmons in graphene Moiré superlattices. Nature Materials. 2015;**14**(12):1217-1222

[24] Kumar A, Malik G, Chandra R, Mulik RS. Bluish emission of economical phosphor h-BN nanoparticle fabricated via mixing annealing route using non-toxic precursor. Journal of Solid State Chemistry. 2020;**288**:121430

[25] Jakšić Z, Obradov M, Jakšić O, Tanasković D, Radović DV. Reviewing MXenes for plasmonic applications: Beyond graphene. In: 2019 IEEE 31st International Conference on Microelectronics (MIEL). Washington, DC, USA: IEEE; 2019. pp. 91-94

[26] Anasori B, Lukatskaya MR, Gogotsi Y. 2D metal carbides and nitrides (MXenes) for energy storage. Nature Reviews Materials. 2017;**2**(2):1-17

[27] Ersan F, Keçik D, Özçelik VO, Kadioglu Y, Aktürk OÜ, Durgun E, et al.

Two-dimensional pnictogens: A review of recent progresses and future research directions. Applied Physics Reviews. 2019;**6**(2):021308

[28] Hulst HC, van de Hulst HC. Light Scattering by Small Particles. New York, USA: Courier Corporation; 1981

[29] Bohren CF, Huffman DR. Absorption and Scattering of Light by Small Particles. Weinhein, Germany: John Wiley and Sons; 2008

[30] Mohammed F, Warmbier R, Quandt A. Computational plasmonics: Theory and applications. In: Recent Trends in Computational Photonics. Cham: Springer; 2017. pp. 315-339

[31] Sen D, Das SK, Basu AN, Sengupta S. Significance of Ehrenfest theorem in quantum–classical relationship. Current Science. 2001;**80**(4):536-541

[32] Iqbal MA, Ashraf N, Shahid W, Afzal D, Idrees F, and Ahmad R. Fundamentals of Density Functional Theory: Recent Developments, Challenges and Future Horizons. UK: IntechOpen; 2021

[33] Taflove A. Review of the formulation and applications of the finite-difference time-domain method for numerical modeling of electromagnetic wave interactions with arbitrary structures. Wave Motion. 1988;**10**(6):547-582

[34] Johnson SG, Joannopoulos JD. Block-iterative frequency-domain methods for Maxwell's equations in a planewave basis. Optics Express. 2001;**8**(3):173-190

[35] Joannopoulos JD, Johnson SG, Winn JN, Meade RD. Photonic Crystals: Molding the Flow of Light. 2nd ed. Princeton, New Jersey: Princeton Univ; 2008

[36] Smith GD, Smith GD, Smith GDS. Numerical Solution of Partial

Differential Equations: Finite Difference Methods. New York, USA: Oxford University Press; 1985

[37] Mohammed F, Warmbier R, Quandt A. Computational plasmonics: Numerical techniques. In: Recent Trends in Computational Photonics. New York City, USA: Springer; 2017. pp. 341-368

[38] Hohenberg P, Kohn W. Inhomogeneous electron gas. Physical Review. 1964;**136**(3B):B864

[39] Becke AD. A new mixing of Hartree–Fock and local density-functional theories. The Journal of Chemical Physics. 1993;**98**(2):1372-1377

[40] Hedin L. New method for calculating the one-particle Green's function with application to the electron-gas problem. Physical Review. 1965;**139**(3A):A796

[41] Warmbier R, Mehay T, Quandt A. Computational plasmonics with applications to bulk and nanosized systems. In: Active Photonic Platforms X. Vol. 10721. San Diego, California, United States: International Society for Optics and Photonics; 2018. p. 107211V

[42] Wei Z, Hai Z, Akbari MK, Qi D, Xing K, Zhao Q, et al. Atomic layer deposition-developed two-dimensional α-MoO3 windows excellent hydrogen peroxide electrochemical sensing capabilities. Sensors and Actuators B: Chemical. 2018;**262**:334-344

[43] Xu H, Akbari MK, Zhuiykov S. 2D Semiconductor nanomaterials and heterostructures: Controlled synthesis and functional applications. Nanoscale Research Letters. 2021;**16**(1):1-38

[44] Kang K, Xie S, Huang L, Han Y, Huang PY, Mak KF, et al. High-mobility three-atom-thick semiconducting films with wafer-scale homogeneity. Nature. 2015;**520**(7549):656-660

[45] Yu J, Li J, Zhang W, Chang H. Synthesis of high quality two-dimensional materials via chemical vapor deposition. Chemical Science. 2015;**6**(12):6705-6716

[46] Xu H, Akbari MK, Kumar S, Verpoort F, Zhuiykov S. Atomic layer deposition–state-of-the-art approach to nanoscale hetero-interfacial engineering of chemical sensors electrodes: A review. Sensors and Actuators B: Chemical. 2021;**331**:129403

[47] Leskelä M, Ritala M. Atomic layer deposition chemistry: Recent developments and future challenges. Angewandte Chemie International Edition. 2003;**42**(45):5548-5554

[48] Brun N, Ungureanu S, Deleuze H, Backov R. Hybrid foams, colloids and beyond: From design to applications. Chemical Society Reviews. 2011;**40**(2):771-788

[49] Daeneke T, Khoshmanesh K, Mahmood N, De Castro IA, Esrafilzadeh D, Barrow SJ, et al. Liquid metals: Fundamentals and applications in chemistry. Chemical Society Reviews. 2018;**47**(11):4073-4111

[50] A de Castro I, Chrimes AF, Zavabeti A, Berean KJ, Carey BJ, Zhuang J, et al. A gallium-based magnetocaloric liquid metal ferrofluid. Nano Letters. 2017;**17**(12):7831-7838

[51] Caturla MJ, Jiang JZ, Louis E, Molina JM. Some Issues in Liquid Metals Research. Switzerland: MDPI; 2015.

[52] Britnell L, Gorbachev RV, Jalil R, Belle BD, Schedin F, Katsnelson MI, et al. Electron tunneling through ultrathin boron nitride crystalline barriers. Nano Letters. 2012;**12**(3):1707-1710

[53] Pham VP, Jang H-S, Whang D, Choi J-Y. Direct growth of graphene on rigid and flexible substrates: Progress, applications, and challenges. Chemical Society Reviews. 2017;**46**: 6276-6300

[54] Pham VP, Kim KH, Jeon MH, Lee SH, Kim KN, Yeom GY. Low damage pre-doping on CVD graphene/Cu using a chlorine inductively coupled plasma. Carbon. 2015;**95**:664-671

[55] Pham VP, Kim KN, Jeon MH, Kim KS, Yeom GY. Cyclic chlorine trap-doping for transparent, conductive, thermally stable and damage-free graphene. Nanoscale. 2014;**6**:15310-15318

[56] Pham VP, Kim, Nguyen MT, Park JW, Kwak SS, et al. Chlorine-trapped CVD bilayer graphene for resistive pressure sensor with high detection limit and high sensitivity. 2D Materials. 2017;**4**:025049

[57] Pham PV. Hexagon flower quantum dot-like Cu pattern formation during low-pressure chemical vapor deposited graphene growth on a liquid Cu/W substrate. ACS Omega. 2018;**3**:8036-8041

[58] Pham PV, Mishra A, Yeom GY. The enhancement of Hall mobility and conductivity of CVD graphene through radical doping and vacuum annealing. RSC Advances. 2017;**7**:16104-16108

[59] Pham PV. Cleaning of graphene surfaces by low-pressure air plasma. Royal Society Open Science. 2018;**6**:172395

[60] Pham VP et al. Low energy BCl_3 plasma doping of few-layer graphene. Science of Advanced Materials. 2016;**8**:884-890

[61] Fang Z, Wang Y, Liu Z, Schlather A, Ajayan PM, Koppens FH, et al. Plasmon-induced doping of graphene. ACS Nano. 2012;**6**(11):10222-10228

[62] Fei Z, Rodin AS, Andreev GO, Bao W, McLeod AS, Wagner M, et al. Gate-tuning of graphene plasmons revealed by infrared nano-imaging. Nature. 2012;**487**(7405):82-85

[63] Zayats AV, Smolyaninov II, Maradudin AA. Nano-optics of surface plasmon polaritons. Physics Reports. 2005;**408**(3-4):131-314

[64] Cui L, Wang J, Sun M. Graphene plasmon for optoelectronics. Reviews in Physics. 2021;**6**:100054

[65] Mokhtar MM, Abo-El-Enein SA, Hassaan MY, Morsy MS, Khalil MH. Mechanical performance, pore structure and micro-structural characteristics of graphene oxide nano platelets reinforced cement. Construction and Building Materials. 2017;**138**:333-339

[66] Rowley-Neale SJ, Randviir EP, Dena ASA, Banks CE. An overview of recent applications of reduced graphene oxide as a basis of electroanalytical sensing platforms. Applied Materials Today. 2018;**10**:218-226

[67] Cao Y, Yang H, Zhao Y, Zhang Y, Ren T, Jin B, et al. Fully suspended reduced graphene oxide photodetector with annealing temperature-dependent broad spectral binary photoresponses. ACS Photonics. 2017;**4**(11):2797-2806

[68] Kretinin AV, Cao Y, Tu JS, Yu GL, Jalil R, Novoselov KS, et al. Electronic properties of graphene encapsulated with different two-dimensional atomic crystals. Nano Letters. 2014;**14**(6): 3270-3276

[69] Kim CJ, Brown L, Graham MW, Hovden R, Havener RW, McEuen PL, et al. Stacking order dependent second harmonic generation and topological defects in h-BN bilayers. Nano Letters. 2013;**13**(11):5660-5665

[70] McCreary A, Kazakova O, Jariwala D, Al Balushi ZY. 2D Materials. 2020;**8**(1):013001-013018

[71] Dillon AD, Ghidiu MJ, Krick AL, Griggs J, May SJ, Gogotsi Y, et al. Highly conductive optical quality solution-processed films of 2D titanium carbide.

Advanced Functional Materials. 2016;**26**(23):4162-4168

[72] Hantanasirisakul K, Gogotsi Y. Electronic and optical properties of 2D transition metal carbides and nitrides (MXenes). Advanced Materials. 2018;**30**(52):1804779

[73] Liu Z, Alshareef HN. MXenes for optoelectronic devices. Advanced Electronic Materials. 2021;**7**(9):2100295

[74] Ersan F, Aktürk E, Ciraci S. Stable single-layer structure of group-V elements. Physical Review B. 2016;**94**(24):245417

[75] Özcelik VO, Aktürk OÜ, Durgun E, Ciraci S. Prediction of a two-dimensional crystalline structure of nitrogen atoms. Physical Review B. 2015;**92**(12):125420

[76] Gusmão R, Sofer Z, Bouša D, Pumera M. Pnictogen (As, Sb, Bi) nanosheets for electrochemical applications are produced by shear exfoliation using kitchen blenders. Angewandte Chemie. 2017;**129**(46):14609-14614

[77] Yua X, Lianga W, Xinga C, Chena K, Chena J, Huangd W, et al. Rising 2D pnictogens for catalytic applications: Status and challenges. Journal of. 2018;**6**(12):4883-5230

[78] Agrawal A, Johns RW, Milliron DJ. Control of localized surface plasmon resonances in metal oxide nanocrystals. Annual Review of Materials Research. 2017;**47**:1-31

[79] Tong L, Huang X, Wang P, Ye L, Peng M, An L, et al. Stable mid-infrared polarization imaging based on quasi-2D tellurium at room temperature. Nature Communications. 2020;**11**(1):1-10

[80] Li F, Jiang X, Zhao J, Zhang S. Graphene oxide: A promising nanomaterial for energy and environmental applications. Nano Energy. 2015;**16**:488-515

[81] Low T, Avouris P. Graphene plasmonics for terahertz to mid-infrared applications. ACS Nano. 2014;**8**(2):1086-1101

[82] Wang H, Li S, Ai R, Huang H, Shao L, Wang J. Plasmonically enabled two-dimensional material-based optoelectronic devices. Nanoscale. 2020;**12**(15):8095-8108

[83] Autore M, Li P, Dolado I, Alfaro-Mozaz FJ, Esteban R, Atxabal A, et al. Boron nitride nanoresonators for phonon-enhanced molecular vibrational spectroscopy at the strong coupling limit. Light: Science and Applications. 2018;**7**(4):17172-17172

[84] Iqbal MA, Ashraf N, Shahid W, Awais M, Durrani AK, Shahzad K, Ikram M. Nanophotonics: Fundamentals, Challenges, Future Prospects and Applied Applications. UK: IntechOpen; 2021

[85] Caldwell JD, Vurgaftman I, Tischler JG, Glembocki OJ, Owrutsky JC, Reinecke TL. Atomic-scale photonic hybrids for mid-infrared and terahertz nanophotonics. Nature Nanotechnology. 2016;**11**(1):9-15

[86] Nguyen M, Kim S, Tran TT, Xu ZQ, Kianinia M, Toth M, et al. Nanoassembly of quantum emitters in hexagonal boron nitride and gold nanospheres. Nanoscale. 2018;**10**(5):2267-2274

[87] Chaudhuri K, Alhabeb M, Wang Z, Shalaev VM, Gogotsi Y, Boltasseva A. Highly broadband absorber using plasmonic titanium carbide (MXene). ACS Photonics. 2018;**5**(3):1115-1122

[88] Aydin K, Ferry VE, Briggs RM, Atwater HA. Broadband polarization-independent resonant light absorption using ultrathin plasmonic super

absorbers. Nature Communications. 2011;**2**(1):1-7

[89] Liu H, Du Y, Deng Y, Peide DY. Semiconducting black phosphorus: Synthesis, transport properties and electronic applications. Chemical Society Reviews. 2015;**44**(9):2732-2743

[90] Chen C, Youngblood N, Peng R, Yoo D, Mohr DA, Johnson TW, et al. Three-dimensional integration of black phosphorus photodetector with silicon photonics and nanoplasmonics. Nano Letters. 2017;**17**(2):985-991

[91] Abbas AN, Liu B, Chen L, Ma Y, Cong S, Aroonyadet N, et al. Black phosphorus gas sensors. ACS Nano. 2015;**9**(5):5618-5624

[92] Shao L, Tao Y, Ruan Q, Wang J, Lin HQ. Comparison of the plasmonic performances between lithographically fabricated and chemically grown gold nanorods. Physical Chemistry Chemical Physics. 2015;**17**(16):10861-10870

Emerging Nanostructured Materials and Applications in Industry and Biomedicine

Chapter 4

Doping and Transfer of High Mobility Graphene Bilayers for Room Temperature Mid-Wave Infrared Photodetectors

Ashok K. Sood, John W. Zeller, Parminder Ghuman,
Sachidananda Babu, Nibir K. Dhar, Randy N. Jacobs,
Latika S. Chaudhary, Harry Efstathiadis, Samiran Ganguly,
Avik W. Ghosh, Sheikh Ziauddin Ahmed
and Farjana Ferdous Tonni

Abstract

High-performance graphene-HgCdTe detector technology has been developed combining the best properties of both materials for mid-wave infrared (MWIR) detection and imaging. The graphene functions as a high mobility channel that whisks away carriers before they can recombine, further contributing to detection performance. Comprehensive modeling on the HgCdTe, graphene, and the HgCdTe-graphene interface has aided the design and development of this MWIR detector technology. Chemical doping of the bilayer graphene lattice has enabled p-type doping levels in graphene for high mobility implementation in high-performance MWIR HgCdTe detectors. Characterization techniques, including SIMS and XPS, confirm high boron doping concentrations. A spin-on doping (SOD) procedure is outlined that has provided a means of doping layers of graphene on native substrates, while subsequently allowing integration of the doped graphene layers with HgCdTe for final implementation in the MWIR photodetection devices. Successful integration of graphene into HgCdTe photodetectors can thus provide higher MWIR detector efficiency and performance compared to HgCdTe-only detectors. New earth observation measurement capabilities are further enabled by the room temperature operational capability of the graphene-enhanced HgCdTe detectors and arrays to benefit and advance space and terrestrial applications.

Keywords: graphene, HgCdTe, photodetectors, MWIR, mobility, doping, transfer

1. Introduction

1.1 Graphene overview

The term *graphene* is a combination of two words—*graphite* and *alkene*. Graphene denotes a two-dimensional (2-D) sheet of graphite of atomic-scale thickness resulting due to intercalation of graphite compounds [1, 2]. Graphene can be divided into three different classifications based on the extent of its layer structure, which are monolayer, bilayer, and multilayer graphene, with the latter designating graphene structures consisting of three or more layers [3].

Graphite, another well-known compound of carbon, basically consists of stacked sheets of graphene held in place by van der Waals forces. A third relatively recently discovered carbon compound that has likewise been envisioned as a catalyst essentially comprises rolled-up sheets of graphene commonly known as *carbon nanotubes*. As thus comprising the basic building block of these key carbon materials of varying dimensionalities, graphene has come to be considered the "mother" of graphitic compounds [4].

Chemically speaking, graphene comprises a two-dimensional hexagonal benzene ring-like structure consisting of sp^2-bonded carbon, packed into honeycomb lattice as shown in **Figure 1**, where the C–C bond length is 0.142 nm. It is one of the first 2-D materials known to be stable at room temperature, and in ambient conditions is crystalline and chemically inert. Although stronger than diamond, graphene yet remains as flexible as rubber [6].

From an electrical standpoint, the valence and conduction bands in the band structure of graphene meet at the corners of the Brillouin zone or Dirac points. The consequence of this is that apart from the influence of thermal excitations, the intrinsic charge carrier concentration is zero, and graphene is consequently characterized as a zero-bandgap semiconductor. Notwithstanding the theoretical implications of this, practical and functioning graphene-based devices still require the existence of charge carriers, as well as control over the quantification of the concentrations and types of charge carriers (i.e., for n- or p-doping) [7, 8].

Intensive research performed over multiple decades into graphene material has further uncovered the remarkable chemical and material properties of this unique and somewhat extraordinary form of carbon. Perhaps most notably are the extremely high charge carrier mobilities in the range of 2000–5000 cm^2/V s, making graphene a choice material for implementation in high-speed electronics, such as flexible ultrafast microelectronics. Graphene is likewise one of the most highly conductive materials known with thermal conductivities reaching 5000 W/m K, facilitating its use for applications such as light-emitting diodes (LEDs).

Furthermore, having Young's modulus reportedly as high as 1 TPa has led to the use of graphene for strength reinforcement in various aerospace and structural/concrete material applications. In offering one of the largest specific surface areas (2630 m^2/g) combined with nearly full optical transparency of 97.7%, graphene has likewise been employed for the advancement of numerous optical and

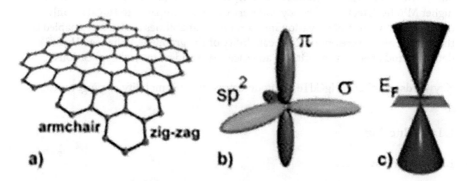

Figure 1.
(a) Graphene geometry; (b) bonding diagram; and (c) associated band diagram [5].

optoelectronic applications [9, 10]. The combined benefits of its high mechanical strength, optical transparency, and mobility of charge carriers have made graphene a choice material for a diverse array of electrical and/or optical applications.

During past decades, starting with the disclosure of the Hummers method in the 1950s followed by the chemical reduction of graphene oxides in 1962, there have been numerous research and studies on graphite oxide synthesis [8]. However, it took the work of A. Geim and K. Novoselov to fully isolate and subsequently characterize pristine graphene by a process of mechanical exfoliation that came to be known as the "Scotch tape" method for the properties of graphene to be more adequately understood, upon which it soon became a primary material of interest for many diverse ongoing research efforts. Later in 2010 Geim and Novoselov were awarded the Nobel Prize in Physics for this research encompassing graphene as a 2-D atomic structure [9].

After the initial success of isolating this graphitic material, many in the scientific community commenced to explore different processes and techniques for the large-scale synthesis and fabrication of graphene. As reported contemporarily in literature, the more common graphene synthesis processes included the oxidation-reduction growth process, chemical vapor deposition (CVD), liquid phase stripping, and epitaxial growth on silicon carbide (SiC) [10]. Among these various methods, the CVD process soon became the most established technique for producing graphene films with the highest quality crystalline and structural integrity, primarily on Cu substrates [11].

However, Cu and other metal substrates are not practical for most applications, as many optoelectronic-, sensor-, and microelectronics-based applications require the placement of graphene films directly on metal oxide or semiconductors. There, therefore, has been and remains a need for more optimized increasingly effective processes through which graphene films can be directly and effectively transferred onto any desired substrate of choice, while also avoiding cracks, wrinkles, and forms of contamination [12].

1.2 Graphene-based technological device development

Infrared detector and focal plane array (FPA) technologies are at the heart of many space-based instruments for NASA and defense missions that provide remote sensing and long-range imaging capabilities [13]. While often considered exotic in comparison to more established detector materials such as HgCdTe, on account of its gapless band structure, strong light-matter interaction, and the relative ease by which heterostructures may be fabricated, graphene can provide numerous capabilities from diverse means for effective broad spectra photodetection.

Graphene detector implementation can likewise further facilitate reduced size, weight, power, and cost (SWaP-C) mid-wave infrared (MWIR) sensors on smaller platforms, a high priority for providing improved measurement and mission capabilities in space. Use of process techniques such as post-growth thermal cycle annealing (TCA) has additionally been reported to enable up to an order of magnitude reduction in the dislocation density down to the saturation limit ($\sim 10^6$ cm^{-2}) for improved high-temperature operability of HgCdTe-on-Si-based MWIR detectors and FPAs [13].

The overall functionality and applicability of a detector device or system are governed primarily by its wavelength range, that is, band, of operation. Of the different infrared (IR) bands spanning the short-wave infrared (SWIR) to very long-wave infrared (VLWIR), the MWIR region is considered of the highest beneficial for long-range imaging and early threat detection [14]. Specifically, the 2–5 μm MWIR spectral band is crucial for NASA Earth Science applications, especially in satellite-based LIDAR systems that require measuring a wide variety of natural features, including cloud aerosol properties, sea surface temperatures, and natural phenomena such as volcano and forest fires.

Though prior scientific reporting of associated experimental results has served to illuminate the key 2-D nanomaterial properties of graphene for enhancement of sensing performance particularly for IR band detection, certain challenges still must be addressed, which are as follows:

1. The existence of a process and technique for doping bilayer graphene with holes and electrons having sufficient charge carrier concentration. The main emphasis here is that the process of doping the graphene should not affect the crystalline structure and that the doped graphene film remains defect-free. This process should moreover be cost-effective and scalable for doping large-area graphene films.

2. The development of a method and process for transferring CVD-grown graphene to the desired substrate that ensures a uniform, clean, and intact transfer required for successful realization in an array of prospective applications.

3. Attainment of graphene-enhanced high mobilities and corresponding high level of photodetection performance in graphene-HgCdTe-based IR, and specificity 2–5 µm MWIR, photodetectors, and optical imaging arrays.

2. Graphene-enhanced MWIR photodetectors

2.1 Motives and objectives of device concept

HgCdTe, or MCT, the most widely used infrared (IR) detector material in military applications, is a direct energy bandgap semiconductor having a bandgap that is tunable from near-infrared (NIR) and SWIR to VLWIR bands through varying the Cd composition [15]. Typically, 0.3:0.7 Cd:Hg ratio results in a detectivity window over the SWIR to MWIR wavelength range.

Additionally, HgCdTe layer growth is highly controllable with certain deposition techniques. Notably among these is molecular-beam epitaxy (MBE), which yields high precision in the deposition of detector material structures leading to excellent control over optical excitation evidenced by the high quantum efficiencies (QE) demonstrated by HgCdTe-based detectors and sensors over the IR.

While adding considerable cost and bulk, cryogenic cooling is commonly utilized for IR detection to minimize thermally generated dark current. Since dark current increases with cutoff wavelength longevity, this requirement becomes even more important for MWIR and long-wave infrared (LWIR) sensors. IR band detector technologies that can operate at or near room temperature and substantially avoid costly and bulky cooling requirements, therefore, offer great practical benefits for many types of applications.

The incorporation of a high mobility graphene channel in HgCdTe-based detectors is a newly discovered means to offer further performance improvements and operational capabilities for MWIR detection. The intrinsic interfacial barrier between the HgCdTe-based absorber and the graphene layers thereby may be designed to effectively reduce the recombination of photogenerated carriers in the detector. The graphene thus functions as a high mobility channel that whisks away carriers before they can recombine, further contributing to the MWIR detection performance compared to in photodetectors only utilizing HgCdTe absorption layers [16].

2.2 Physical graphene-enhanced detector structure

The graphene-enhanced HgCdTe MWIR detector structure fabricated on a silicon substrate comprises three principal layers. First, a layer of CdTe is grown to act as a buffer layer functioning as the gate terminal (1). This layer provides an electrical field in the "vertical" direction into the detector heterostructure that aids in carrier transport in that direction. **Figure 2** shows a schematic of this detector structure as shown in a study by Srivastava et al. [12].

The HgCdTe absorber layer (2) is grown above the silicon substrate and the CdTe buffer layer acts as the active optical layer where photogeneration of carriers takes place. The HgCdTe absorber material and its physical properties, such as bandgap, determine the sensitivity of the absorber layer to the detection wavelength window. In addition, the absorber material governs the photogeneration rate, quantum efficiency, and carrier lifetime, which collectively contribute to overall detection performance.

Finally, the graphene layer (3) incorporates the role of high mobility, low noise channel that quickly whisks away the photogenerated carriers in the absorber into the contacts, and subsequently into the readout integrated circuit (ROIC) for electrical readout. This layer, therefore, is directly contacted to the ROIC.

2.3 Graphene-HgCdTe detector operating principle

The general operating principle of the graphene-HgCdTe MWIR photodetector may be described in terms of the life cycle of the photogenerated carriers [17]. Incident IR photons transmitted through the Si substrate and CdTe layers into the HgCdTe region are absorbed and produce electron-hole pairs, or excitons (**Figure 3(a)**). The vertical electric field in the absorber applied through modulation of the gate voltage effectively separates the electron-hole pairs due to the consequent opposing forces on electrons and holes. This separation of the carriers physically isolates the two photogenerated carrier species and suppresses the Auger recombination within the absorber, minimizing the loss of photogenerated carriers and is thus critical to the ultimate performance of the detector.

After separation, the carriers are transported through the absorber film toward the graphene interface and then injected into it (**Figure 3(b)**). Modulation of the gate voltage bias to preferentially inject only one of the photogenerated species into the graphene in a rectifier-like action enables dynamic control of the interface properties. As this process involves the injection of both species, it further prevents any Auger recombinations from taking place in the graphene.

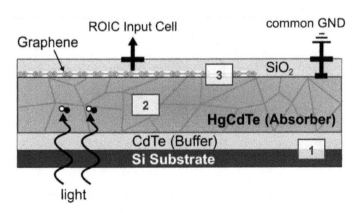

Figure 2.
Heterostructure layer structure of HgCdTe-graphene-based IR photodetector [12].

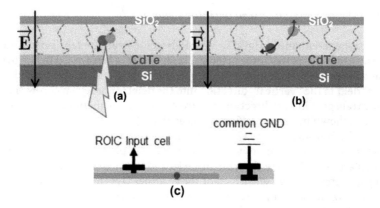

Figure 3.
(a) Generation of excitons from incident photons separation of electrons and holes due to applied gate electric field. (b) Photogenerated carrier transport and injection into graphene. (c) Horizontal transport of the photogenerated carriers in the graphene.

The carriers injected from the absorber into the graphene are transported laterally to the ROIC terminal and subsequently collected into it (**Figure 3(c)**). The establishment of a separate high mobility channel enabled by the graphene allows faster modulation frequencies with reduced **1/f** noise and consequently higher performance metrics. Dynamic gating is additionally provided through the electrical control of carriers injected into the graphene.

3. Graphene-HgCdTe detector modeling effort

3.1 Modeling approach

The modeling effort was individually built upon various elements that combine to form a comprehensive model for this detector material technology. The overall goal of this modeling approach was to determine through simulations an accurate determination of electrical detector device behavior, including I-V characteristics, noise, responsivity, and other performance metrics. In addition, the directivity D* and noise equivalent temperature difference (NETD) may be derived from basic material parameters and device design and operating specifications to allow and guide further design optimizations.

The modeling effort, depicted schematically in **Figure 4**, entailed modular construction of the complete detector simulation platform from the individual models as data were made available from experiments and device characterizations. These have involved specific material modeling of the HgCdTe, graphene, and HgCdTe-graphene interface.

3.2 Modeled performance parameters in HgCdTe

This graphene-HgCdTe MWIR detector fundamentally functions as a photo-controlled current source rather than a light/heat-dependent resistance, characteristic of the typical operating mode for bolometers. **Figure 5** compares the theoretical dark current and photocurrent, film resistance, and detectivity (D*) performance parameters in the HgCdTe for a conventional photoconductive detector (**Figure 5(a)**) with that for this type of HgCdTe-based MWIR detector (**Figure 5(b)**). It is here noted that D* does not change appreciably because ultimately the material properties are

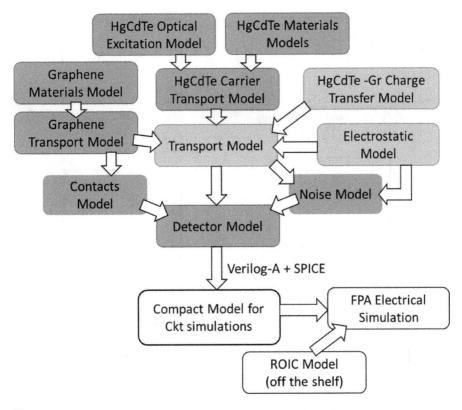

Figure 4.
Flowchart diagram illustrating the modeling approach and relationship between the different models utilized.

the same; altering the area of the current collection does not significantly change this fundamental property. The current and resistance, however, are each significantly lower for this latter detector design in view (**Figure 5(b)**). The incorporation of high mobility graphene in this detector can further enable higher responsivity and greater D*.

3.3 Graphene-HgCdTe interface band structure models

Figure 6 shows the *E-k* dispersion relation and density of states (DOS) determined for the HgCdTe/graphene interface. The contribution of individual atoms to the DOS is likewise computed. The carbon contributes maximally to the conduction band, while HgCdTe species contribute to the valence band.

Bandgap engineering of the HgCdTe detector material is additionally possible through adaptive control of the epitaxial growth process parameters. This provides the capability to optimize the performance to achieve desired spectral range and operating temperature specifications for the development of graphene-enhanced MWIR detectors and FPAs.

The work function of $Hg_{0.73}Cd_{0.27}Te$, ΦMCT, is determined (5.52 eV) based on the following relation:

$$\Phi_{MCT} = x\Phi_{CdTe} + (1-x)\Phi_{HgTe} \tag{1}$$

where x is the CdTe concentration in $Hg_{1-x}Cd_xTe$, and Φ_{CdTe} and Φ_{HgTe}, the work functions of CdTe and HgTe, are 4.5 eV and 5.9 eV, respectively. As shown in

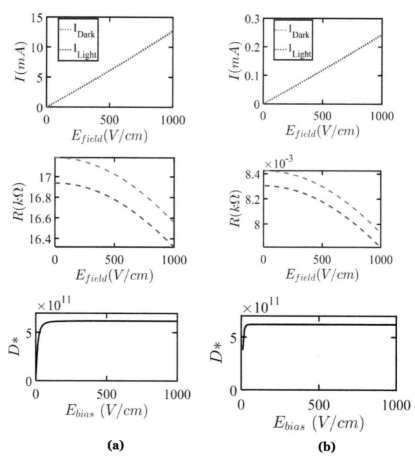

Figure 5.
Modeled dark current and photocurrent, film resistance, and detectivity (D*) (from top to bottom) in HgCdTe for (a) conventional photoconductive detector design, and (b) design of this HgCdTe MWIR photoconductive detector.

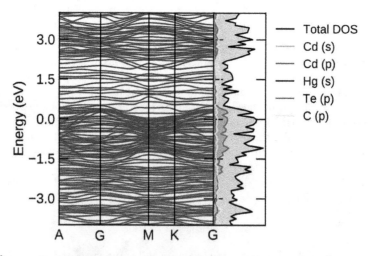

Figure 6.
E-k dispersion relation and density of states for HgCdTe/graphene interface.

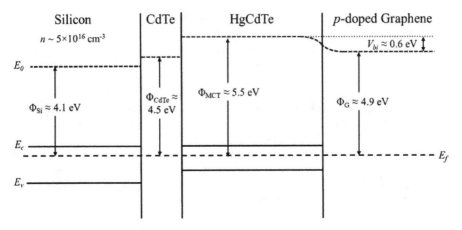

Figure 7.
Band diagram for graphene/HgCdTe/Si detector heterostructure.

Figure 7, $Hg_{0.7}Cd_{0.3}Te$ produces a built-in V_{bi} potential with p-doped graphene of ~0.6 eV. (With n-doped graphene having a work function 4.25, V_{bi} becomes as high as 1.27 eV.) Given its intermediate work function between that of HgCdTe (5.5 eV) and n-doped Si (4.1 eV), the use of the CdTe buffer layer facilitates band matching of the HgCdTe/CdTe/Si layers.

4. Doping of graphene bilayers

4.1 Historical development of graphene doping techniques

Recently a significant amount of research has been dedicated toward the manipulation of the physicochemical and electrical properties of graphene to specifically tailor it for various applications. One means to achieve this is through chemically functionalizing the graphene, involving modification of its carbon sp^2 honeycomb structure [18]. This chemical functionalization in turn necessitates chemically doping the atomic lattice of graphene with atoms from other compatible elements of the periodic table, essentially modifying graphene lattice originally undoped into a heteroatomically doped one.

The technique of doping through inducing charge carriers comprising either holes or electrons may be divided into two broad categories, which are as follows:

a. *Electrical doping*: In this process, charge carriers are induced by the application of an electric field. This can take place using a graphene-based field-effect transistor (FET), wherein the charge carriers are induced by an electric field produced by the gate structure. For example, with a Si^+/SiO_2 substrate varying the gate voltage V_g and consequently the concentration of electron/holes charge carriers enables the concentration of the induced carriers to be controlled by way of the applied gate voltage. If V_g is positive excitation of electrons and n-type doping will result, while on the other hand, an applied negative voltage will lead to induction of holes and a p-doped material. The concentration of charge carriers induced by this method can be as high as 10^{13} cm^{-2} [19].

b. *Chemical doping*: This technique involves the association of other chemical species with graphene and is further subdivided into two additional classifications, which are *substitutional doping* and *surface transfer*. *Substitutional doping* is a process whereby carbon atoms in the graphene lattice are substituted with other atoms, leading to either *p*-type or *n*-type conductivity [20]. Likewise, *surface transfer* describes a nondestructive technique for inducing charge carriers, in this case within the graphene lattice involving charge transfer between surface adsorbates and the graphene [21].

Doping by surface transfer may occur as the result of two different mechanisms—*electronical doping* and *electrochemical doping* [22]. Electronic doping is due to the direct transfer of charge between the graphene and adsorbate. In the presence of a differing electronic chemical potential, the doping type is controlled by the position of the graphene Fermi level relative to the highest occupied (HOMO) and lowest occupied (LUMO) level molecular orbitals of the adsorbate. While graphene is usually *n*-doped when the adsorbate HOMO lies above graphene Fermi level, *p*-type doping occurs when the LUMO of the adsorbate is found below the graphene Fermi levels [7]. The representation of molecular orbitals levels to the graphene Fermi levels for (a) *p*-type and (b) *n*-type doped graphene is shown in **Figure 8**. In contrast to electrical doping, electrochemical doping is a time-dependent process influenced by various factors that include the reaction rate and diffusion rate of molecular species [23].

The focus here is on inducing *p*-type doping in graphene through chemical doping. Chemical heteroatom doping of graphene is generally performed using either a one-step or two-step synthesis method. The one-step method involves employing CVD to introduce both carbon and boron sources into the chamber while heating the copper foil at high temperatures [24].

The alternative two-step synthesis process for boron doping includes thermal annealing [25] and rapid Wurtz-type reactive coupling [26] techniques, among others. Nevertheless, such two-step methods generally involve more complex experimental setups, tend to result in defects present in the doped graphene films, and require the use of toxic chemicals as the source/precursor as well as relatively high temperatures. These factors clearly limit the types of substrates that may be practically used [27].

However, a recently developed technique known as the spin-on dopant (SOD) process has made it possible to avoid these shortcomings in large part [28]. This

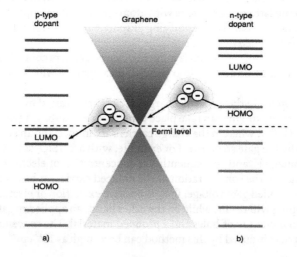

Figure 8.
Relative position of highest occupied (HOMO) and lowest unoccupied (LUMO) molecular orbitals of an adsorbate to the Fermi level of graphene for (a) p-type and (b) n-type dopants [6].

method, which we have adopted for producing p-type doping in graphene and shall subsequently be described in more comprehensive detail, requires only a relatively basic experimental setup without the need for toxic precursor gases.

4.2 Boron doping of bilayer graphene

Dopants used for chemically doping graphene include, but are not limited to, S, N, B, P, I, Se, O, and I [20, 21]. Among these dopant elements, the two most common and notable are nitrogen (N) and boron (B), for inducing n-type conductivity p^+-type conductivity, respectively, in graphene [22].

As the MWIR photodetector device under consideration requires p-doped graphene for optimal performance, the boron doping method is here in view. Boron is one of the most natural choices for doping among the atomic elements, having valence atoms that differ in number by only a single atom compared to the number of its carbon ones [29]. Atoms of boron also have a similar atomic size (0.088 nm atomic radius) to those of carbon (atomic radius of 0.077 nm), a factor that further facilitates p-type conduction in graphene [30].

When a dopant atom such as in this instance boron is bonded within a carbon framework, a defect is introduced into the neighboring site. Since boron only contains three valence electrons compared to four in carbon atoms, this can cause uneven charge distribution resulting in charge transfer between nearby carbon atoms, further expounding their electrochemical behavior [31]. The incorporation of a relatively small number of boron atoms thereby effectively lowers the Fermi level and formation of an acceptor level in the doped graphene. The introduction of boron likewise contributes to improved stabilization of the extremities of the graphene material and similarly aids in mitigating the termination of its layers, thereby promoting layer-by-layer growth of larger portions or sections of graphene [32].

4.3 Graphene spin-on doping process

We have developed and implemented a distinct spin-on dopant (SOD) process to produce highly boron-doped bilayer graphene. The SOD process involves spin-coating a dopant solution onto a source substrate and annealing the latter in conjunction with a target substrate in a tube furnace [31]. The bilayer graphene sheets doped using this process were deposited on SiO_2/Si substrates (300 nm SiO_2, p-type doped) by CVD acquired from Graphenea, Inc.

In the high-temperature environment and inert gas (e.g., argon) atmosphere, diffusion of the dopant (boron) from the source substrate into the target sample (bilayer graphene) occurs when the B atoms replace the C atoms to form p-doped graphene. The main advantages of this technique are its low cost and simplistic setup, combined with the capability to provide uniform and consistent doping profiles [33]. **Figure 9** depicts a schematic representation for substitutional doping of boron in the honeycomb lattice of graphene by the SOD process.

For the SOD procedure schematically illustrated in **Figure 9**, we start with the CVD-deposited graphene on Si/SiO_2 substrates. The spin-on diffusant used is Filmtronics B-155 (4% boron conc.). This boron source is spin-coated onto a Si wafer at 2300 rpm for 30 s using a CEE vacuum coater tool [34].

This boron-solution-coated source wafer is then placed in a custom-designed silica boat approximately 10 mm apart from and facing a target graphene sample. These are each inserted into a tube furnace that is pumped down to 10 Torr vacuum pressure and annealed in the presence of flowing Ar gas.

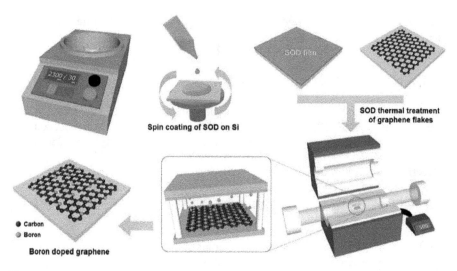

Figure 9.
Schematic of processing steps for the fabrication of boron-doped GFs through a spin-on dopant (SOD) method [32].

Parameter	Value
Temperature	500, 600°C
Flow gas	Argon
Flow rate	550 sccm
Ramp rate	15°C/min
Vacuum pressure	10 Torr
Boron source	Filmtronics B155 (4%)
Spin speed	2300 rpm
Spin time	30 s

Table 1.
Annealing and spin coating parameters for the SOD process.

Experimental parameters for the boron doping of graphene are given in **Table 1**. The goal of this process is to achieve required high doping levels while maintaining graphene surface features and quality for graphene-enhanced HgCdTe MWIR photodetectors.

4.4 Graphene boron-doping concentration analysis

To determine the structural properties, chemical bonding states, and doping concentration changes in the doped bilayer graphene, Raman spectroscopy and time-of-flight (ToF) secondary-ion mass spectroscopy (SIMS) techniques were performed as along with X-ray photoelectron spectroscopy (XPS).

Doping concentration vs. depth profile results for different boron-doped graphene bilayers using SIMS are presented in **Figure 10**. This graphene sample on Si/SiO$_2$ was doped with the SOD process using 15 min. Annealing duration. The SIMS analysis indicates boron doping levels of ~1.8 × 10^{20} cm^{-3} of boron in the graphene bilayers further confirmed by XPS.

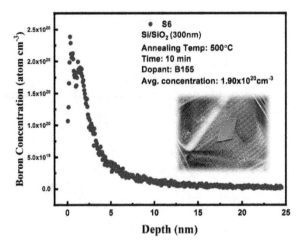

Figure 10.
SIMS atomic dopant concentrations vs. depth profiles for two different p-doped graphene samples on Si/SiO$_2$ substrates.

5. Substrate transfer of doped bilayer graphene

5.1 Evolution of substrate transfer techniques

Considerable research has been undertaken to synthesize graphene on metal substrates using the CVD process to produce high-quality large-area graphene films. Sufficiently high quality of graphene films is demonstrated by a single-crystalline structure free of wrinkles, contamination, and cracks [35]. However, various and often critical applications can require that the CVD-grown graphene films be transferred onto other more suitable substrates.

The earliest form of transfer for graphene can be traced back to the synthesis of graphene by "Scotch tape method," whereby graphene flakes were exfoliated and isolated from the graphite substrate onto another target substrate [34]. The development of an efficient process for transfer of graphene from its native substrate onto another foreign nevertheless has since proved relatively challenging, especially for large-area and intact graphene films [36].

While CVD-grown graphene on metal substrates is typically of high quality and purity, the deposited graphene when transferred off onto other types of substrates can easily degrade and suffer from contamination and structural damage [37]. Common sources of contamination during the process of graphene transfer include residues from the source substrate, etchant solutions used to dissolve the source substrates, and unwanted organic contamination due to the adherence of polymer compounds to the graphene following completion of the transfer [38]. These factors can lead to the formation of more charge carrier scattering centers that affect the electrical properties and mechanical stability of that graphene films, generally resulting in undesired doping of the graphene [39].

Additionally, the extreme thinness of graphene (single-layer atomic thickness for monolayer graphene) makes it inherently more vulnerable to altercation and impairment [40]. During cleaning and repeated transfer, mechanical strains can arise in the graphene that potentially can cause irreversible damage [41]. Hence, the need to maintain structural integrity and uniformity of the graphene for various applications can be very essential, for instance for optoelectronic devices or sensors requiring charge injection between the active functional layer and highly pure and conductive graphene [42].

Furthermore, implementation of an optimized transfer process is critical to boost yield and reproducibility for low cost and scalable production of large-area graphene films [43]. Consequently, significant research is being undertaken to further optimize the process of transferring graphene to attain intact and dislocation- and defect-free graphene films.

The procedure required for preparing graphene for transfer may be characterized by the more relevant process steps involved, which typically include—(a) graphene layer removal from substrate utilizing liquid etchant, bubble transfer, or thermal peel off; (b) use of supportive layers (e.g., polymers such as PMMA and camphor) to prevent cracks, creases, and other structural damage; and (c) cleaning and removal/transfer of the grown layer from the substrate and protective layers [35].

The major graphene transfer methods reported in the literature are as follows:

a. *Bubble-mediated transfer*: In this process, H_2 and O_2 bubbles are produced due to electrochemical reactions, that is, by the graphene when CVD-grown on a metal substrate such as Cu/Ni acting as an electrode (either anode or cathode). These bubbles when generated apply a pealing-inducing force on the substrate surface, eventually leading to delamination of the graphene from the growth substrate.

Although this method is challenging in certain respects and more limited in that conductive substrates are necessary for the actualization of the electrode-initiated electrochemical reactions [30], considerable improvements have been discovered and incorporated over time. Goa et al. developed a nondestructive bubble-mediated transfer process that enabled repeated use of the growth (Pt) substrate, whereby the transferred graphene was found to have high carrier mobility along with minimal wrinkles [42]. Another study examined the use of PMMA/graphene/Cu acting as both an anode and cathode to remove a graphene sheet by bubble delamination, resulting in the reportedly high-quality transfer of the graphene films [40].

b. *Wet transfer*: This transfer method involves the use of ionic etchants to dissolve the growth substrate, after which the graphene is washed with a liquid cleaning agent and transferred to desired target substrate without drying [43]. The commonly used liquid etchant consists of ammonium persulfate aqueous solution and ferric chloride solution for dissolving Cu/Ni foil [44].

c. *Dry transfer*: While techniques for transferring graphene have traditionally relied on liquid etchants and cleaning solutions, this renders the growth substrates unusable consequently making these processes less economically viable. To circumvent in part these limitations, the dry transfer method was developed involving the incorporation of an inorganic metal oxide lifting layer (e.g., of MoO_3), which due to its low binding energy with the graphene films may be subsequently washed away completely. This process leads to high-quality graphene films without further contamination [45].

Through this experimentation toward the achievement of clean, smooth, and reduced-residue transfer, a PMMA-based resist-assisted transfer process was identified and established. In contrast to more conventional PMMA transfer processed known to leave residues during the graphene transfer, the method we have adapted and further developed features a more straightforward process that uses different polymeric supportive layers for residue-free and clean transfer of graphene to avoid cleaning and support removal steps. This new method for transferring graphene from Si/SiO$_2$ to HgCdTe substrates combines both wet transfer and nonelectrochemical reaction-based transfer methods.

5.2 Experimental bilayer graphene transfer

Boron p^+ doping of bilayer graphene on Si/SiO$_2$ substrates has been accomplished to provide required electrical performance characteristics for a high mobility graphene channel in MWIR HgCdTe photodetector devices. The final step as discussed in this process involves transferring the sheets of highly p-doped bilayer graphene from the original Si/SiO$_2$ onto HgCdTe substrates for incorporation in the MWIR photodetector and FPA devices. This subsequently p-doped bilayer graphene on Si/SiO$_2$ is transferred using a PMMA-assisted wet transfer process [44]. **Figure 11** shows schematically this experimental procedure for transferring the graphene onto HgCdTe [18].

Through this relatively straightforward procedure, the successful transfer of doped bilayer graphene sheets deposited on SiO$_2$/Si onto HgCdTe has been demonstrated. The graphene bilayers are preserved, and no morphological changes were observed, following the transfer process with their relocation where the spatial configurations of the bilayers were maintained across macroscopic regions.

5.3 Characterization of graphene transferred onto HgCdTe

Following the transfer of the p-doped graphene onto HgCdTe substrates, the doped bilayer graphene on HgCdTe was measured using optical microscopy and Raman spectroscopy to determine if any significant changes had occurred in its properties through the process. **Figure 12** presents optical microscopy images of

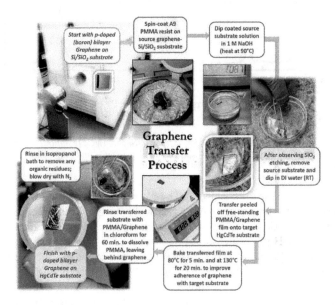

Figure 11.
Schematic outline of the experimental process enabling the removable transfer of bilayer graphene from SiO$_2$/Si onto HgCdTe substrates.

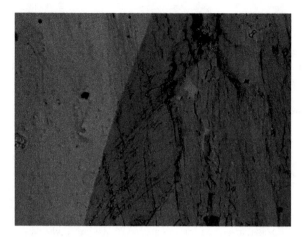

Figure 12.
Optical microscopy image of graphene deposited on HgCdTe, where darker areas represent graphene on HgCdTe and lighter area portions the bare HgCdTe substrate.

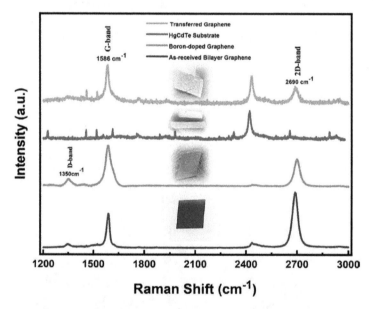

Figure 13.
Raman spectroscopy analysis of boron-doped graphene transferred onto HgCdTe substrate, compared to spectra of bare HgCdTe substrate, that of the pristine bilayer graphene on Si/SiO₂, and the graphene on Si/SiO₂ following the boron doping but before transfer.

the transferred graphene onto HgCdTe. The darker areas represent the part of the HgCdTe covered with graphene (having relatively marginal but practically observable differences in optical absorption), while the lighter areas indicate uncovered portions of the bare HgCdTe substrate.

Figure 13 shows Raman spectra of the transferred doped graphene on HgCdTe, in comparison to the as-received graphene on Si/SiO₂; the graphene following the boron doping; and a bare HgCdTe substrate.

The Raman spectroscopy analysis shows the G-band peak resulting from in-plane vibrations of sp^2-bonded carbon atoms, and the D-band peak due to out-of-plane vibrations attributed to the presence of structural defects. The associated

D/G ratio relates to the sp^3/sp^2 carbon ratio. The 2D-band, the second order of the D-band, is the result of a two-phonon lattice vibrational process.

The ratio of 2D/G intensities provides insight into the properties of the graphene layers. For example, a 2D/G band ratio in this case found in the range of 1–2 indicates a bilayer graphene structure. In addition, the same D-band and 2G-band graphene peaks present in the bilayer graphene prior to doping as well as in graphene samples doped on Si/SiO$_2$ have likewise observed in graphene transferred onto the HgCdTe substrate, thus demonstrating preservation of the structural integrity in the transferred bilayers of doped graphene.

6. Conclusions

The material and electrical properties of high-performance graphene-HgCdTe detector technology, where the graphene layer functions as a high mobility channel, developed for MWIR sensing and imaging for NASA Earth Science applications have been assessed. Comprehensive modeling of HgCdTe, graphene, and the HgCdTe-graphene interface has aided in the design and development of this MWIR detector technology.

By using a SOD process, we have achieved boron doping of the bilayer graphene. SIMS, XPS, and Raman spectroscopy-based characterization of the doping levels and properties have confirmed higher boron doping concentrations >10^{20} cm^{-3} in the graphene layers. The p-doped graphene bilayers originally on Si/SiO$_2$ substrates have been furthermore transferred onto HgCdTe substrates, and the structural integrity of the transferred doped layers confirmed through various methods of characterization for implementation as high mobility channels in uncooled MWIR graphene-enhanced HgCdTe detection devices.

Successful integration of enhanced graphene into HgCdTe photodetectors can thereby provide higher MWIR detector performance as compared to HgCdTe detectors alone. Combined with the room temperature operational capability of the graphene-HgCdTe detectors and arrays, the fulfillment of the objective of attaining new earth observation measurement capabilities is a step closer to benefit and advancing critical NASA Earth Science applications.

Acknowledgements

This research is and has been funded by the National Aeronautics and Space Administration (NASA), Contract No. 80NSSC18C0024. The views and conclusions contained in this document are those of the authors and should not be interpreted as representing the official policies, either express or implied, of NASA or the U.S. Government.

Author details

Ashok K. Sood[1,2*], John W. Zeller[1,2], Parminder Ghuman[3], Sachidananda Babu[3],
Nibir K. Dhar[4], Randy N. Jacobs[5], Latika S. Chaudhary[6], Harry Efstathiadis[6],
Samiran Ganguly[7], Avik W. Ghosh[7], Sheikh Ziauddin Ahmed[7]
and Farjana Ferdous Tonni[7]

1 Magnolia Optical Technologies, Inc., Woburn, MA, United States

2 Magnolia Optical Technologies Inc., Albany, NY, United States

3 NASA Earth Science Technology Office, Greenbelt, MD, USA

4 Department of Electrical and Computer Engineering, Virginia Commonwealth
University, Richmond, VA, USA

5 U.S. Army Night Vision and Electronic Sensors Directorate, Fort Belvoir, VA, USA

6 College of Nanoscale Science and Engineering, State University of New York
Polytechnic Institute, Albany, NY, USA

7 Department of Electrical and Computer Engineering, University of Virginia,
Charlottesville, VA, USA

*Address all correspondence to: aksood@magnoliaoptical.com

IntechOpen

References

[1] Zhang X, Rajaraman BR, Liu H, Ramakrishna S. Graphene's potential in materials science and engineering. RSC Advances. 2014;**4**:28987-29011. DOI: 10.1039/C4RA02817A

[2] Pham VP. Direct Growth of Graphene on Flexible Substrates Towards Flexible Electronics: A Promising Perspective. London: IntechOpen; 2018. DOI: 10.5772/intechopen.73171

[3] Jariwala D, Srivastava A, Ajayan PM. Graphene synthesis and band gap opening. Journal of Nanoscience and Nanotechnology. 2011;**11**:6621-6641. DOI: 10.1166/jnn.2011.5001

[4] Randviir EP, Brownson DA, Banks CE. A decade of graphene research: Production, applications and outlook. Materials Today. 2014;**17**:426-432. DOI: 10.1016/j.mattod.2014.06.001

[5] Sood AK, Lund I, Puri YR, Efstathiadis H, Haldar P, Dhar NK, et al. Review of graphene technology and its applications for electronic devices. In: Farzad Ebrahimi F, editor. Graphene—New Trends and Developments. London: IntechOpen; 2015. pp. 59-89. DOI: 10.5772/61316

[6] Pinto H, Markevich A. Electronic and electrochemical doping of graphene by surface adsorbates. Beilstein Journal of Nanotechnology. 2014;**5**:1842-1848. DOI: 10.3762/bjnano.5.195

[7] Novoselov KS, Jiang D, Schedin F, Booth TJ, Khotkevich VV, Morozov SV, et al. Two-dimensional atomic crystals. Proceedings of the National Academy of Sciences. 2005;**102**:10451-10453. DOI: 10.1073/pnas.0502848102

[8] Novoselov KS, Fal VI, Colombo L, Gellert PR, Schwab MG, Kim K. A roadmap for graphene. Nature. 2012;**490**:192-200. DOI: 10.1038/nature11458

[9] Geim A, Novoselov K. The rise of graphene. Nature Materials. 2007;**6**:183-191. DOI: 10.1038/nmat1849

[10] Zhu Y, Murali S, Cai W, Li X, Suk JW, Potts JR, et al. Graphene and graphene oxide: Synthesis, properties, and applications. Advanced Materials. 2010;**22**:3906-3924. DOI: 10.1002/adma.201001068

[11] Dasari BL, Nouri JM, Brabazon D, Naher S. Graphene and derivatives—Synthesis techniques, properties and their energy applications. Energy. 2017;**140**:766-768. DOI: 10.1016/j.energy.2017.08.048

[12] Srivastava S, Jain SK, Gupta G, Senguttuvan TD, Gupta BK. Boron-doped few-layer graphene nanosheet gas sensor for enhanced ammonia sensing at room temperature. RSC Advances. 2020;**10**:1007-1014. DOI: 10.1039/C9RA08707A

[13] Sood AK, Zeller JW, Ghuman P, Babu S, Dhar NK, Jacobs R, et al. Development of high-performance graphene-HgCdTe detector technology for mid-wave infrared applications. In: Proceedings of SPIE. Vol. 11831. Bellingham: SPIE; 2021. p. 1183103

[14] Jacobs RN, Benson JD, Stoltz AJ, Almeida LA, Farrell S, Brill G, et al. Analysis of thermal cycle-induced dislocation reduction in HgCdTe/CdTe/Si(211) by scanning transmission electron microscopy. Journal of Crystal Growth. 2013;**366**:88-94. DOI: 10.1016/j.jcrysgro.2012.12.007

[15] Sood AK, Zeller JW, Pethuraja GG, Welser RE, Dhar NK, Wijewarnasuriya PS. Nanostructure technology for EO/IR detector applications. In: Ghamsari MS, Dhara S, editors. Nanorods and Nanocomposites. London: IntechOpen; 2019. pp. 69-93. DOI: 10.5772/intechopen.85741

[16] Dhar NK, Dat R, Sood AK. Advances in infrared detector array technology. In: Pyshkin SL, Ballato JM, editors. Optoelectronics—Advanced Materials and Devices. London: IntechOpen; 2013. pp. 165-208. DOI: 10.5772/51665

[17] Sood AK, Zeller JW, Ghuman P, Babu S, Dhar NK. Development of high-performance detector technology for UV and IR applications. In: IGARSS 2019-2019 IEEE International Geoscience and Remote Sensing Symposium; Yokohama; 28 July-2 August 2019. New York: IEEE; 2019. p. 20200000463

[18] Sood AK, Zeller JW, Ghuman P, Babu S, Dhar NK, Jacobs R, et al. Development of high-performance graphene-HgCdTe detector technology for mid-wave infrared applications. In: Proceedings of SPIE. Vol. 11723. Bellingham: SPIE; 2021. p. 117230I

[19] Agnoli S, Favaro M. Doping graphene with boron: A review of synthesis methods, physicochemical characterization, and emerging applications. Journal of Materials Chemistry A. 2016;**4**:5002-5025. DOI: 10.1039/C5TA10599D

[20] Bae S, Kim H, Lee Y, Xu X, Park JS, Zheng Y, Balakrishnan J, et al. Roll-to-roll production of 30-inch graphene films for transparent electrodes. Nature Nanotechnology. 2020;**5**:574-578. DOI: 10.1038/nnano.2010.132

[21] Mousavi H, Moradian R. Nitrogen and boron doping effects on the electrical conductivity of graphene and nanotube. Solid State Sciences. 2011;**13**:1459-1464. DOI: 10.1016/j.solidstatesciences.2011.03.008

[22] Yu X, Han P, Wei Z, Huang L, Gu Z, Peng S, et al. Boron-doped graphene for electrocatalytic N_2 reduction. Joule. 2018;**2**:1610-1622. DOI: 10.1016/j.joule.2018.06.007

[23] Nankya R, Lee J, Opar DO, Jung H. Electrochemical behavior of boron-doped mesoporous graphene depending on its boron configuration. Applied Surface Science. 2019;**489**:552-559. DOI: 10.1016/j.apsusc.2019.06.015

[24] Tennyson WD, Tian M, Papandrew AB, Rouleau CM, Puretzky AA, Sneed BT, et al. Bottom up synthesis of boron-doped graphene for stable intermediate temperature fuel cell electrodes. Carbon. 2017;**123**:605-615. DOI: 10.1016/j.carbon.2017.08.002

[25] Panchakarla LS, Subrahmanyam KS, Saha SK, Govindaraj A, Krishnamurthy HR, Waghmare UV, et al. Synthesis, structure, and properties of boron-and nitrogen-doped graphene. Advanced Materials. 2009;**21**:4726-4730. DOI: 10.1002/adma.200901285

[26] Fujisawa K, Hayashi T, Endo M, Terrones M, Kim JH, Kim YA. Effect of boron doping on the electrical conductivity of metallicity-separated single-walled carbon nanotubes. Nanoscale. 2018;**10**:12723-12733. DOI: 10.1039/C8NR02323A

[27] Feng L, Qin Z, Huang Y, Peng K, Wang F, Yan Y, et al. Boron-, sulfur-, and phosphorus-doped graphene for environmental applications. Science of the Total Environment. 2020;**698**: 134239. DOI: 10.1016/j.scitotenv.2019.134239

[28] Santhosh R, Raman SS, Krishna SM, Sai Ravuri S, Sandhya V, Ghosh S, et al. Heteroatom doped graphene based hybrid electrode materials for supercapacitor applications. Electrochimica Acta. 2018;**276**:284-292. DOI: 10.1016/j.electacta.2018.04.142

[29] Usachov DY, Fedorov AV, Vilkov OY, Petukhov AE, Rybkin AG, Ernst A, et al. Large-scale sublattice asymmetry in pure and boron-doped graphene. Nano Letters. 2016;**16**:4535-4543. DOI: 10.1021/acs.nanolett.6b0179

[30] Sin DY, Park IK, Ahn HJ. Enhanced electrochemical performance of phosphorus incorporated carbon nanofibers by the spin-on dopant method. RSC Advances. 2016;**6**:58823-58830. DOI: 10.1039/C6RA06782D

[31] Wu Y, Han Z, Younas W, Zhu Y, Ma X, Cao C. P-Type boron-doped monolayer graphene with tunable bandgap for enhanced photocatalytic H_2 evolution under visible-light irradiation. ChemCatChem. 2019;**11**:5145-5153. DOI: 10.1002/cctc.201901258

[32] Sahoo M, Sreena KP, Vinayan BP, Ramaprabhu S. Green synthesis of boron doped graphene and its application as high performance anode material in Li ion battery. Materials Research Bulletin. 2015;**61**:383-930. DOI: 10.1016/j.materresbull.2014.10.049

[33] Ren S, Rong P, Yu Q. Preparations, properties and applications of graphene in functional devices: A concise review. Ceramics International. 2018;**44**:11940-11955. DOI: 10.1016/j.ceramint.2018.04.089

[34] Jang AR, Lee YW, Lee SS, Hong J, Beak SH, Pak S, et al. Electrochemical and electrocatalytic reaction characteristics of boron-incorporated graphene via a simple spin-on dopant process. Journal of Materials Chemistry A. 2018;**6**:7351-7356. DOI: 10.1039/C7TA09517A

[35] Zagozdzon-Wosik W, Grabiec PB, Lux G. Silicon doping from phosphorus spin-on dopant sources in proximity rapid thermal diffusion. Journal of Applied Physics. 1994;**75**:337-344. DOI: 10.1063/1.355855

[36] Lee HC, Liu WW, Chai SP, Mohamed AR, Aziz A, Khe CS, et al. Review of the synthesis, transfer, characterization and growth mechanisms of single and multilayer graphene. RSC Advances. 2017;7:15644-15693. DOI: 10.1039/C7RA00392G

[37] Chen Y, Gong XL, Gai JG. Progress and challenges in transfer of large-area graphene films. Advanced Science. 2016;**3**:1500343. DOI: 10.1002/advs.201500343

[38] Kang S, Yoon T, Kim S, Kim TS. Role of crack deflection on rate dependent mechanical transfer of multilayer graphene and its application to transparent electrodes. ACS Applied Nano Materials. 2019;**2**:1980-1985. DOI: 10.1021/acsanm.9b00014

[39] Ullah S, Yang X, Ta HQ, Hasan M, Bachmatiuk A, Tokarska K, et al. Graphene transfer methods: A review. Nano Research. 2021;**5**:1-7. DOI: 10.1007/s12274-021-3345-8

[40] Ma LP, Ren W, Cheng HM. Transfer methods of graphene from metal substrates: A review. Small Methods. 2019;**3**:1900049. DOI: 10.1002/smtd.201900049

[41] Kostogrud IA, Boyko EV, Smovzh DV. The main sources of graphene damage at transfer from copper to PET/EVA polymer. Materials Chemistry and Physics. 2018;**219**:67-73. DOI: 10.1016/j.matchemphys.2018.08.001

[42] Gao L, Ren W, Xu H, Jin L, Wang Z, Ma T, et al. Repeated growth and bubbling transfer of graphene with millimetre-size single-crystal grains using platinum. Nature Communications. 2012;**3**:1-7. DOI: 10.1038/ncomms1702

[43] Leong WS, Wang H, Yeo J, Martin-Martinez FJ, Zubair A, Shen PC, et al. Paraffin-enabled graphene transfer. Nature Communications. 2019;**10**:1-8. DOI: 10.1038/s41467-019-08813-x

[44] Gorantla S, Bachmatiuk A, Hwang J, Alsalman HA, Kwak JY, Seyller T, et al. A universal transfer route for graphene.

Nanoscale. 2014;6:889-896.
DOI: 10.1039/C3NR04739C

[45] Park H, Lim C, Lee CJ, Kang J,
Kim J, Choi M, et al. Optimized poly
(methyl methacrylate)-mediated
graphene-transfer process for
fabrication of high-quality graphene
layer. Nanotechnology. 2018;29(41):
415303. DOI: 10.1088/1361-6528/aad4d9

Chapter 5

Analysis of Heat Transfer in Non-Coaxial Rotation of Newtonian Carbon Nanofluid Flow with Magnetohydrodynamics and Porosity Effects

Wan Nura'in Nabilah Noranuar,

Ahmad Qushairi Mohamad, Sharidan Shafie, Ilyas Khan,

Mohd Rijal Ilias and Lim Yeou Jiann

Abstract

The study analyzed the heat transfer of water-based carbon nanotubes in non-coaxial rotation flow affected by magnetohydrodynamics and porosity. Two types of CNTs have been considered; single-walled carbon nanotubes (SWCNTs) and multi-walled carbon nanotubes (MWCNTs). Partial differential equations are used to model the problem subjected to the initial and moving boundary conditions. Employing dimensionless variables transformed the system of equations into ordinary differential equations form. The resulting dimensionless equations are analytically solved for the closed form of temperature and velocity distributions. The obtained solutions are expressed in terms of a complementary function error. The impacts of the embedded parameters are graphically plotted in different graphs and are discussed in detail. The Nusselt number and skin friction are also evaluated. The temperature and velocity profiles have been determined to meet the initial and boundary conditions. An augment in the CNTs' volume fraction increases both temperature and velocity of the nanofluid as well as enhances the rate of heat transport. SWCNTs provides high values of Nusselt number compared to MWCNTs. For verification, a comparison between the present solutions and a past study is conducted and achieved excellent agreement.

Keywords: Nanofluids, Carbon nanotubes, Newtonian fluid, Magnetohydrodynamics, Heat transfer

1. Introduction

The growing demand in manufacturing has led to a significant process of heat energy transfer in industry applications such as nuclear reactors, heat exchangers, radiators in automobiles, solar water heaters, refrigeration units and the electronic cooling devices. Enhancing the heating and cooling processes in industries will save energy, reduce the processing time, enhances thermal rate and increase the equipment's lifespan. Sivashanmugam [1] found that nanofluid emergence has improved heat transfer capabilities for processes in industries. Choi and Eastman [2] established the nanofluid by synthesizing nanoparticles in the conventional base fluid. To be specific, nanofluid is created by suspending nano-sized particles with commonly less than 100 nm into the ordinary fluids such as ethylene glycol, propylene glycol, water and oils [3]. Various materials from different groups can be used as the nanoparticles such as Al_2O_3 and CuO from metalic oxide, Cu, Ag, Au from metals, SiC and TiC from carbide ceramics, as well as TiO_3 from semiconductors [4]. In addition, immersion of nanoparticles is a new way of enhancing thermal conductivity of ordinary fluids which directly improves their ability in heat transportation [5]. In line with nanofluid's contribution in many crucial applications, a number of research has been carried out to discover the impacts of various nanofluid suspension on the flow features and heat transfer with several effects including Sulochana et al. [6] considering CuO-water and TiO-water, Sandeep and Reddy [7] using Cu-water, and Abbas and Magdy [8] choosing Al_2O_3-water as their nanofluid.

Magnetohydrodynamics (MHD) is known as the resultant effect due to mutual interaction of magnetic field and moving electrical conducting fluid. Their great applications such as power generation system, MHD energy conversion, pumps, motors, solar collectors have drawn significant attention of several researcher for MHD nanofluid in convective boundary layer flow [9]. Benos and Sarris [10] studied the impacts of MHD flow of nanofluid in a horizontal cavity. Hussanan et al. [11] analyzed the transportation of mass and heat for MHD nanofluid flow restricted to an accelerated plate in a porous media. In this study, water-based oxide and non-oxide had been considered as the nanofluids. Prasad et al. [12] performed similar work as [11] concerning the radiative flow of nanofluid over a vertical moving plate. Anwar et al. [13] conducted the MHD nanofluid flow in a porous material with heat source/sink and radiation effects. Cao et al. [14] analyzed the heat transfer and flow regimes for a Maxwell nanofluid under MHD effect. While, Ramzan et al. [15] investigated for a radiative Jeffery nanofluid and Khan et al. [16] carried out for a Casson nanofluid with Newtonian heating.

One of the greatest discoveries in material science history is carbon nanotubes (CNTs), which was discovered by a Japanese researcher in the beginning of the 1990s. Since the discovery, due to the unique electronic structural and mechanical characteristics, CNTs are found as valuable nanoparticles, especially in nanotechnology field. CNTs are great conductance which is highly sought in medical applications. They have been used as drug carriers and have benefited cancer therapy treatments [17]. The high thermal conductivity of CNTs has attracted significant attention from many researchers, including Xue [18], Khan et al. [19] and Saba et al. [20]. CNTs are hollow cylinders of carbon atoms in the forms of metals or semiconductors. CNTs are folded tubes of graphene sheet made up of hexagonal carbon rings, and their bundles are formed. CNTs are classified into two types with respectively differ in the graphene cylinder arrangement which are single-walled carbon nanotubes (SWCNTs) and multi-walled carbon nanotubes (MWCNTs). SWCNTs has one layer [21], while MWCNTs consist of more than one graphene cylinder layers [22]. Khalid et al. [23] studied the characteristics of flow and heat transfer for CNTs nanofluid affected by MHD and porosity effects. Acharya et al. [24] discussed a comparative study on the properties of MWCNTs and SWCNTs suspended in water with the imposition of magnetic field. The CNTs nanofluid flow induced by a moving plate was investigated by Anuar et al. [25] and a prominent effect on heat transfer and skin friction by SWCNTs was observed. Ebaid et al. [26] analyzed convective boundary layer for CNTs nanofluid under magnetic field effect. The closed form solution was derived using Laplace transform method and the findings showed increasing magnetic strength and volume fraction of CNTs had deteriorated the rate of heat transport. Aman et al. [27] improved heat transfer for a Maxwell CNTs nanofluid moving over a vertical state plate with constant wall temperature. The investigation of velocity slip of carbon nanotubes flow with diffusion species was conducted by Hayat et al. [28]. Recently, the heat transmission analysis for water-based CNTs was discussed by Berrehal and Makinde [29], considering the flow over non-parallel plates and Ellahi et al. [30] considering flow past a truncated wavy cone.

Due to its broad range of uses such as car brake system, manufacturing of glass and plastic films, gas turbines, and medical equipment's, numerous researchers have effectively studied heat transfer and fluid flow in a rotation system [31]. The impact of MHD and porosity on rotating nanofluid flow with double diffusion by using regular nanoparticles was discussed by Krishna and Chamkha [32]. More features of heat transfer affected by porosity and magnetic field for a rotating fluid flow were referred in Das et al. [33] and Krishna et al. [34, 35]. Kumam et al. [36] implemented CNTs in analyzing the flow behavior for a rotating nanofluid. The nanofluid was considered as an electrical conducting fluid moving in a channel under heat source/sink and radiation effects. More study on the heat propagation for a convective flow of nanofluid in a rotating system affected by CNTs with several effects and different geometries were presented by Imtiaz et al. [37], Mosayebidorcheh and Hatami [38] and Acharya et al. [39]. Interestingly, several researchers had recently concentrated their study on the non-coaxial rotation flow. Mixer machines in food processing industry, cooling pad of electronic devices and rotating propellers for aircraft have become great application to exemplify the non-coaxial rotating phenomenon in various industries. Mohamad et al. [40] presented the mathematical expression for heat transfer in non-coaxial rotation of viscous fluid flow. As the extension of the previous study, the heat and mass transfer effects (double diffusion) were considered by Mohamad et al. [41] and followed by Mohamad et al. [42] investigating porosity effect in double diffusion flow of MHD fluid. Ersoy [43] imposed a disk with non-torsional oscillation to study the convective non-coaxial rotating flow for a Newtonian fluid. Mohamad et al. [44, 45] worked on similar study considering the second grade fluid and Rafiq et al. [46] concerning the Casson fluid model. The time dependent flow of an incompressible fluid with MHD, chemical reaction and radiation effects under non-coaxial rotation was investigated by Rana et al. [47]. Subjecting to the same type of rotation, Mohamad et al. [48] studied the porosity and MHD consequences in mixed convection flow influenced by an accelerated disk. The study was improved by Noranuar et al. [49] including the effects of double diffusive flow. According to the review of non-coaxial rotation, it is clear that most of the study are subjected to the ordinary fluid. However, the study of non-coaxial rotation for nanofluid by using regular nanoparticles had been performed by Das et al. [50] and Ashlin and Mahanthesh [51] but then the study reporting the implementation of CNTs in non-coaxial rotation flow remains limited.

Inspiring from the above literature, new study is essential to explore more findings on non-coaxial rotation of CNTs nanofluid. Therefore, the investigation of MHD non-coaxial rotating flow of CNTs nanofluid due to free convection in a porous medium become the primary focus of the current study. Water base fluid is chosen to suspend nanoparticle of SWCNTs and MWCNTs. The exact solutions for velocity and temperature distributions are attained by solving the problem analytically using the Laplace transform method. The results are illustrated in several graphs and tables for further analysis of various embedded parameters.

2. Problem formulation

The incompressible time-dependent carbon nanofluid instigated by non-coaxial rotation past a vertical disk with an impulsive motion is considered as illustrated in **Figure 1**, where x and z are the Cartesian coordinates with x-axis is chosen as the upward direction and z-axis is the normal of it. The semi-finite space $z > 0$ is occupied by nanofluid that composed by constant kinematic viscosity v_{nf} of SWCNTs and MWCNTs suspended in water and acts as an electrically conducting fluid flowing through a porous medium. The disk is placed vertically along the x-axis with forward motion and a uniform transverse magnetic field of strength B_0 is applied orthogonal to it. The plane $x = 0$ is considered as rotation axes for both disk and fluid. Initially, at $t = 0$, the fluid and disk are retained at temperature T_∞ and rotate about z'-axis with the same angular velocity Ω. After time $t > 0$, the fluid remains rotating at z'-axis while the disk begins to move with velocity U_0 and rotates at z-axis. Both rotations have a uniform angular velocity Ω. The temperature of the disk raises to T_w and the distance between the two axes of rotation is equal to ℓ. With above assumptions, the usual Boussinesq approximation is applied, and the nanofluid model proposed by Tiwari and Das [52] is used to represent the problem in the governing equations, express as

$$\rho_{nf} \frac{\partial F}{\partial t} + \left(\rho_{nf} \Omega i + \sigma_{nf} B_0{}^2 + \frac{\mu_{nf}}{k_1} \right) = \mu_{nf} \frac{\partial^2 F}{\partial z^2} + (\rho \beta_T)_{nf} g_x (T - T_\infty)$$
$$+ \left(\rho_{nf} \Omega i + \sigma_{nf} B_0{}^2 + \frac{\mu_{nf}}{k_1} \right) \Omega \ell, \qquad (1)$$

$$(\rho C_p)_{nf} \frac{\partial T}{\partial t} = k_{nf} \frac{\partial^2 T}{\partial z^2}. \qquad (2)$$

The corresponding initial and boundary conditions are

$$F(z, 0) = \Omega \ell; T(z, 0) = T_\infty; z > 0,$$
$$F(0, t) = U_0; T(0, t) = T_w; t > 0, \qquad (3)$$
$$F(\infty, t) = \Omega \ell; T(\infty, t) = T_\infty; t > 0,$$

in which $F = f + ig$ is the complex velocity; f and g are (real) primary and (imaginary) secondary velocities respectively, T is the temperature of nanofluid and U_0 is the characteristic velocity. The following nanofluid constant for dynamic

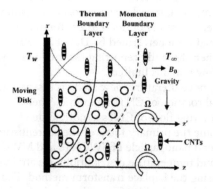

Figure 1.
Physical model of the problem.

viscosity μ_{nf}, density ρ_{nf}, heat capacitance $\left(\rho C_p\right)_{nf}$, electrical conductivity σ_{nf}, thermal expansion coefficient $(\beta_T)_{nf}$ and thermal conductivity k_{nf} can be used as

$$\mu_{nf} = \frac{\mu_f}{(1-\phi)^{2.5}}, \rho_{nf} = (1-\phi)\rho_f + \phi\rho_{CNTs},$$

$$\left(\rho C_p\right)_{nf} = (1-\phi)\left(\rho C_p\right)_f + \phi\left(\rho C_p\right)_{CNTs},$$

$$\frac{\sigma_{nf}}{\sigma_f} = 1 + \frac{3\left(\dfrac{\sigma_{CNTs}}{\sigma_f} - 1\right)\phi}{\left(\dfrac{\sigma_{CNTs}}{\sigma_f} + 2\right) - \phi\left(\dfrac{\sigma_{CNTs}}{\sigma_f} - 1\right)}, \tag{4}$$

$$(\beta_T)_{nf} = \frac{(1-\phi)(\rho\beta_T)_f + \phi(\rho\beta_T)_{CNTs}}{\rho_{nf}},$$

$$\frac{k_{nf}}{k_f} = \frac{1 - \phi + 2\phi\dfrac{k_{CNTs}}{k_{CNTs} - k_f}\ln\left(\dfrac{k_{CNTs} + k_f}{2k_f}\right)}{1 - \phi + 2\phi\dfrac{k_f}{k_{CNTs} - k_f}\ln\left(\dfrac{k_{CNTs} + k_f}{2k_f}\right)},$$

where the subscripts f is for fluid and $CNTs$ is for carbon nanotubes. Meanwhile, ϕ is the solid volume fraction of nanofluid. The constants in Eq. (4) are used based on the thermophysical features in **Table 1**.

Introducing following dimensionless variables

$$F^* = \frac{F}{\Omega\ell} - 1, z^* = \sqrt{\frac{\Omega}{v}}z, t^* = \Omega t, T^* = \frac{T - T_\infty}{T_w - T_\infty}. \tag{5}$$

Using Eqs. (4) and (5), the governing equations in Eqs. (1)–(3) reduce to (excluding the * notation to simplify the equations)

$$\frac{\partial F}{\partial t} + d_1 F = \frac{1}{\phi_1}\frac{\partial^2 F}{\partial z^2} + \phi_3 GrT, \tag{6}$$

$$\frac{\partial T}{\partial t} = \frac{1}{a_1}\frac{\partial^2 T}{\partial z^2} \tag{7}$$

and the conditions take the form

$$F(z, 0) = 0, T(z, 0) = 0; z > 0,$$

$$F(0, t) = U - 1, T(0, t) = 1; t > 0, \tag{8}$$

$$F(\infty, t) = 0, T(\infty, t) = 0; t > 0,$$

Material	Properties				
	$\rho\ (Kgm^{-3})$	$C_p\ (JKg^{-1}K^{-1})$	$k\ (Wm^{-1}K^{-1})$	$\beta \times 10^{-5}\ (K^{-1})$	$\sigma\ (Sm^{-1})$
Water	997.1	4179	0.613	21	0.05
SWCNTs	2600	425	6600	27	$10^6 - 10^7$
MWCNTs	1600	796	3000	44	1.9×10^{-4}

Table 1.
Thermophysical features of water, SWCNTs, and MWCNTs.

where

$$d_1 = \left(i + M^2 \phi_2 + \frac{1}{\phi_1 K} \right), a_1 = \frac{Pr\phi_4}{\lambda}, M = \frac{\sigma_f B_0^2}{\Omega \rho_f} \frac{1}{K} = \frac{\upsilon_f}{k_1 \Omega},$$

(9)

$$Pr = \frac{\upsilon_f (\rho C_p)_f}{k_f}, Gr = \frac{g_x \beta_{Tf} (T_w - T_\infty)}{\Omega^2 \ell}, U = \frac{U_0}{\Omega \ell}.$$

At this point, d_1 and a_1 are constant parameters, M is the magnetic parameter (magnetic field), K is the porosity parameter, Pr is Prandtl number, Gr is Grashof number and U is the amplitude of disk. Besides that, the other constant parameters are

$$\lambda = \frac{k_{nf}}{k_f}, \phi_1 = (1 - \phi)^{2.5} \left((1 - \phi) + \frac{\phi \rho_{CNTs}}{\rho_f} \right),$$

$$\phi_2 = \left(1 + \frac{3 \left(\frac{\sigma_{CNTs}}{\sigma_f} - 1 \right) \phi}{\left(\frac{\sigma_{CNTs}}{\sigma_f} + 2 \right) - \phi \left(\frac{\sigma_{CNTs}}{\sigma_f} - 1 \right)} \right) \frac{1}{(1 - \phi) + \frac{\phi \rho_{CNTs}}{\rho_f}},$$

(10)

$$\phi_3 = \frac{(1 - \phi) + \frac{\phi (\rho \beta)_{CNTs}}{(\rho \beta)_f}}{(1 - \phi) + \frac{\phi \rho_{CNTs}}{\rho_f}}, \phi_4 = (1 - \phi) + \frac{\phi (\rho C_p)_{CNTs}}{(\rho C_p)_f}.$$

3. Exact solution

Next, the system of equations in Eqs. (6)–(8) after applying Laplace transform yield to the following form

$$\frac{d^2}{dz^2} \overline{F}(z, q) - (\phi_1 q + d_2) \overline{F}(z, q) = -d_3 Gr \overline{T}(z, q),$$

(11)

$$\overline{F}(0, q) = (U - 1) \frac{1}{q}, \overline{F}(\infty, q) = 0,$$

(12)

$$\frac{d^2}{dz^2} \overline{T}(z, q) - (a_1 q) \overline{T}(z, q) = 0,$$

(13)

$$\overline{T}(0, q) = \frac{1}{q}, \overline{T}(\infty, q) = 0.$$

(14)

Then, Eqs. (11) and (13) are solved by using the boundary conditions, Eqs. (12) and (14). After taking some manipulations on the resultant solutions, the following Laplace solutions form

$$\overline{F}(z, q) = \overline{F}_1(z, q) - \overline{F}_2(z, q) - \overline{F}_3(z, q) + \overline{F}_4(z, q) + \overline{F}_5(z, q) - \overline{F}_6(z, q),$$

(15)

$$\overline{T}(z, q) = \frac{1}{q} exp \left(-z \sqrt{a_1 q} \right),$$

(16)

where

$$\overline{F}_1(z,q) = \frac{U}{q} \exp\left(-z\sqrt{\phi_1 q + d_2}\right), \overline{F}_2(z,q) = \frac{1}{q} \exp\left(-z\sqrt{\phi_1 q + d_2}\right),$$

$$\overline{F}_3(z,q) = \frac{a_4}{q} \exp\left(-z\sqrt{\phi_1 q + d_2}\right), \overline{F}_4(z,q) = \frac{a_4}{q - a_3} \exp\left(-z\sqrt{\phi_1 q + d_2}\right),$$

$$\overline{F}_5(z,q) = \frac{a_4}{q} \exp\left(-z\sqrt{a_1 q}\right), \overline{F}_6(z,q) = \frac{a_4}{q - a_3} \exp\left(-z\sqrt{a_1 q}\right)$$

$$(17)$$

are defined, respectively. The exact solutions for the temperature and velocity are finally generated by utilizing the inverse Laplace transform on Eqs. (15) and (16). Hence, it results

$$F(z,t) = F_1(z,t) - F_2(z,t) - F_3(z,t) + F_4(z,t) + F_5(z,t) - F_6(z,t) \tag{18}$$

$$T(z,t) = erfc\left(\frac{z}{2}\sqrt{\frac{a_1}{t}}\right) \tag{19}$$

with

$$F_1(z,t) = \frac{U}{2} \exp\left(z\sqrt{\phi_1 d_4}\right) erfc\left(\frac{z}{2}\sqrt{\frac{\phi_1}{t}} + \sqrt{d_4 t}\right)$$

$$+ \frac{U}{2} \exp\left(-z\sqrt{\phi_1 d_4}\right) erfc\left(\frac{z}{2}\sqrt{\frac{\phi_1}{t}} - \sqrt{d_4 t}\right),$$

$$F_2(z,t) = \frac{1}{2} \exp\left(z\sqrt{\phi_1 d_4}\right) erfc\left(\frac{z}{2}\sqrt{\frac{\phi_1}{t}} + \sqrt{d_4 t}\right)$$

$$+ \frac{1}{2} \exp\left(-z\sqrt{\phi_1 d_4}\right) erfc\left(\frac{z}{2}\sqrt{\frac{\phi_1}{t}} - \sqrt{d_4 t}\right),$$

$$F_3(z,t) = \frac{a_4}{2} \exp\left(z\sqrt{\phi_1 d_4}\right) erfc\left(\frac{z}{2}\sqrt{\frac{\phi_1}{t}} + \sqrt{d_4 t}\right)$$

$$+ \frac{a_4}{2} \exp\left(-z\sqrt{\phi_1 d_4}\right) erfc\left(\frac{z}{2}\sqrt{\frac{\phi_1}{t}} - \sqrt{d_4 t}\right), \tag{20}$$

$$F_4(z,t) = \frac{a_4}{2} \exp\left(a_3 t + z\sqrt{\phi_1(a_3 + d_4)}\right) erfc\left(\frac{z}{2}\sqrt{\frac{\phi_1}{t}} + \sqrt{(a_3 + d_4)t}\right)$$

$$+ \frac{a_4}{2} \exp\left(a_3 t - z\sqrt{\phi_1(a_3 + d_4)}\right) erfc\left(\frac{z}{2}\sqrt{\frac{\phi_1}{t}} - \sqrt{(a_3 + d_4)t}\right),$$

$$F_5(z,t) = a_4 erfc\left(\frac{z}{2}\sqrt{\frac{a_1}{t}}\right),$$

$$F_6(z,t) = \frac{a_4}{2} \exp\left(a_3 t + z\sqrt{a_1 a_3}\right) erfc\left(\frac{z}{2}\sqrt{\frac{a_1}{t}} + \sqrt{a_3 t}\right)$$

$$+ \frac{a_4}{2} \exp\left(a_3 t - z\sqrt{a_1 a_3}\right) erfc\left(\frac{z}{2}\sqrt{\frac{a_1}{t}} - \sqrt{a_3 t}\right),$$

where

$$d_2 = \phi_1 d_1, d_3 = \phi_1 \phi_3, d_4 = \frac{d_2}{\phi_1}, a_2 = a_1 - \phi_1, a_3 = \frac{d_2}{a_2}, a_4 = \frac{d_3 Gr}{a_2 a_3}. \tag{21}$$

4. Physical quantities

In this study, the skin friction $\tau(t)$ and Nusselt number Nu for the flow of Newtonian nanofluid in non-coaxal rotation are also analyzed. Their dimensional form is expressed as

$$\tau(t) = -\mu_{nf} \frac{\partial F}{\partial z}\Big|_{z=0} \tag{22}$$

$$Nu = -k_{nf} \frac{\partial T}{\partial z}\Big|_{z=0} \tag{23}$$

Incorporating Eqs. (22) and (23) with the nanofluid model Eq. (4), dimensionless variables Eq. (5) and solutions Eqs. (18) and (19), the following dimensionless skin friction and Nusselt number form as

$$\tau(t) = -\frac{1}{(1-\phi)^{2.5}} \frac{\partial F}{\partial z}\Big|_{z=0},$$
$$= -\frac{1}{(1-\phi)^{2.5}} (\tau_1(t) - \tau_2(t) - \tau_3(t) + \tau_4(t) - \tau_5(t) + \tau_6(t)), \tag{24}$$

$$Nu = -\frac{k_{nf}}{k_f} \frac{\partial T}{\partial z}\Big|_{z=0} = \lambda \sqrt{\frac{a_1}{\pi t}}, \tag{25}$$

where

$$\tau_1(t) = U\sqrt{\phi_1 d_4}\,\text{erfc}\left(\sqrt{d_4 t}\right) - U\sqrt{\phi_1 d_4} - \frac{U}{2}\sqrt{\frac{\phi_1}{\pi t}}\exp\left(-d_4 t\right),$$

$$\tau_2(t) = \sqrt{\phi_1 d_4}\,\text{erfc}\left(\sqrt{d_4 t}\right) - \sqrt{\phi_1 d_4} - \sqrt{\frac{\phi_1}{\pi t}}\exp\left(-d_4 t\right),$$

$$\tau_3(t) = a_4\sqrt{\phi_1 d_4}\,\text{erfc}\left(\sqrt{d_4 t}\right) - a_4\sqrt{\phi_1 d_4} - a_4\sqrt{\frac{\phi_1}{\pi t}}\exp\left(-d_4 t\right),$$

$$\tau_4(t) = a_4\sqrt{\phi_1(a_3 + d_4)}\,\exp\left(a_3 t\right)\text{erfc}\left(\sqrt{(a_3 + d_4)t}\right) - a_4\sqrt{\frac{\phi_1}{\pi t}}\exp\left(-d_4 t\right) \tag{26}$$
$$\quad - a_4\sqrt{\phi_1(a_3 + d_4)}\,\exp\left(a_3 t\right),$$

$$\tau_5(t) = -a_4\sqrt{\frac{a_1}{\pi t}},$$

$$\tau_6(t) = a_4\sqrt{a_1 a_3}\,\exp\left(a_3 t\right)\text{erfc}(\sqrt{a_3 t}) - a_4\sqrt{a_1 a_3}\,\exp\left(a_3 t\right) - a_4\sqrt{\frac{a_1}{\pi t}},$$

with $\tau^* = \tau\sqrt{\nu_f}/\mu_f \Omega^{\frac{3}{2}}\ell.$

5. Analysis of results

The dimensionless differential equations of non-coaxial rotating nanofluid flow with associated boundary and initial conditions are analytically solved using the method of Laplace transform to obtain the closed form solutions of heat transfer. Further analysis for the role of dimensionless time t, Grashof number Gr, volume fraction of nanoparticles ϕ, porosity parameter K, magnetic field parameter M and amplitude of disk U on velocity and temperature distributions as well as Nusselt number and skin friction are presented in figures and tables. The profiles are plotted with the physical value of parameters as $Pr = 6.2, Gr = 0.5, M = 0.2, K = 2.0, \phi = 0.02, U = 2.0$ and $t = 0.2$. The values are same unless for the investigated parameter of the profile. Since the rotating nanofluid is part of the problem, the results are discussed by presenting the graph of velocity profile in real and imaginary parts, specifically describes the primary f and secondary g velocities. The velocity profiles are demonstrated in **Figures 2–7** and the temperature profiles are illustrated in **Figures 8** and **9**. From these profiles, it is found that all the obtained results satisfy both boundary and initial conditions. SWCNTs and MWCNTs have an identical nature of fluid flow and heat transfer.

Figure 2 depicts the plotting of f and g profiles with varying t values. Overall, the velocity of both SWCNTs and MWCNTs rises over time. As t increases, the buoyancy force becomes more effective and functions as an external source of energy to the flow, causing the velocity of fluid to increase. **Figure 3** illustrates the variation of f and g profiles for SWCNTs and MWCNTs cases under the effect of Gr. It is essential to note that Gr is an approximation of the buoyancy force to the viscous force exerting on the flow. Hence, an increase of Gr suggests to the domination of

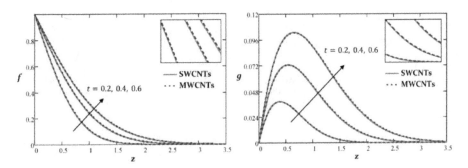

Figure 2.
Profile of f and g for varied values of t.

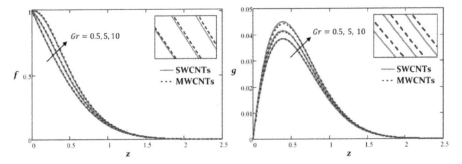

Figure 3.
Profile of f and g for varied values of Gr.

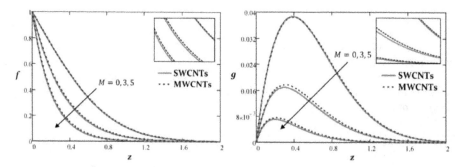

Figure 4.
Profile of f and g for varied values of M.

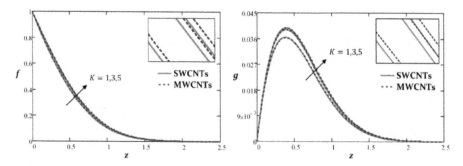

Figure 5.
Profile of f and g for varied values of K.

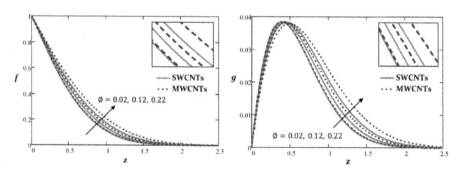

Figure 6.
Profile of f and g for varied values of φ.

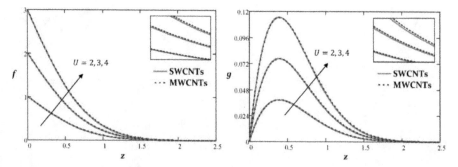

Figure 7.
Profile of f and g for varied values of U.

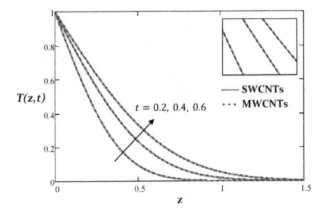

Figure 8.
Profile of $T(z,t)$ for varied values of t.

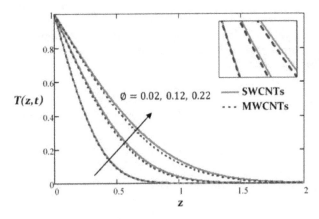

Figure 9.
Profile of $T(z,t)$ for varied values of ϕ.

buoyancy force and reduces the viscosity of fluid. Thus, growing Gr leads to an augment of fluid velocity.

On the other hand, **Figure 4** discloses the nature of fluid flow in response to M. For both SWCNTs and MWCNTs cases, the figure suggests that amplifying M decreases f and g profiles. This impact is owing to the fact that a greater M value increases the frictional forces acting on the fluid, commonly known as the Lorentz force. Consequently, the fluid encounters substantial resistance along the flow and its velocity decreases. Next, the contribution of K in SWCNTs and MWCNTs nanofluids for both f and g profiles are displayed in **Figure 5**. It suggests that K value increases linearly with the velocities for both SWCNTs and MWCNTs. Noting that porosity is also greatly affected by the permeability of a medium, where it determines the ability of a medium to enable the fluid to flow through it. Then, the increasing values of K cause the medium to be more permeable and the fluid can easily pass through the medium. Therefore, it increases both f and g profiles.

Figure 6 reveals the consequences of ϕ on f and g profiles in the cases of SWCNTs and MWCNTs. It shows that increasing ϕ values result in the increment of f profiles and fluctuating trend of g profiles. This suggests significant advantages of non-coaxial rotation in CNTs, especially in industrial and medical applications. In line with a general finding, an analysis proceeding in cancer treatment has reported that the CNTs with higher velocity have been used to reach the tumor's site. Besides, referring to **Figure 7**, it is noticed that ascending U also has a positive impact on velocity profiles for both CNTs suspensions, where the velocity ascends linearly with the values of U. As U increases, this proposes to the creation of external sources, which are used to enhance the thrust force acting in the fluid flow. Thus, the velocity fluid elevates with increasing U.

Furthermore, the temperature profiles $T(z,t)$ under the impacts of t and ϕ are displayed graphically in **Figures 8** and **9**. It reveals that increment of t and ϕ contributes to a rise in nanofluid temperature for both types of CNTs case and followed by the magnification of thermal boundary layer. Physically, the addition of sufficient ϕ of CNTs can improve nanofluid's thermal conductivity. The more CNTs being inserted, the higher the thermal conductivity, which unsurprisingly improves the ability of fluid to conduct heat. Therefore, a growth of temperature profile is exhibited for increasing ϕ. The comparison of physical behavior for SWCNTs and MWCNTs are clearer when referring to the zooming box of each graph. Overall, **Figures 2–7** reveal that the velocity profile of MWCNTs case is more significant compared to the velocity of SWCNTs. This behavior is agreed to the thermophysical features in **Table 1**, where MWCNTs have low density, which also being a key factor for the increase of velocity profiles. Meanwhile, from **Figures 8** and **9**, SWCNTs have provided a prominent effect on temperature profiles as it is affected by a high thermal conductivity property.

t	Gr	M	K	ϕ	U	SWCNTs		MWCNTs	
						τ_p	τ_s	τ_p	τ_s
0.2	0.5	0.2	2	0.02	2	1.3811	−0.2550	1.3691	−0.2523
0.4	0.5	0.2	2	0.02	2	1.0318	−0.3492	1.0236	−0.3455
0.2	5	0.2	2	0.02	2	0.6276	−0.2705	0.6195	−0.2676
0.2	0.5	3	2	0.02	2	3.2171	−0.1596	3.0871	−0.1620
0.2	0.5	0.2	3	0.02	2	1.3377	−0.2578	1.3252	−0.2552
0.2	0.5	0.2	2	0.12	2	1.6820	−0.3138	1.6016	−0.2966
0.2	0.5	0.2	2	0.02	3	2.8459	−0.5082	2.8214	−0.5030

*The significance of bold emphasis used in **Table 2** is for the comparison of the effects for varied values of the particular parameters. For each parameter, the changes of skin friction values are compared among the bold values of parameters.*

Table 2.
Values of primary τ_p and secondary τ_s skin friction for SWCNTs and MWCNTs.

t	ϕ	Nu	
		SWCNTs	MWCNTs
0.2	0.02	3.6238	3.5818
0.4	0.02	2.5624	2.5327
0.2	0.12	5.4840	5.3185

*The significance of bold emphasis used in **Table 3** is for the comparison of the effects for varied values of the particular parameters. For each parameter, the changes of Nusselt number values are compared among the bold values of parameters.*

Table 3.
Values of Nusselt number Nu for SWCNTs and MWCNTs.

Tables **2** and **3** show the results of skin friction (τ_p and τ_s) and Nusselt number *Nu* for various parameters on both cases SWCNTs and MWCNTs. According to **Table 2**, it shows that both τ_p and τ_s of SWCNTs and MWCNTs rise when the strength of *M* higher. These effects cause the surface to produce high friction drag due to the maximization of wall shear stress. On the contrary, as *Gr*, *K* and *t* increase, both suspension of SWCNTs and MWCNTs report a diminution in τ_p and τ_s. This shows that augmentation of *Gr*, *K* and *t* have reduced the friction between fluid and surfaces which lead the velocity to increase. Meanwhile, as ϕ and *U* increase, both suspension of SWCNTs and MWCNTs report a growth of τ_p and a diminution in τ_s. From **Table 3**, it shows that *Nu* for both CNTs cases decrease as the values of *t* increase. However, when involving high ϕ, both SWCNTs and MWCNTs have large *Nu* which also implies to have a great of heat transfer rate. This effect is also directly affected by the reduction of nanofluid heat capacitance as ϕ increases. Overall, for **Table 3**, it is found that SWCNTs case have high value of *Nu* compared to MWCNTs, due to its reduction of heat capacitance. This effect also signifies for a better heat transfer process that can be used in several engineering and industrial system.

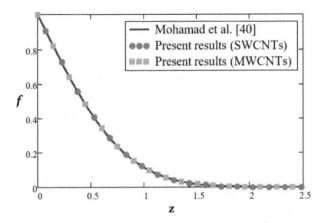

Figure 10.
Comparison of f profiles from present results in Eq. (18) with the published work by Mohamad et al. [40] in Eq. (53).

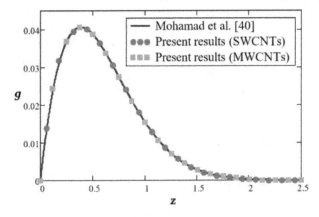

Figure 11.
Comparison of g profiles from present results in Eq. (18) with the published work by Mohamad et al. [40] in Eq. (53).

z	Exact Eq. (18)		Numerical Laplace Eq. (15)	
	SWCNTs	MWCNTs	SWCNTs	MWCNTs
0	1.0000	1.0000	1.0000	1.0000
0.5	0.4165	0.4206	0.4165	0.4206
1.0	0.1089	0.1122	0.1089	0.1121
1.5	0.0171	0.0182	0.0172	0.0182
2.0	0.0016	0.0017	0.0015	0.0017

Table 4.
Comparison of exact and numerical solution of f profiles for SWCNTs and MWCNTs with $t = 0.2, Gr = 0.5, M = 0.2, K = 2, \phi = 0.02, U = 2, Pr = 6.2$.

z	Exact Eq. (18)		Numerical Laplace Eq. (15)	
	SWCNTs	MWCNTs	SWCNTs	SWCNTs
0	0.0000	0.0000	0.0000	0.0000
0.5	0.0366	0.0367	0.0366	0.0367
1.0	0.0146	0.0149	0.0146	0.0150
1.5	0.0027	0.0029	0.0027	0.0029
2.0	0.0003	0.0003	0.0003	0.0003

Table 5.
Comparison of exact and numerical solution of g profiles for SWCNTs and MWCNTs with $t = 0.2, Gr = 0.5, M = 0.2, K = 2, \phi = 0.02, U = 2, Pr = 6.2$.

The accuracy of the obtained solution is verified by comparing solution in Eq. (18) with the solution obtained by Mohamad et al. [40] in Eq. (53). The comparison is conducted by letting magnetic parameter and nanoparticle volume fraction $M = \phi = 0$, and porosity parameter $K \rightarrow \infty$ in the present solution for both types of CNTs and letting phase angle $\omega = 0$ and amplitude of disk oscillation $U = 2$ in the published work. This comparison shows that f and g profiles for both present and previous works are identical to each other as clearly presented in **Figures 10** and **11**, which thus proves that the accuracy of obtained solution is verified. Meanwhile, another verification is also carried out to verify the validity of present solution by comparing the values of velocity profiles from the present work with the numerical values solved by numerical Gaver-Stehfest algorithm [53, 54]. **Tables 4** and **5** observe that the results of f and g profiles from the exact solution in Eq. (18) and the results from numerical solution are in excellent agreement.

6. Summary with conclusion

The unsteady non-coaxial rotation of water-CNTs nanofluid flow in a porous medium with MHD effect is analytically solved for the exact solutions by applying the Laplace transform method. The temperature and velocity profiles with various values of parameter for the immersion of SWCNTs and MWCNTs are plotted graphically and analyzed for their effects. From the discussion, significant findings emerge:

1. Both primary and secondary velocities for SWCNTs and MWCNTs suspension increase as the values of $t, Gr, K,$ and U increase while decrease as the values of M increase.

2. The insertion of higher ϕ of SWCNTs and MWCNTs increases the primary velocity profiles while for secondary velocity profiles, fluctuating trend is reported for both cases.

3. The temperature of nanofluid increases when ϕ and t increase for both SWCNTs and MWCNTs cases.

4. MWCNTs have higher primary and secondary velocity profiles compared to SWCNTs because of their low-density property

5. SWCNTs have higher temperature profile than MWCNTs owing to their high thermal conductivity property.

6. The increasing values of t, Gr and K decrease both primary and secondary skin friction for both types of CNTs while the increase of M gives opposite effect on both skin friction.

7. Nusselt number for both CNTs cases reduce as t increases and amplify as ϕ increases.

8. The findings in present work are in accordance to findings in Mohamad et al. [40] and numerical values obtained by numerical Gaver-Stehfest algorithm.

Acknowledgements

The authors would like to acknowledge the Ministry of Higher Education Malaysia and Research Management Centre-UTM, Universiti Teknologi Malaysia (UTM) for the financial support through vote number 17 J98, FRGS/1/2019/STG06/UTM/02/22 and 08G33.

Conflict of interest

The authors declare that they have no conflicts of interest to report regarding the present study.

Nomenclature

β_T	Thermal expansion coefficient
C_p	Specific heat
ρ	Density
σ	Electrical conductivity
μ	Dynamic viscosity
g_x	Acceleration due to gravity
k	Thermal conductivity
T	Temperature of nanofluid

T_∞	Free stream temperature
T_w	Wall temperature
B_0	Magnetic field
k_1	Permeability
U_0	Characteristic of velocity
Nu	Nusselt number
τ	Skin friction
τ_p	Primary skin friction
τ_s	Secondary skin friction
F	Complex velocity
f	Primary velocity
g	Secondary velocity
ϕ	Volume fraction nanoparticles
Ω	Angular velocity
t	Time
i	Imaginary unit
Pr	Prandtl number
Gr	Grashof number
K	Porosity

Subscripts

$CNTs$	Carbon nanotubes
nf	Nanofluid
f	Fluid

Author details

Wan Nura'in Nabilah Noranuar[1], Ahmad Qushairi Mohamad[1*], Sharidan Shafie[1], Ilyas Khan[2], Mohd Rijal Ilias[3] and Lim Yeou Jiann[1]

1 Faculty of Science, Department of Mathematical Sciences, Universiti Teknologi Malaysia, Johor Bahru, Malaysia

2 Faculty of Mathematics and Statistics, Ton Duc Thang University, Ho Chi Minh City, Vietnam

3 Faculty of Computer and Mathematical Sciences, Centre of Mathematics Studies, Universiti Teknologi MARA (UiTM), Shah Alam, Selangor, Malaysia

*Address all correspondence to: ahmadqushairi@utm.my

IntechOpen

References

[1] Sivashanmugam P. Application of nanofluids in heat transfer, In: S. N. Kazi (Ed.), An Overview of Heat Transfer, INTECH Publications, Croatia, 2012, 411-440.

[2] Choi SUS, Eastman JA. Enhancing thermal conductivity of fluids with nanoparticles, United States, 1995.

[3] Azhar WA, Vieru D, Fetecau C. Free convection flow of some fractional nanofluids over a moving vertical plate with uniform heat flux and heat source. Physics of Fluids. 2017:29(9):1-13. DOI: http://dx.doi.org/10.1063/1.4996034

[4] Gbadeyan JA, Titiloye EO, Adeosun AT. Effect of variable thermal conductivity and viscosity on Casson nanofluid flow with convective heating and velocity slip. Heliyon. 2020:6: e03076. DOI: https://doi.org/10.1016/j.heliyon.2019.e03076

[5] Rehman F, Khan MI, Sadiq M, Malook A. MHD flow of carbon in micropolar nanofluid with convective heat transfer in the rotating frame. Journal of Molecular Liquids. 2017:231:353-263. DOI: 10.1016/j.molliq.2017.02.022

[6] Sulochana C, Ashwinkumar GP, Sandeep N. Effect of frictional heating on mixed convection flow of chemically reacting radiative Casson nanofluid over an inclined porous plate. Alexandria Engineering Journal. 2018:57(4): 2573-2584. DOI: http://dx.doi.org/10.1016/j.aej.2017.08.006

[7] Sandeep N, Reddy MG. Heat transfer of nonlinear radiative magnetohydrodynamic Cu-water nanofluid flow over two different geometries. Journal of Molecular Liquids. 2017:225:87-94. DOI: http://dx.doi.org/10.1016/j.molliq.2016.11.026

[8] Abbas W, Magdy MM. Heat and mass transfer analysis of nanofluid flow based on Cu, Al2O3, and TiO2 over a moving rotating plate and impact of various nanoparticle shapes. Mathematical Problems in Engineering. 2020:2020:1-12. DOI: https://doi.org/10.1155/2020/9606382

[9] Anwar T, Kumam P, Watthayu W. An exact analysis of unsteady MHD free convection flow of some nanofluids with ramped wall velocity and ramped wall temperature accounting heat radiation and injection/consumption. Scientific Reports. 2020:10(1):1-19. DOI: https://doi.org/10.1038/s41598-020-74739-w

[10] Benos L, Sarris IE. Analytical study of the magnetohydrodynamic natural convection of a nanofluid filled horizontal shallow cavity with internal heat generation. International Journal of Heat and Mass Transfer. 2019:130: 862-873. DOI: https://doi.org/10.1016/j.ijheatmasstransfer.2018.11.004

[11] Hussanan A, Salleh MZ, Khan I, Chen ZM. Unsteady water functionalized oxide and non-oxide nanofluids flow over an infinite accelerated plate. Chinese Journal of Physics. 2019:62:115-131. DOI: https://doi.org/10.1016/j.cjph.2019.09.020

[12] Prasad PD, Kumar RVMSSK, Varma SVK. Heat and mass transfer analysis for the MHD flow of nanofluid with radiation absorption. Ain Shams Engineering Journal. 2018:9(4):801-813. DOI: http://dx.doi.org/10.1016/j.asej.2016.04.016

[13] Anwar T, Kumam P, Shah Z, Watthayu W, Thounthong P. Unsteady radiative natural convective MHD nanofluid flow past a porous moving vertical plate with heat source/sink. Molecules. 2020:25(4):1-21. DOI: 10.3390/molecules25040854

[14] Coa Z, Zhoa J, Wang Z, Liu F, Zheng L. MHD flow and heat transfer of

fractional Maxwell viscoelastic nanofluid over a moving plate. Journal of Molecular Liquids. 2016:222: 1121-1127. DOI: 10.1016/j. molliq.2016.08.012

[15] Ramzan M, Bilal M, Chung JD, Mann AB. On MHD radiative Jeffery nanofluid flow with convective heat and mass boundary conditions. Neural Computing and Applications. 2017:30 (9):2739-2748. DOI: https://doi.org/ 10.1007/s00521-017-2852-8

[16] Khan A, Khan D, Khan I, Ali F, Karim FU, Imran M. MHD flow of sodium alginate-based Casson type nanofluid passing through a porous medium with Newtonian heating, Scientific Reports. 2018:8(1):1-18. DOI: 10.1038/s41598-018-26994-1

[17] Saleh H, Alali E, Ebaid A. Medical applications for the flow of carbon-nanotubes suspended nanofluids in the presence of convective condition using Laplace transform. Journal of the Association of Arab Universities for Basic and Applied Sciences. 2017:24: 206-212. DOI: http://dx.doi.org/ 10.1016/j.jaubas.2016.12.001

[18] Xue QZ. Model for thermal conductivity of carbon nanotubes-based composites. Physica B: Condensed Matter. 2005:368(1–4):302-307. DOI: 10.1016/j.physb.2005.07.024

[19] Khan ZH, Khan WA, Haq RU, Usman M, Hamid M. Effects of volume fraction on water-based carbon nanotubes flow in a right-angle trapezoidal cavity: FEM based analysis. International Communications in Heat and Mass Transfer. 2020:116:1-10. DOI: https://doi.org/10.1016/j.icheatmasstra nsfer.2020.104640

[20] Saba F, Ahmed N, Hussain S, Khan U, Mohyud-Din ST, Darus M. Thermal analysis of nanofluid flow over a curved stretching surface suspended by carbon nanotubes with internal heat

generation. Applied Sciences. 2018:8(3): 395. DOI: https://doi.org/10.3390/a pp8030395

[21] Pham VP, Jo YW, Oh JS, Kim SM, Park JW, Kim SH, Jhon MS, Yeom GY. Effect of plasma-nitric acid treatment on the electrical conductivity of flexible transparent conductive films. Japanese Journal of Applied Physics. 2013:52(7R): 075102. DOI: http://dx.doi.org/10.7567/ JJAP.52.075102

[22] Ellahi R, Hassan M, Zeeshan A. Study of natural convection MHD nanofluid by means of single and multi-walled carbon nanotubes suspended in a salt water solution. IEEE Transactions on Nanotechnology. 2015:14(4):726-734. DOI: 10.1109/TNANO.2015.2435899

[23] Khalid A, Khan I, Khan A, Shafie S, Tlili I. Case study of MHD blood flow in a porous medium with CNTS and thermal analysis. Case Studies in Thermal Engineering. 2018:12:374-380. DOI: https://doi.org/10.1016/j.csite .2018.04.004

[24] Acharya N, Bag R, Kundu PK. On the mixed convective carbon nanotube flow over a convectively heated curved surface Heat Transfer. 2020:49(4): 1713-1735. DOI: 10.1002/htj.21687

[25] Anuar N, Bachok N, Pop I. A stability analysis of solutions in boundary layer flow and heat transfer of carbon nanotubes over a moving plate with slip effect. Energies. 2018:11(12): 1-20. DOI: 10.3390/en11123243

[26] Ebaid A, Al Sharif MA. Application of Laplace transform for the exact effect of a magnetic field on heat transfer of carbon nanotubes-suspended nanofluids. Zeitschrift für Naturforschung A. 2015:70(6):471-475. DOI: 10.1515/zna-2015-0125

[27] Aman S, Khan I, Ismail Z, Salleh MZ, Al-Mdallal QM. Heat transfer enhancement in free

convection flow of CNTS Maxwell nanofluids with four different types of molecular liquids. Scientific Reports. 2017:7(1):2445. DOI: 10.1038/s41598-017-01358-3

[28] Hayat T, Hussain Z, Alsaedi A, Hobiny A. Computational analysis for velocity slip and diffusion species with carbon nanotubes. Results in Physics. 2017:7(1):3049-3058. DOI: http://dx.doi.org/10.1016/j.rinp.2017.07.070

[29] Berrehal H, Makinde OD. Heat transfer analysis of CNTs-water nanofluid flow between nonparallel plates: Approximate solutions, Heat Transfer. 2021:1-15. DOI: https://doi.org/10.1002/htj.22112

[30] Ellahi R, Zeeshan A, Waheed A, Shehzad N, Sait SM. Natural convection nanofluid flow with heat transfer analysis of carbon nanotubes–water nanofluid inside a vertical truncated wavy cone. Mathematical Methods in the Applied Sciences. 2021:1-19. DOI: https://doi.org/10.1002/mma.7281

[31] Jayadevamurthy PGR, Rangaswamy NK, Prasannakumara BC, Nisar KS. Emphasis on unsteady dynamics of bioconvective hybrid nanofluid flow over an upward–downward moving rotating disk. Numerical Methods for Partial Differential Equations. 2020:1-22. DOI: 10.1002/num.22680

[32] Krishna MV, Chamkha AJ. Hall and ion slip effects on MHD rotating boundary layer flow of nanofluid past an infinite vertical plate embedded in a porous medium. Results in Physics. 2019:15:1-10. DOI: https://doi.org/10.1016/j.rinp.2019.102652

[33] Das S, Tarafdar B, Jana RN. Hall effects on unsteady MHD rotating flow past a periodically accelerated porous plate with slippage. European Journal of Mechanics / B Fluids. 2018:72:135-143. DOI: https://doi.org/10.1016/j.euromechflu.2018.04.010

[34] Krishna MV, Ahamad NA, Chamkha AJ. Hall and ion slip effects on unsteady MHD free convective rotating flow through a saturated porous medium over an exponential accelerated plate. Alexandria Engineering Journal. 2020:59(2):565-577. DOI: https://doi.org/10.1016/j.aej.2020.01.043

[35] Krishna MV, Ahamad NA, Chamkha AJ. Numerical investigation on unsteady MHD convective rotating flow past an infinite vertical moving porous surface, Ain Shams Engineering Journal. 2020:1-11. DOI: https://doi.org/10.1016/j.asej.2020.10.013

[36] Kumam P, Shah Z, Dawar A, Rasheed HU, Islam S. Entropy generation in MHD radiative flow of CNTs Casson nanofluid in rotating channels with heat source/sink, Mathematical Problems in Engineering. 2019:2019(9158093):1-14. DOI: https://doi.org/10.1155/2019/9158093

[37] Imtiaz M, Hayat T, Alsaedi A, Ahmad B. Convective flow of carbon nanotubes between rotating stretchable disks with thermal radiation effects. International Journal of Heat and Mass Transfer. 2016:101:948-957. DOI: http://dx.doi.org/10.1016/j.ijheatmasstransfer.2016.05.114

[38] Mosayebidorcheh S, Hatami M. Heat transfer analysis in carbon nanotube-water between rotating disks under thermal radiation conditions. Journal of Molecular Liquids. 2017:240:258-267. DOI: http://dx.doi.org/10.1016/j.molliq.2017.05.085

[39] Acharya N, Das K, Kundu PK. Rotating flow of carbon nanotube over a stretching surface in the presence of magnetic field: a comparative study. Applied Nanoscience. 2018:8(3):369-378. DOI: https://doi.org/10.1007/s13204-018-0794-9

[40] Mohamad AQ, Khan I, Ismail Z, Shafie S. Exact solutions for unsteady

free convection flow over an oscillating plate due to non-coaxial rotation. Springerplus. 2016:5(1):1-22. DOI: 10.1186/s40064-016-3748-2

[41] Mohamad AQ, Khan I, Shafie S, Isa ZM, Ismail Z. Non-coaxial rotating flow of viscous fluid with heat and mass transfer. Neural Computing and Applications. 2017:30(9):1-17. DOI: https://doi.org/10.1007/s00521-017-2854-6

[42] Mohamad AQ, Khan I, Jiann LY, Shafie S, Isa ZM, Ismail Z. Double convection of unsteady MHD non-coaxial rotation viscous fluid in a porous medium, Bulletin of the Malaysian Mathematical Sciences Society. 2018:41(4):2117-2139. DOI: https://doi.org/10.1007/s40840-018-0627-8

[43] Ersoy HV. Unsteady flow due to a disk executing non-torsional oscillation and a Newtonian fluid at infinity rotating about non-coaxial axes. Sādhanā. 2017:42(3):307-315. DOI: 10.1007/s12046-017-0600-5

[44] Mohamad AQ, Jiann LY, Khan I, Zin NAM, Shafie S, Ismail Z. Analytical solution for unsteady second grade fluid in presence of non-coaxial rotation. Journal of Physics: Conference Series. 2017:890(1):1-7. DOI: 10.1088/1742-6596/890/1/012040

[45] Mohamad AQ, Khan I, Zin NAM, Isa ZM, Shafie S, Ismail Z. Mixed convection flow on MHD non-coaxial rotation of second grade fluid in a porous medium, AIP Conference Proceedings. 2017:1830(1):020044. DOI: https://doi.org/10.1063/1.4980907

[46] Rafiq SH, Nawaz M, Mustahsan M. Casson fluid flow due to non-coaxial rotation of a porous disk and the fluid at infinity through a porous medium, Journal of Applied Mechanics and Technical Physics. 2018:59(4):601-607. DOI: 10.1134/S0021894418040053

[47] Rana S, Iqbal MZ, Nawaz M, Khan HZI, Alebraheem J, Elmoasry A. Influence of chemical reaction on heat and mass transfer in MHD radiative flow due to non-coaxial rotations of disk and fluid at infinity. Theoretical Foundations of Chemical Engineering. 2020:54(4):664-674. DOI: 10.1134/S0040579520040247

[48] Mohamad AQ, Ismail Z, Omar NFM, Qasim M, Zakaria MN, Shafie S, Jiann LY. Exact solutions on mixed convection flow of accelerated non-coaxial rotation of MHD viscous fluid with porosity effect, Defect and Diffusion Forum. 2020:399:26-37. DOI: 10.4028/www.scientific.net/DDF.399.26

[49] Noranuar WNN, Mohamad AQ, Shafie S, Khan I. Accelerated non-coaxial rotating flow of MHD viscous fluid with heat and mass transfer, IOP Conference Series: Materials Science and Engineering. 2021:1051(1):012044. DOI: 10.1088/1757-899X/1051/1/012044

[50] Das S, Tarafdar B, Jana RN. Hall effects on magnetohydrodynamics flow of nanofluids due to non-coaxial rotation of a porous disk and a fluid at infinity. Journal of Nanofluids. 2018:7(6):1172-1186. DOI: https://doi.org/10.1166/jon.2018.1527

[51] Ashlin TS, Mahanthesh B. Exact solution of non-coaxial rotating and non-linear convective flow of Cu–Al2O3–H2O hybrid nanofluids over an infinite vertical plate subjected to heat source and radiative heat. Journal of Nanofluids. 2019:8(4):781-794. DOI: https://doi.org/10.1166/jon.2019.1633

[52] Tiwari RK, Das MK. Heat transfer augmentation in a two-sided lid-driven differentially heated square cavity utilizing nanofluids. International Journal of Heat and Mass Transfer. 2007:50:2002-2018. DOI: https://doi.org/10.1016/j.ijheatmasstransfer.2006.09.034

[53] Stehfest H. Algorithm 368: Numerical inversion of Laplace transforms [D5]. Communications of the ACM. 1970:13(1):47-49. DOI: https://doi.org/10.1145/361953.361969

[54] Villinger H. Solving cylindrical geothermal problems using the Gaver-Stehfest inverse Laplace transform. Geophysics. 1985:50(10):1581-1587. DOI: https://doi.org/10.1190/1.1441848

Chapter 6

Semiconductor Epitaxial Crystal Growth: Silicon Nanowires

Maha M. Khayyat

Abstract

The topic of nanowires is one of the subjects of technological rapid-progress research. This chapter reviews the experimental work and the advancement of nanowires technology since the past decade, with more focus on the recent work. Nanowires can be grown from several materials including semiconductors, such as silicon. Silicon is a semiconductor material with a very technological importance, reflected by the huge number of publications. Nanowires made of silicon are of particular technological importance, in addition to their nanomorphology-related applications. A detailed description of the first successfully reported Vapor–Liquid–Solid (VLS) 1-D growth of silicon crystals is presented. The bottom-up approach, the supersaturation in a three-phase system, and the nucleation at the Chemical Vapor Deposition (CVD) processes are discussed with more focus on silicon. Positional assembly of nanowires using the current available techniques, including Nanoscale Chemical Templating (NCT), can be considered as the key part of this chapter for advanced applications. Several applied and conceptional methods of developing the available technologies using nanowires are included, such as Atomic Force Microscopy (AFM) and photovoltaic (PV) cells, and more are explained. The final section of this chapter is devoted to the future trend in nanowires research, where it is anticipated that the effort behind nanowires research will proceed further to be implemented in daily electronic tools satisfying the demand of low-weight and small-size electronic devices.

Keywords: nanowires, semiconductors, silicon, CVD, catalyst, PV, AFM

1. Introduction

The topic of semiconductor nanowires is timely developing research. A legitimate question is then: what makes a material in a nanowire different from a bulk one? The direct answer is the extremely large surface-to-volume ratio. Any application occurs at the surface, such as chemical reactions; it will speed up at a medium of high surface area. Indeed, there are more features of integrating nanowires with the current available technologies (such as PV, AFM, and Raman spectroscopy) [1–5] and as stand-alone applications, such as sensors [6, 7]. Moreover, semiconductor nanowires can be functionalized and tailored in accordance with different requirements. For example, we can dope them with particular elements in the growth stage to change electrical properties or change the growth conditions to vary their shape or size.

There are several techniques of growing semiconductor nanowires. These fabrication methods are based on the semiconductor industrial capabilities, mainly top-down and bottom-up approaches. Photoresist patterning on top of a silicon-on-insulator layer followed by etching silicon and creating vertical silicon columns is explained as a top-down approach. Techniques based on the direct epitaxial growth of nanowires from a seeding material on a substrate are called bottom-up growth techniques, which is the main technique discussed in this chapter (see **Figure 1**).

Studies on silicon nanowires (Si-NWs) started with the pioneer work of Wagner and Ellis in 1965 [1]. The Vapor–Liquid–Solid (VLS) growth method uses metallic droplets or particles as a catalyst to nucleate the growth and absorb gaseous precursors and precipitate them into a solid form to permit crystal growth. The classic example is the VLS growth of Si-NWs on a Si substrate using gold (Au) eutectic droplets. Studying the structural properties of NWs is particularly important so that a reliable procedure of fabrication based on the desired functionality can be designed. Due to the enhanced surface-to-volume ratio in nanowires, their properties may depend critically on their surface condition and geometrical configuration. Even nanowires made of the same material may possess dissimilar properties due to differences in their crystal phase, crystalline size, surface conditions, and aspect ratios, which depend on the synthesis methods and conditions used in their fabrication. Moreover, the temperature of growing NWs is of critical importance, when we integrate them with other electronic devices so that the rest of the components of the electronic circuit do not get damaged at high temperatures. Si-NWs have been thus tried to be grown at low temperatures by utilizing metal catalysts, such as gold (Au) and Aluminum (Al), whose alloys with Si have low eutectic temperatures. Generally speaking, Si-NWs have been shown to provide a promising framework for applying the bottom-up approach (Feynman, 1959) for the design of nanostructures for nanoscience investigations and for potential nanotechnology applications. We are progressing in accordance with the predictions of Moore's Law, which suggests that the number of transistors in a dense integrated circuit doubles every two years [8]. Electronic devices are getting smaller and smaller, and the capabilities of these devices are becoming more leading-edge with the integration with the NWs technology.

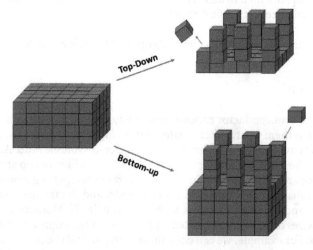

Figure 1.
Schematic representation shows the original substrate, and the top main approaches of creating nanowires: Bottom-up and top-down approaches. Notice that the building blocks of materials (atoms) are moving toward (deposition) the substrate in the bottom-up process, while atoms are moving away (etching) from the substrate in the top-down mechanism.

2. Semiconductor nanowires

The procedure of the bottom-up growth process of semiconductor NWs can be described as follows (see **Figure 2**); a semiconductor substrate, which could a bulk semiconductor substrate or an epitaxial layer of a semiconductor materials on a glass, for instance, (see **Figure 2**, step 1). A thin continuous layer of few nanometers-thick metal (step 2) evaporated on the surface of the semiconductor epitaxial layer, which segregates in isolated droplets during annealing (step 3). Precursor gas flows in the Chemical Vapor Deposition (CVD) reactor, where semiconductor atoms react at the metal-droplet surfaces, depositing semiconductor vapor atoms into solution within the metal droplets (step 4). The catalyst droplets supersaturate, inducing precipitation of crystalline semiconductor vapor atoms upon the substrate. As precipitation occurs only at the droplet metal (liquid)–semiconductor (solid) interfaces, the semiconductor atoms crystallite in wire structures with diameters comparable to the diameter of the metal droplet (step 5). This growth protocol has been called by Wagner and Ellis as VLS growth after the three coexisting phases: the vaporous precursors (such as Si_v), liquid catalyst droplets (such as Au_l), and solid silicon substrate (Si_s). Notice the possible incorporation of some of the metal atoms (Au) which

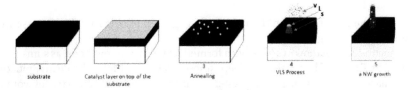

Figure 2.
A schematic representation shows the sequence of the VLS process in five main steps. The substrate, depicted 1, can be bulk semiconductor materials or a relatively thin film of a semiconductor on a cheap substrate such as glass or polycarbonate (PC) or polymethyl methacrylate (PMMA) sheets; 2: Catalyst thin layer; 3: Catalyst after annealing where it balls up; 4: The sample was placed at the CVD reactor, allowing the precursor gas to flow; temperature reaches the eutectic; three phases coexist; and precipitation begins; and 5: Growth continues forming NWs.

Semiconductor materials	Growth techniques	Catalyst	References
TiO_2/In_2O_3	Electron Beam Evaporation	Catalytic free glancing angle deposition technique	Guney et al. [9]
GaAs	Metal–organic chemical vapor deposition (MOCVD). A vapor–liquid–solid (VLS) mechanism	Au	Zeng et al. [10]
SnO$_2$ nanowires	A solvothermal process	Pd	Lu et al. [11]
p-type α-Bi$_2$O$_3$	Vapor–Liquid–Solid (VLS)	different catalysts (Au, Pt and Cu) as seed layers but the highest aspect ratio was obtained using Au	Moumen et al. [7]
β-Ga$_2$O$_3$	The chemical vapor deposition (CVD) method	Au catalysts	Miao et al. [12]

Semiconductor materials	Growth techniques	Catalyst	References
Si	CVD reactor	employing Sn nanospheres as catalyst	Mazzetta et al. [13]
InAs/InP	Molecular Beam Epitaxy (MBE)	gold catalyst	Helmi et al. [14]
Poly(3-hexylthiophene) (P3HT) nanowires	N/A	N/A	Jeong et al. [15]
Si-doped GaAs nanowires (NWs)	VS selective area growth patterned with SiO2 MBE	No catalysts	Ruhstorfer et al. [16]
II-VI semiconductors CdTe, CdS, ZnSe, and ZnS	Vapor–liquid–solid (VLS) process	Bismuth and tin	Yang et al. [17]
GaAs, InAs, and InGaAs nanowires	Molecular Beam Epitaxy (MBE)	Gold as the growth catalyst	Jabeen et al. [18]
Si	VLS-CVD	Al	Wacaser et al. [3]
Si NWs On Si(100) and Si(111)	VLS-CVD	Gold catalyst	Lindner et al. [19]

Table 1.
Published experimental research articles of semiconductor nanowires including the techniques and the catalysts.

catalyzed the growth within the frame of the grown NW (Si-NW), as presented schematically in **Figure 2**.

Semiconductor nanowires have been formed using various methods, as summarized in **Table 1**. Chemical Vapor Deposition (CVD) and Molecular Beam Epitaxy (MBE) have been the main growing systems since the past decade up to recent work for growing various semiconductor nanowires using several catalysts or without catalysts.

3. Epitaxial growth: Silicon nanowires

The nanowires growth is usually performed in a chemical vapor deposition (CVD) reactor or can be at the Molecular Beam Epitaxy (MBE); see **Figure 3**. The CVD growth mechanism involves the absorption of source material from the vapor phase into a liquid droplet of catalyst above the solid substrate as explained in the original work of Wagner and Ellis, in 1969 [20, 21]. The original proposed VLS mechanism of growing Si-NWs using Au as a catalyst is based on three critical parameters; the presence of the arriving Si **vapor** atoms to the metal droplet in a **liquid** state acting as a preferred position on the **solid** substrate. The detailed growth conditions at the CVD such as the pressure, flow rate, and temperature are placed accordingly. On the other hand, the motivation for using molecular beam epitaxy (MBE) to grow nanowires is that although MBE growth is both complicated and challenging, its high precision and flexibility can give good control over the growth of thin layers and abrupt junctions, which may be an advantage in future nanostructure devices [22].

According to the binary phase diagram of Si and Au, as shown in **Figure 4**, the lowest melting temperature for the Au–Si eutectic is approximately 363°C obtained for a composition of Si and Au. The eutectic is lower than the melting point of Au (1064°C) and Si (1414°C) [21]. Considering that the liquid phase is thermodynamically equilibrated with the solid one, the lowering of the melting point, with the size of the droplet, is given by Eq. (1), as follows [23]:

(a)

(b)

Figure 3.
Photographs of (a) chemical vapor deposition (CVD) and (b) molecular beam epitaxy (MBE) systems (pictures were taken with permission from nanoscience Center, Cambridge, UK).

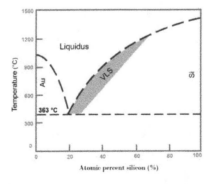

Figure 4.
Phase diagram for the Au–Si system. The shaded zone represents the range of temperatures and alloy compositions at which VLS growth might occur [22].

$$\delta T = 2\sigma.T_0 / (\rho.L.r) \tag{1}$$

where δT is the lowering of the melting point, σ is the interfacial energy, T_0 is the melting point of the bulk metal, ρ is the density of the material,

L is the latent heat, and r is the radius of the circle of the catalysts. Thus, heating Au film deposited on a Si substrate to a temperature of 363°C results in the formation of liquid Au–Si eutectic. The eutectic is simply a mixture of two elements at such proportions that its melting point is at the lowest possible temperature, much lower than the melting point of either of the two elements that make it up. If these Au–Si melted alloys are placed in an environment containing a gaseous silicon precursor such as silane (SiH_4), the precursor molecules decompose into Si and H_2 at the outer surface of the metal droplets, thereby supplying additional Si to the Au–Si alloy, and precipitate at the interface between the liquid alloy droplet and the solid substrate.

It has been shown that Si-NWs grow perpendicularly on Si(111), as represented in **Figure 5**.

A variety of derivatives of CVD methods exist, which can be classified by parameters such as the base and operation pressure or the treatment of the precursor. Since Si is known to oxidize easily, it is a key parameter for a successful epitaxial growth of Si-NWs to reduce the oxygen background pressure. In particular, when oxygen-sensitive catalyst materials are used, it turns out to be useful to combine catalyst deposition and nanowire growth in one system, so that growth experiments can be performed without breaking the vacuum in between [24, 25].

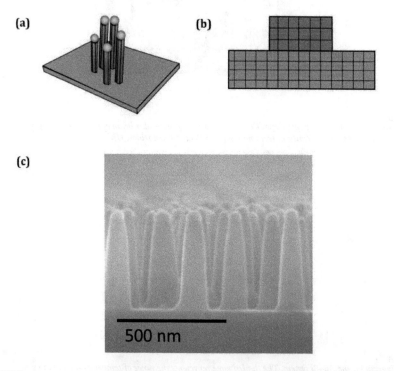

(a)

(b)

(c)

500 nm

Figure 5.
Schematic representation of epitaxial growth of Si-NWs (a), where the grown nanowires copy the crystal structure of the substrate, (c) epitaxial grown Si-NWs on Si(111) substrate catalyzed with Al. Detailed growth conditions can be found in Khayyat et al. [24].

It is often noted in VLS wire growth that the radius of the catalyst droplet exceeds the radius of the nanowire Eq. (2) [22].

$$R = r\sqrt{1/\left(1-\left(\sigma_{ls}/\sigma_l\right)^2\right)} \qquad (2)$$

where R is radius of the catalyst droplet, r is the radius of the nanowire, σ_l is the surface tension of the liquid catalyst, and σ_{ls} is the surface tension of the liquid–catalyst interface. Based on this, one can estimate the growth conditions and deduce the diameters of the catalyst droplet of the resulted growth of NWs with a certain average diameter.

4. Si-NWs as building blocks for bottom-up nanotechnology

Controlling the growth position of an NW is important for fabricating devices, especially when involving a large array of nanowires. The growth reproducibility is critically a key parameter in the progress of implementing nanowires in advance applications.

Free-standing nanowires can be yielded and their position on the wafer can be determined by predefining the position of the seed on the wafer using lithography. There are several research groups working on optimizing the growth of positional Si-NWs [1, 26–28]. Most of the studies till date have used Au due to the convenience of handling that arises from its resistance to oxidation. The current technique is a new method of controlling the position of the grown Si-NWs seeded with oxygen-reactive materials such as Al, which is a standard metal in silicon industrial process line. The technique is based on electron beam lithography for patterning the Si substrate and then forming a Si alloy with Al during a subsequent annealing step. Moreover, it does not require removal of the patterned compound oxide layer [25].

4.1 Nanoscale chemical templating (NCT) technique

It is an innovative technique that arises as a solution of the issue of the defective planar growth between the grown Si-NWs seeded with Al (or any other chemically active elements). The technique is called Nanoscale Chemical Templating (NCT) of oxygen-reactive elements. Now, what makes NCT an innovative solution? [25].

I. Does not require Al removal for selective growth.

II. Does not require any lithography steps.

III. Multiple application space.

As explained in **Figure 6** (I), the process that does not require Al removal (I-a) shows the patterned SiO_2 layer after photolithography, etching, and resist removal. (I-b) After Al deposition and annealing, notice the agglomerated Al:Si feature where it balls up in the openings forming the NW seeds, while the Al in contact with SiO_2 has reacted forming Al_2O_3. (I-c) After NW growth. The NWs are epitaxial and appear as bright spots in the plan view. In the cross sectional view, tapering is visible, due to a thin, non-seeded, Si layer approximately thinner than 1/100 of the length of the nanowire. Notice that a single NW per opening is achieved. (I-c3) and

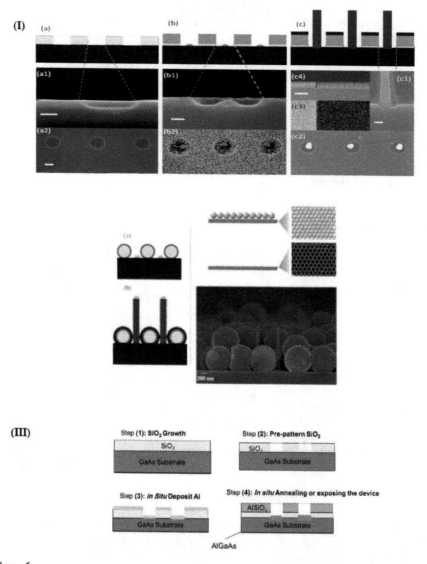

Figure 6.
(I) (a)–(c) schematic illustration of NCT of NWs with corresponding SEM images, cross sectional ((a1), (b1), (c1) and (c3)) and plan views ((a2), (b2), (c2) and (c4)). The scale bars are: (a1) and (b1) 100 nm; (c1) 300 nm; (a2), (b2), and (c2) 1 μm; and (c3) and (c4) 20 μm. (II) representation of NCT using silica microspheres (a&b) and the corresponding SEM original proof of the concept. (III) schematic representation of the selective growth of AlGaAs for further applications.

(I-c4) show a larger area containing both a patterned area on the right and an area with no oxide on the left where random growth occurs.

The position control of Si-NWs can be achieved using silica microsphere, as described in 6 (II). The schematic representation of a spinning silica microsphere on the Si substrate, followed by thin-layer evaporation of Al and the subsequent annealing, is shown in (a), where (b) shows the Si-NWs growth. Moreover, patterning III-V semiconductors selectively is considered as one of the possible multiple applications schemes (III).

4.2 Applications on NCT: functional devices of Si-NWs

It is of great interest to find applications for Si-NWs, which could be as stand-alone innovative structures such as in photovoltaic (PV) cells (**Figure 7**) or integrating with conventional structures such as Atomic Force Microscopy (AFM) (**Figure 8**) [26], and MOSFET (**Figure 9**), for the purpose of developing and miniaturizing.

PV Cells made of Si-NWs have several potential benefits over conventional bulk Si one- or thin-film devices related primarily to cost reduction. It is possible to form the p–n core–shell junctions in high-density arrays, which have the advantages of decoupling the absorption of light from charge transport by allowing lateral diffusion of minority carriers to the p–n junction which is at most 50–500 nm away rather than many microns away as in Si conventional bulk photovoltaic cells. Based on this, the potential cost benefits come from lowering the purity standard and the amount of semiconductor material needed to obtain sophisticated efficiencies, increasing the defect tolerance, and lattice-matched substrates [6]. The concept of

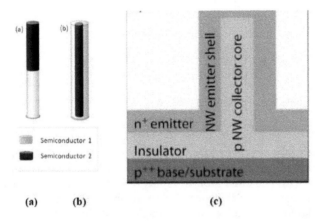

Figure 7.
Schematic illustration of the main types of heterojunction nanowires. (a) Axial heterojunction. (b) Radial heterojunction (core-shell). (c) a core–shell PV cell [3].

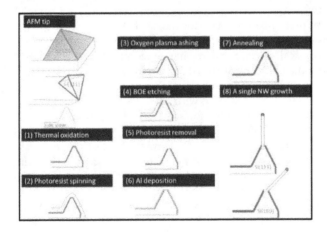

Figure 8.
Schematic demonstration of detailed steps involved in Si-NWs integration with AFM tip [26].

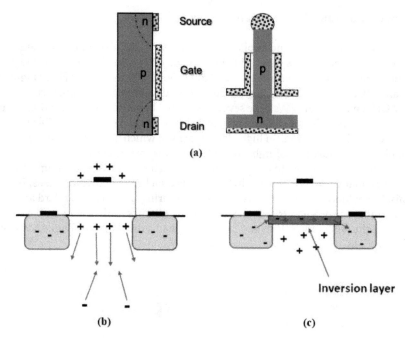

Figure 9.
(a) Shows a schematic representation of an (npn) MOSFET, the conventional one in parallel with the innovative one of NWs; (b) shows the migration of charges based on the applied voltages; and (c) presents the formation of the inversion layer, the channel across the diameter of the NW.

nanowire-based solar cells has attracted significant attention because of their potential benefits in carrier transport, charge separation, and light absorption. The Lieber [1] and Atwater [28] and other groups [29–31] have developed core–shell growth and contact strategy for their silicon p–i–n nanowire solar cells, with sophisticated efficiencies. Moreover, the ability to make single-crystalline nanowires on low-cost substrates, such as Al foil, and to relax strain in subsequent epitaxial layers removes two more major cost hurdles associated with high-efficiency planar solar cells. A schematic representation is shown in **Figure 7** where SiO_2 has been used as a separation layer between the planar defective growth, which occurs during NWs growth, and the substrate to enhance the performance of the PV core–shell junctions [3].

AFM was invented in 1968, which has opened new perspectives for various micro- and nanoscale surface imaging in science and industry. Nanotechnology has benefited from the invention of the AFM, and in turn AFM is developing based on the progress of Si-NWs growth techniques. Based on the NCT technique that is based on catalyzing the growth of Si-NWs with Al, it has been proposed to improve the resolution of AFM tips in a production scale [26]. The concept of "Production Scale Fabrication Method for High Resolution AFM Tips" is demonstrated in **Figure 8**, along with the various steps of the Si-NW growth on the tip of the available Si(100) or Si(111) AFM tips. The grown Si-NW on the squared-base Si(100) tip is 45° tilted, while Si-NW grow perpendicularly on the AFM tip of the triangular base of Si(111) [32–35] where further reduction of the average wire diameter to the nanometer scale can be done via hydrogen annealing or oxidation [8, 36–41]. As the diameter decreased, the tensile strength tended to increase from 4.4 to 11.3 GPa. Under bending, the Si-NWs demonstrated considerable plasticity [42].

The proposed structure in **Figure 8** has not yet been experimentally demonstrated [26]; the fabrication of the structure could be achieved by the described processes where it illustrates the potential mass scalability of this technique. A strategy has been presented to equip microcantilever beams with single Si-NW scanning tips that were directly grown by Au-catalyzed VLS synthesis. It was evident from AFM measurements evidently that the assembled Si-NW scanning tips are suitable for topography reconstruction as well as for overall comparison with conventional pyramidal scanning tips besides their high aspect-ratio nature and a superior durability [39, 43–45].

MOSFET can be designed in the form of NWs, as shown in **Figure 9** [8]. The channel can be altered using NWs as shown when a positive voltage is applied to the gate. The holes in the p-type Si are repelled from the surface, and minority carrier conduction electrons are attracted to the surface. If the gate voltage exceeds the threshold value, then an inversion layer is created near the surface. In this layer, the material behaves as an n-type and provides a conducting channel between the source and the drain. The width of the conduction channel is dominated by the diameter of the NW.

Because of the enhanced surface-to-volume ratio of NWs, their transport behavior may be modified by changing their surface conditions, and this property may be utilized for sensor applications to provide improved sensitivity compared to conventional sensors based on bulk material. Si-NWs sensors will potentially be smaller, more sensitive, demand less power, and react faster than their macroscopic counterparts [42, 43, 46].

5. Future remarks of nanowires research

In this chapter, we attempt to summarize progresses made in this field during the last several years, ranging from nanowire growth with precise control at the atomic level [41]. Probing novel properties in 1D systems using a stand-alone innovative novel device was presented, in addition to integration and assembly methods of large numbers of NWs for practical applications.

We conclude this chapter with some outlooks for future research. Will nanowires research lead to new science or discovery of new phenomena? Will it lead to new applications? [47–50]. The answer is clearly yes based on the research activity done on the topic of nanowires. Studies are among the most potential in the topic of nanoscience, as shown in **Figure 10**. The cumulative published studies starting from 2010 up to 2020 on the topic of nanowires have been increased, thus markedly reflecting the technological importance of this topic.

The ever-growing demand for smaller electronic devices is prompting the scientific community to produce circuits whose components satisfy the size and weight requirements. The well-controlled NW growth process, with distinct chemical composition, structure, size, and morphology, implies that semiconductor nanowires can be integrated within the process of the development of nanodevices. Control of the synthesis and the surface properties of Si-NWs may open new opportunities in the field of silicon nanoelectronics and use them as nanocomponents to build nanocircuits and nanobiosensors. Moreover, Si-NWs possess the combined attributes of cost effectiveness and mature manufacturing infrastructures [51–55].

The conventional thin-film technologies grown at MBE have technical limitations, mainly the interfacial lattice mismatch issues that often result in highly defective optical materials. In this regard, Si-NWs growth provides a natural mechanism for relaxing the lattice strain at the interface and enables dislocation-free semiconductor growth on lattice-mismatched substrates, where radial strain relaxation allows for uncharted combinations of semiconductor materials (III–V on Si). In this regard, efforts must be made to break new grounds in this promising research field to stimulate more creative ideas about nanowire research and applications. Many

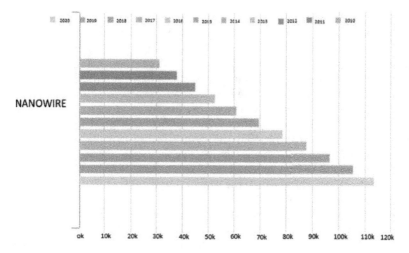

Figure 10.
Cumulative nanowires publications in 11 years 2010–2020 (there are no data available for 2021, where the x-axis represents the number of articles in kilo. In 2020, there were more than 100,000 articles on the topic on nanowires, according to the number of nanotechnology-related articles indexed in web of science (WoS) (ISI web of knowledge). https://statnano.com/report/s29/3.

promising applications are now at the early demonstration stage but are moving ahead rapidly because of their promise for new functionality, not previously available, to the fields of electronics, optoelectronics, biotechnology, magnetics, and energy conversion and generation, among others [56, 57].

Integration of nanoelectronic units, such as Si-NWs, and biosystems is a multidisciplinary field that has the potential for multilateral impact on various scientific fields including biotechnology. The combination of these multidisciplinary research backgrounds promises to yield revolutionary advances in our everyday life through, for example, the creation of new and powerful tools that enable direct, sensitive, and rapid analysis of biological and chemical species [49, 50].

To simulate future research on NWs, one would look at the progress achieved by the industries of semiconductors which have produced devices and systems that are part of our daily lives, including transistors, sensors, lasers, light-emitting diodes, solar panels, computers, and cell phones [51–54, 58]. Then, imagine changing the morphology of semiconductors from the bulk to the nanowire form; one might wonder how much fundamental difference there is. Where, sometimes the intersection of top-down and bottom-up approaches toward building nanostructures for practical functionality is also possible.

Acknowledgements

I would like to thank Materials Science Research Institute, KACST, for the kind professional support and fruitful discussion.

Conflict of interest

The author declares no conflict of interest.

Author details

Maha M. Khayyat
Nanotechnology and Semiconductors National Center, Materials Science Research
Institute, King Abdulaziz City of Science and Technology, Riyadh, Saudi Arabia

*Address all correspondence to: mkhayyat@kacst.edu.sa

IntechOpen

References

[1] Lieber C, Wang Z. Functional nanowires. MRS Bulletin. 2007;**32**

[2] Jihun O, Yuan H, Branz H. An 18.2%-efficient black-silicon solar cell achieved through control of carrier recombination in nanostructures. Nature Nanotechnology. vol. 7. 2012. DOI: 10.1038/NNANO.2012.166

[3] Wacaser B, Reuter M, Khayyat M, Haight R, Guha S, Ross F. The role of microanalysis in micro/nanowire-based future generation photovoltaic devices. Microscopy and Microanalysis. 2010;**16** (Supplement S2)

[4] Sahoo M, Paresh Kale P. Integration of silicon nanowires in solar cell structure for efficiency enhancement: A review. Journal of Materiomics. 2019; 5:34-48

[5] Zamfira M, Nguyenab H, Moyena E, Leeac Y, Pribata D. Silicon nanowires for Li-based battery anodes: A review. Journal of Materials Chemistry A. 2013;**1**:9566

[6] Tien L, Shih Y, Chen R. Broadband photodetectors based on layered 1D GaTe nanowires and 2D GaTe nanosheets. Journal of Alloys and Compounds. 2021;**876** (Cover date: 25 September 2021) Article 160195

[7] Moumen A, Zappa D, Comini E. Catalyst–Assisted vapor liquid solid growth of α-Bi2O3 nanowires for acetone and ethanol detection. Sensors and Actuators B: Chemical. 2021;**346** (Cover date: 1 November 2021) Article 130432)

[8] Colinge J, Greer J. Nanowire Transistors: Physics of Devices and Materials in One Dimension. Cambridge, UK: Cambridge University Press; 2016 ISBN 978-1-107-05240-6 Hardback

[9] Güney H, İskenderoğlu D. CdO:Ag semiconductor nanowires grown by spray method. Journal of Alloys and Compounds. 2021;**865** (Cover date: 5 June 2021) Article 158924

[10] Zeng L, Li L, Liu G. Morphology characterization and growth of GaAs nanowires on selective-area substrates. Chemical Physics Letters. 2021;**779**; Article 138887

[11] Lu S, Zhang Y, Liu H. Sensitive H_2 gas sensors based on SnO_2 nanowires. Sensors and Actuators B: Chemical. 2021;**345** (Cover date: 15 October 2021) Article 130334

[12] Miao Y, Liang B, Chen C. Epitaxial growth of β-Ga2O3 nanowires from horizontal to obliquely upward evolution. Vacuum. 2021;**192** (Cover date: October 2021) Article 110444

[13] Mazzetta I, Rigoni F, Palma F. Large-scale CMOS-compatible process for silicon nanowires growth and BC8 phase formation. Solid-State Electronics. 2021;**186**

[14] Helmi M, Alouane H, Nasr O, Chauvin N. Temperature dependence of optical properties of InAs/InP quantum rod-nanowires grown on Si substrate. Journal of Luminescence. 2020;**231** (Cover date: March 2021) Article 117814

[15] Jeong G, Choi S, Chang M. Dyes and Pigments. 2020;**185**(Part B):108962

[16] Ruhstorfer D, Mejia S, Ramsteiner M, Doblinger M, Riedl H, Jonathan, et al. Demonstration of n-type behavior in catalyst-free Si-doped GaAs nanowires grown by molecular beam epitaxy. Applied Physics Letters. 2020;**116**:052101

[17] Yang L, Wang W, Jia J. Novel route to scalable synthesis of II–VI semiconductor nanowires:

Catalyst-assisted vacuum thermal evaporation. Journal of Crystal Growth. 2010; **312**(20):2852-2856

[18] Jabeen F, Rubini S, Martelli F. Growth of III–V semiconductor nanowires by molecular beam epitaxy. Microelectronics Journal. 2009;**40**(3): 442-445

[19] Lindner J, Bahloul D, Kraus D, Weinl M, Melin T, Strizker B. TEM Characterization of Si nanowires grown by CVD on Si pre-structures by nanosphere lithography. Materials Science in Semiconductor Processing. 2008;**11**:169-174

[20] Wagner RS, Ellis WC. The VLS mechanism of whisker growth. Transactions of the Metallurgical Society of AIME 1965;**233**:1053

[21] Wagner RS. Whisker Technology. New York: Wiley; 1970

[22] Haakenaasen R, and Selvig E, Molecular Beam Epitaxy Growth of Nanowires in the $Hg_{1-x}Cd_xTe$ Material System, Open access peer-reviewed chapter, 2010

[23] Wagner RS, Ellis WC. vapor-liquid-solid mechanism of single crystal growth. Applied Physics Letters. 1964;**4**:89-91

[24] Khayyat M, Wacaser B, Reuter M, Ross F, Sadana D, Chin T. Nanoscale chemical templating of Si-NWs seeded with Al. Nanotechnology. 2013;**24**: 235301

[25] Khayyat M, Wacaser B, Sadana D. Nanoscale chemical templating with oxygen reactive materials. USPTO Patent number 8349715

[26] Cohen G., Reuter M., Wacaser B., Khayyat M., Production scale fabrication method for high resolution

AFM tips, USPTO Patent number 8321961

[27] Kim H, Bae H, Chang T, (2021) Huffaker D. III-V nanowires on silicon (100) as plasmonic-photonic hybrid meta-absorber. Scientific Reports 11:13813

[28] Kayes BM, Filler MA, Putnam MC, Kelzenberg MD, Lewis NS, Atwater HA. Growth of vertically aligned Si wire arrays over large areas (>1 cm2) with Au and Cu catalysts. Applied Physics Letters. 2007;**91**:103110

[29] Erik C, Garnett E, Mark L, Brongersma M, Cui Y, McGehee M. Nanowire solar cells. Annual Review of Materials Research. 2011;**41**:269-295

[30] McIntosh, K., Cudzinovic R., Michael J, Smith David D, Mulligan William P, Swanson Richard M. The choice of silicon wafer for the production of low-cost rear-contact solar cells. In Photovoltaic Energy Conversion, 2003. Proceedings of 3rd World Conference on. 2003. IEEE

[31] Jana T, Mukhopadhyay Sumita M, Swati R. Low temperature silicon oxide and nitride for surface passivation of silicon solar cells. Solar Energy Materials and Solar Cells. 2002;**71**(2):197-211

[32] Behroudj A, Salimitari P, Nilsen M, Strehle S. Exploring nanowire regrowth for the integration of bottom-up grown silicon nanowires into AFM scanning probes. Journal of Micromechanics and Microengineering. 2021;**31**:055010 (11pp)

[33] Mansoori G. Principles of Nanotechnology: Molecular-Based Study of Condensed Matter in Small Systems. British Library-in-Publication Data: Singapore: World Scientific Publishing Co. Pte. Ltd; 2005. p. 341. ISBN 981-256-154-4

[34] Bououdina M, Davim J, editors. Handbook of Research on Nanoscience,

Nanotechnology & Advanced Materials. INSPEC: SCOPUS; 2004. p. 617 DOI: 10.4018/978-1-4666-5824-0

[35] Khayyat M, Wacaser B, Reuter M, Sadana D. Templating silicon nanowires seeded with oxygen reactive materials. Saudi International Electronics Communications and Photonics Conference (SIECPC). IEEE; 2011

[36] Wacaser B. Nanoscale Crystal Growth: The Importance of Interfaces and Phase Boundries. Sweden: Lund University; 2007

[37] Dicka KA, Hansena AE, Martenssona T, Paneva N, Perssona AI, Seiferta W, et al. Semiconductor nanowires for 0D and 1D physics and applications. Physica E. 2004;**25**: 313-318

[38] Suzuki H, Araki H, Tosa M, Noda T. Formation of silicon nanowires by CVD using gold catalysts at low temperatures. Materials Transactions. 2007;**48**(8): 2202-2206

[39] Sutter P, Wimer S, Sutter E. Chiral twisted van der Waals nanowires. Nature. 2019;**569**:1-4

[40] Christiansen S, Becker M, Fahlbusch S, Michler J, Sivakov V, Andra G, et al. Signal enhancement in nano-Raman spectroscopy by gold caps on silicon nanowires obtained by vapour–liquid–solid growth. Nanotechnology. 2007;**18**:035503 (6pp)

[41] Wacaser B, Reuter M, Khayyat M, Wen C, Haight R, Guha S, et al. Growth systems, structure, and doping of aluminum seeded epitaxial silicon nanowires. Nano Letters. 2009; 9:3291-3301

[42] Paulo A, Paulo Ä, Arellano N, Plaza J, He R, Carraro C, et al. Suspended mechanical structures based on elastic silicon nanowire arrays. Nano Letters. 2007;7(4):1100-1104.

Publication Date: 22 March 2007. DOI: 10.1021/nl062877n

[43] Haraguchi K, Katsuyama T, Hiruma K. Polarization dependence of light emitted from GaAs p-n junctions in quantum wire crystals. Journal of Applied Physics. 1994;**75**:4220 https://doi.org/10.1063/1.356009

[44] Tang D, Ren C, Wang M, Wei X, Kawamoto N, Liu, Bando Y, Mitome M, Fukata N, and Golberg D, Mechanical properties of Si nanowires as revealed by in situ transmission electron microscopy and molecular dynamics simulations, dx.doi.org/10.1021/nl204282y | Nano Letters. 2012, 12, 1898–1904

[45] Nebol'sin VA, Shchetinin AA. Role of surface energy in the vapor–liquid–solid growth of silicon. Inorganic Materials. 2003;**39**(9):899-903

[46] Fleischmann M, Hendra P, McQuillan A. Raman spectra of pyridzine adsorbed at silver electrode. Chemical Physics Letters. 1974;**26**(2)

[47] Kim J, Hong A, Nah J, Shin B, Ross FM, Sadana DK. Three-dimensional a-Si:H solar cells on glass nanocone arrays patterned by self-assembled Sn nanospheres. ACS Nano. 2012;**6**(1):265-271

[48] Fogel K., Kim J., Nah J; Sadana D; Shiu K; Nanowires formed by employing soled nanodots; Patent No.: US9.231,133B2, 2016

[49] Yang L, Huh D, Ning R, Rapp V, Zeng Y, Liu Y, et al. High thermoelectric figure of merit of porous Si nanowires from 300 to 700 K. Nature Communications. 2021;**12**:3926

[50] Melosh N, Boukai A, Diana F, Gerardot B, Badolato A, Petroff P, et al. Ultrahigh-density nanowire lattices and circuits. Science. 2003;**300**

[51] Tsivion D, Chvartzman M, Popovitz-Biro R, Huth P, Joselevich E.

Guided growth of millimeter-long
horizontal nanowires with controlled
orientations. Science. 2011;**333**(19)

[52] Costa I, Cunha T, Chiquito A.
Investigation on the optical and
electrical properties of undoped and
Sb-doped SnO_2 nanowires obtained by
the VLS method. Physica E: Low-
dimensional Systems and
Nanostructures. 2021;**134**;
Article 114856

[53] Zhang K, Chen T, Gong J. Atomic
arrangement matters: band-gap
variation in composition-tunable
(Ga1–xZnx)(N1–xOx) nanowires.
Matter. 2021;**4**:3

[54] Jabeen F, Rubini S, Martelli F.
Growth of III–V semiconductor
nanowires by molecular beam epitaxy.
Microelectronics Journal.
2009;**40**(3):442-445

[55] Lieber CM. Nanoscale science and
technology: Building a big future from
small things. MRS Bulletin. 2003

[56] Marcel TM, Pachauri V, Sven IS,
Vu X. Process variability in top-Ddwn
fabrication of silicon nanowire-based
biosensor arrays. Sensors. 2021;**21**:5153

[57] Lu W, Lieber C. Topical review:
Semiconductor nanowires. Journal of
Physics D: Applied Physics.
2006;**39**:R387-R406

[58] Gates B. Self-Assemply: Nanowires
find their place, nature
nanotechnology. 2010;**5**

Chapter 7

Ultrathin Metal Hydroxide/Oxide Nanowires: Crystal Growth, Self-Assembly, and Fabrication for Optoelectronic Applications

Gayani Pathiraja and Hemali Rathnayake

Abstract

The fundamental understanding of transition metal oxides nanowires' crystal growth to control their anisotropy is critical for their applications in miniature devices. However, such studies are still in the premature stage. From an industrial point of view, the most exciting and challenging area of devices today is having the balance between the performance and the cost. Accordingly, it is essential to pay attention to the controlled cost-effective and greener synthesis of ultrathin TMOS NWs for industrial optoelectronic applications. This chapter provides a comprehensive summary of fundamental principles on the preperation methods to make dimensionality controlled anisotropic nanowires, their crystal growth studies, and optical and electrical properties. The chapter particularly addresses the governing theories of crystal growth processes and kinetics that controls the anisotropy and dimensions of nanowires. Focusing on the oriented attachment (OA) mechanism, the chapter describes the OA mechanism, nanocrystal's self-assembly, interparticle interactions, and OA-directed crystal growth to improve the state-of-the art kinetic models. Finally, we provide the future perspective of ultrathin TMOS NWs by addressing their current challenges in optoelectronic applications. It is our understanding that the dimension, and single crystallinity of nanowires are the main contributors for building all functional properties, which arise from quasi-1-D confinement of nanowire growth.

Keywords: transition metal oxides, ultrathin nanowires, optoelectronics, oriented attachment, kinetics

1. Introduction

1.1 Transition metal oxides nanowires

Transition metal oxides nanowires are known as an important class of materials with a rich collection of physical and chemical properties due to their superior performances based on quantum confinement effects for various general applications, including electronic devices, optical devices, gas sensors, photovoltaics, photonic devices, energy storage devices, and catalysts [1–5]. The fabrication of one-dimensional (1 D) nanostructures of transition metal oxides semiconductor

IntechOpen

(TMOS) nanowires and nanorods has been recently fascinating for the next-generation high-performance "trillion sensor electronics" era for Internet of things (IoT) applications, mainly due to their high surface to volume ratio, high crystallinity, and low power consumption [6–10]. The ultrathin nanowires that have below 10 nm diameter offer interesting characteristics including new surface determined structures by tuning the surface chemistry of surfaces [11, 12]. Furthermore, high surface area and increased colloidal stability are also inherited to ultrathin nanowires that are related to the decreased diameter of nanowires [11]. These improved overall properties of ultrathin TMOS nanowires as an ideal building block will continue the miniaturization and functional scaling of integrated circuits (ICs), nanoelectronics and optoelectronic devices by achieving the limitations of Moore's law.

A significant research endeavor has been devoted to the fabrication of 1D metal hydroxide/oxide nanowires. However, their controlled fabrication to tailor the shape, size, crystallinity, and anisotropy is remaining a challenge. Tremendous efforts are needed to devote to the development of effective, greener fabrication methods that has scalability, reproducibility, and stability of nanowires for the successful commercialization and integration of devices. The fundamental in-depth understanding of guiding principles such as crystal growth mechanisms, kinetics, and phase transformation during the fabrication of metal-hydroxide/oxide nanowires with different strategies is crucial to accomplish this goal for promoting TMOS nanowire-based optoelectronic devices. In this review, first, we give a comprehensive overview of different synthesis strategies of transition metal-oxide/hydroxide nanowires and their electrical and optical properties. Then we provide the fundamentals of crystal growth mechanisms and a detailed overview of oriented attachment (OA), crystal growth mechanism that makes anisotropic nanowires. We subsequently discuss the state-of-the-art kinetic models that explain the OA crystal growth mechanism. Then we present a greener facile synthesis approach for the fabrication of ultrathin copper hydroxide/oxide nanowires and their optoelectronic properties. Finally, we provide an outlook of the challenges of future prospective to fabricate ultrathin transition metal hydroxide/oxide nanowires for optoelectronic applications.

1.2 Synthesis strategies of transition metal oxide nanowires

The bottom-up techniques are prominent to produce transition metal oxide NWs due to the high purity of the product, low-cost fabrication, and dimension controllability [13]. The controlled fabrication of transition metal hydroxide/oxide nanowires can be done with either vapor or solution phase growth strategies. Shen and coworkers have summarized different 1-D metal oxide nanostructures including nanowires, nanobelts, nanorods and nanotubes synthesized using both vapor or solution phase growth strategies [14]. However, the controlled pressure of the inert atmosphere and high-temperature vapor-phase approaches including vapor phase growth such as vapor–liquid–solid (VLS), solution–liquid–solid (SLS), vapor–solid–solid (VSS) or vapor–solid (VS) process, physical vapor deposition (PVD) and chemical vapor deposition (CVD) are expensive and need sophisticated instruments [15–18]. Therefore, solution-based wet chemical strategies such as hydro-thermal method, thermal decomposition, electrochemical methods, solvothermal methods, sol–gel routes became popular as they are inexpensive, energy savers, excellent control over size and morphologies, with ease of larger-scale production [19].

The wet chemical routes can be performed through precipitation or oxidation of the precursor by the aid of catalyst or surfactants and the aid of heating in an oxygen-rich environment to make metal oxide/hydroxide NWs [20]. In a hydro-thermal route, heating the precursor solution/substrate and then annealing in an oxygen environment is required to form metal hydroxide/oxide nanowires [19].

Solvothermal methods are performed by heating a metal precursor solution to a high temperature with the presence of a solvent [21]. Microwave-assisted methods are another powerful approach that require heating to a higher temperature in a microwave [22]. However, these most wet chemical growth strategies require expensive chemicals, heating procedures, longer reaction times or templates or impurities that needs to remove after the procedure [20].

The sol–gel method is a green and low-temperature method, which is widely employed to make homogeneous, highly stoichiometric and high-quality metal oxide/hydroxide nanowires in a larger scale production [23, 24]. It is often used to fabricate size and shape-controlled nanostructures starting from a metal salt as the precursor and catalyzed by base or acid to form an integrated network or gel. This method is popular to make different solid networks such as inorganic hydroxides/oxides, organic polymers, organic–inorganic hybrids, and composites due to its advantages such as high yield, better reproducibility, low operation temperature and low-cost method of highly stoichiometric and homogeneous products [25–27]. The sol–gel process can be defined as the transition of a liquid solution "sol" into a solid "gel" phase, involving both physical and chemical reactions such as hydrolysis, condensation, drying, and densification, as shown in **Figure 1** [28]. When preparing metal hydroxides/oxides, first, the sol is formed from the hydrolysis of metal precursors. Sol is a stable dispersion of colloidal particles in a liquid solution and particles can be amorphous or crystalline. Then the gel is formed through condensation, polycondensation and aging to form a gel network using metal–oxo–metal or metal–hydroxy–metal coordinate bonds. Drying and densification involve densifying the gel by collapsing the porous gel network.

The morphology is influenced either due to different surface energies of the crystal faces or the external growth environment, which combines with different factors such as the precursor to base concentration, solvent polarity, temperature, and crystal growth mechanism. Different types of metal oxides have been synthesized using sol–gel route, exhibiting different 1D morphologies such as nanorods, nanoplatelets, and wires [29–32]. The conventional sol–gel process is hydrolytic, as shown in **Figure 2** and oxo ions are originated from water in the reaction medium. When using organic solvents as reagents in the medium, they are nonhydrolytic sol–gel process pathways and the oxygen atoms are originated from the organic O-donor [25, 33–35]. However, both sol–gel processes have their limitations with different metal precursors [25, 35].

1.3 Electronic and optical properties of transition metal oxides

Transition metal oxides offer very diverse and fascinating electrical and optical properties of materials, which arise from the outer d electrons of the transition metal ions. Many transition metals can form binary oxides of the formula M_xO_y. The range of electrical conductivity of transition metal oxides is wide and varied from metals to semiconductors and insulators. A few examples of transition metal oxide insulators are CaO, NiO, TiO_2 and the semiconducting materials are FeO, ZnO, and CuO, while TiO, NbO, CrO_2, ReO_3 show the metallic properties

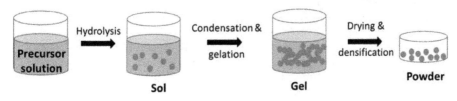

Figure 1.
Steps of a typical sol–gel process.

Hydrolytic sol-gel formation

$\equiv$M-OR + H$_2$O $\quad\rightarrow\quad$ $\equiv$M-OH + ROH $\quad$ Hydrolysis

$\equiv$M-OH + $\equiv$M-OR $\quad\rightarrow\quad$ $\equiv$M-O-M$\equiv$ + ROH

$\equiv$M-OH + $\equiv$M-OH $\quad\rightarrow\quad$ $\equiv$M-O-M$\equiv$ + H$_2$O $\quad$ $\left.\right\}$ Condensation

Nonhydrolytic sol-gel formation

$\equiv$M-OR + M-X $\quad\rightarrow\quad$ $\equiv$M-O-M$\equiv$ + R-X $\quad$ (X =OR, Cl, OOCR)

$\equiv$M-OR + RO-M $\rightarrow$ $\equiv$M-O-M$\equiv$ + R-O-R

Here, M denotes for metal.

Figure 2.
The formation of metal hydroxides/oxides via hydrolytic and nonhydrolytic sol–gel process.

respectively [36]. Every $3d$ transition metals monoxides are widely used in different applications due to their higher abundance and low cost compared to $4d$ and $5d$ transition metal oxides.

Among transition metal oxides, copper oxide has received considerable attention in recent years as an alternative element for expensive silver and gold due to its second-highest electrical conductivity and higher abundance [37]. Cupric oxide (CuO) is a p-type semiconductor with a narrow and indirect energy bandgap of ~1.2 eV [36]. Copper hydroxide (Cu(OH)$_2$ is the hydroxide form, having an indirect bandgap of 1.97 eV [38, 39]. Zinc oxide (ZnO) is also a widely used and studied material, which is composed of the next element to Cu in the periodic table. It is an n-type semiconductor with a direct wide bandgap of 3.3 eV [40]. The electronic properties have been widely studied for CuO and ZnO nanostructures [41–43]. By fabricating smoother surfaces with minimum defects of metal oxide dielectrics can utilize their good electrical insulation without compensating its high k [44]. Since metal oxides are typically wide bandgap materials, they show excellent optical properties. For example, ZnO nanostructures have been mainly reported for laser diodes and LEDs that is the potential to operate at room temperature, owing to their higher exciton binding energy [45–47]. Ye Zhao et al. reported the optical properties of MoO$_3$, exhibiting a wide optical band gap of ~3.05 eV [48]. The controlled hydrothermal synthesis of ZrO$_2$ 1D nanostructures has shown optical properties, which are suitable in light-emitting devices [49]. In addition, CuO, NiO, SnO, and Ta$_2$O$_5$ are also have shown optoelectronic properties [10, 50–53].

2. Crystal growth mechanisms and kinetics

2.1 Oriented attachment (OA) mechanism

The nonclassical mechanism, named oriented attachment (OA) is the most common crystal growth mechanism to understand the aggregation-based crystal growth of materials at the nanoscale. It is the self-assembly of adjacent nanocrystals to form a secondary crystal through Brownian motion by sharing a common crystallographic orientation (**Figure 3(a)**) [54, 55]. This OA-based crystal growth was first described in 1998 for hydrothermally synthesized TiO$_2$ nanocrystals. Penn and coworkers observed the anisotropic chain of TiO$_2$ anatase crystals attachment across the {112} facets using a high-resolution transmission electron microscope (HR-TEM) [54, 56–58]. The OA-based crystal growth is governed by thermodynamics as the result of the reduction of total crystal surface energy [54, 59]. Highly ordered monocrystalline materials can be formed through OA, which is a versatile approach for the preparation of anisotropic 1D nanowires and nanorods.

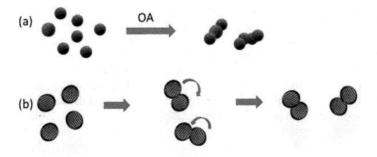

Figure 3.
Schematic representation of (a) OA-based crystal growth and (b) grain-rotation-induced grain coalescence mechanism.

As described by Moldovan et al., the primary nanocrystal colloids in a solution rotate for a crystal match to achieve a perfect coherent grain–grain interface in nearby crystals and start coalescence of nanocrystals, eliminating the common grain boundaries to form a single nanocrystal, shown in **Figure 3(b)** [60, 61]. This model is named as grain-rotation-induced grain coalescence (GRIGC) mechanism to describe the crystal growth process of OA-based nanomaterials [62]. This mechanism is based on the reduction of the crystal surface energy by minimizing the area of high-energy surfaces. Leite et al. have observed the OA mechanism experimentally in the growth process of SnO_2 nanocrystals at room temperature [63]. With the recent advancement of liquid-phase high-resolution transmission electron microscopy (liquid phase HR-TEM), Li and coworkers directly observed the OA mechanism of iron oxyhydroxide nanoparticles [64].

The ultrathin nanowires produced by the OA process provide unique features such as constant nanowire diameter during the growth by direct attachment of nanocrystals to the tip of the growing nanowire similar to polymerization reactions [65]. Therefore, the diameter of the nanowire can be predetermined by the diameter of the nanocrystals, which are monodispersed. However, the disadvantages of OA-based nanowire synthesis methods are poor yield and having residues of ligands and solvents attached to the nanowire [11].

2.2 Crystal growth kinetic models and prior arts

The crystal growth kinetics is mainly depending on the nature of the material, interface of crystal facets, working temperature, the type of surrounding solution and the concentration of the solution [66]. In a colloidal solution, the crystal growth mechanism for the formation of metal and metal hydroxide/oxide microstructures is often explained by Ostwald ripening (OR) theory. The OR crystal growth is controlled by diffusion, where larger particles grow at the expense of smaller particles [56, 57, 67]. The kinetic model of the OR mechanism is LSW kinetic model, which is attributed to the first-order chemical reactions, as in Eq. (1) [68, 69].

$$D^n - D_0^n = k(t - t_0)$$
(1)

where D and D_0 are the mean particle sizes at time t and t_0, k is a temperature-dependent rate constant, n is an exponent relevant to the coarsening mechanism.

Three kinetic models were developed to explain the OA-based crystal growth of nanocrystals based on their diameter [69, 70]. These models can be categorized based on the collision between two primary nanoparticles ($A_1 + A_1$ model) or a primary particle and a multilevel particle ($A_1 + A_i$ model) or two multilevel particles (($A_i + A_j$) multistep kinetic model), as shown in **Figure 4(a)–(c)**, respectively. The aggregation of nanocrystals using the OA mechanism is described in these models using the time evolution size distribution of nanocrystals, based on the Smoluchowski Equation [69]. These population growth matrixes of OA kinetic models account for nanostructures with different diameter sizes. The simplest growth model is $A_1 + A_1$ primary particle model and it can be stated in Eq. (2) [71].

$$d = \frac{d_0\left(\sqrt[3]{2}k_1 t + 1\right)}{\left(k_1 t + 1\right)} \tag{2}$$

where d_0 is the mean diameter at time $t = 0$ (primary nanoparticle), d is the mean diameter at time t (secondary nanoparticle).

The prior arts of these existing kinetic models that were used to understand the crystal growth mechanism by fitting the experimental observation of synthesis of different nanocrystals have summarized in the following **Table 1**. As we can see in these studies, most often, both OR and OA mechanisms occur simultaneously or coexist in the same model. Zhang and coworkers found that the solo OA mechanism causes to grow the nanocrystals by hindering OR mechanism at the initial stages by introducing surfactants to strongly adsorb them onto crystal surfaces [66, 74]. Moreover, Zhuang et al. observed that the OA mechanism become dominated under unsaturated solutions by suggesting that OR mechanism can be thermodynamically prohibited without having enough concentration gradient to dissolve nanoparticles in a surfactant-free hydrothermal synthesis route [76].

These state of the art of OA-based nanostructures show that all the existing kinetic studies have been performed based on the nanoparticle's diameter growth using the modified Smoluchowski equation. In addition, two reports have attempted to explain the kinetic rate for the elongation of 1 D nanorods by considering dipole attraction for their alignment [77, 84]. The in-depth understanding of the kinetics of crystal growth mechanisms for the fabrication of 1 D nanostructures is still in its preliminary stage although there is rapid progress in fabricating semiconductor nanowires. Therefore, it is vital to develop new kinetic models for

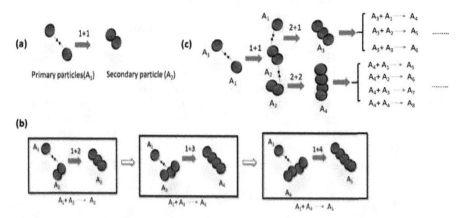

Figure 4.
The illustration of (a) $A_1 + A_1$ model, (b) $A_1 + A_i$ model, and (c) $A_i + A_j$ model (A_1 is a primary particle and A_i and A_j are multilevel particles; n = 1, 2, 3, 4, ...).

Growth system	Capping ligands	Model	Published year
Nano ZnS	Ligand free (water)	"1 + 1" mixed OA + OR	2003 [72]
Nano ZnS	Mercaptoethanol	"1 + 1" sequential OA + OR	2003 [70]
Nano ZnS	NaOH	"i + j" sequential OA + OR	2006 [66], 2007 [73]
Thiol-capped PbS NPs	Ligand free (water)	"i + j" OA and "1 + 1" OA + OR	2007 [74]
SnO$_2$ NPs	Ligand free (water)	"i + j" OA	2006 [75], 2009 [76]
CdS nanorods	Amine	"i + j" OA	2010 [77]
TiO$_2$ anatase NPs	Succinic acid	"1 + 1" OA, shrinkage, and OR	2010 [78]
TGA-capped CdTe NPs	Ligand free (water)	"1 + 1" mixed OA + OR	2011 [79]
ZnO QDs	Ethanol	"1 + 1" sequential OA + OR	2012 [80]
CdS QDs	TGA	"1 + 1" mixed OA + OR	2013 [81]
CdTe QDs	Mercaptopropionic acid	"1 + 1" mixed OA + OR	2014 [82]
Co$_2$FeO$_4$	Polypeptide (c25-mms6)	"1 + 1" sequential OA + OR	2014 [83]
Gd$_2$O$_3$ nanorods	Ligand free (water)	"1 + 1" OA	2016 [84]
ZnO QDs	Ligand free (water)	"1 + 1" mixed OA + OR	2019 [85]

Table 1.
Summary of crystal growth kinetics of different metal/metal oxide nanocrystals and nanostructures studied from 2003 to 2019.

1 D nanostructures with directing their length for the elevation of the controlled fabrication of ultrathin metal hydroxide/oxide nanowires.

3. Greener synthesis approaches for fabricating ultrathin metal hydroxide/oxide nanowires for optoelectronic applications

Due to the remarkable physical properties of 1 D transition metal hydroxide/oxide nanostructures, greener fabrication of anisotropic metal hydroxide/oxide nanowires has attracted increasing attention in many applications. Yang et al. summarized the recent efforts of controlled synthesis of metal oxide and hydroxide 1 D nanostructures such as NiO nanorods and Co$_3$O$_4$ nanowires via hydrothermal route for high-performance electrochemical electrodes and catalysts [19]. A wet chemical synthesis method was demonstrated the fabrication of ZnO nanorods, with a diameter of ~15 nm by oriented attachment mechanism [86]. Very recently, our group introduced a versatile sol–gel synthesis combined with the solvothermal process to make ZnO nanorods for optoelectronic devices [87]. Ultrathin ZnO nanorods with a diameter of ~7 nm were fabricated using a modified solvothermal route [88]. These nanorods show a strong UV band edge emission suggesting the applications of photoelectric nanodevices. Furthermore, Chaurasiya et al. synthesized TiO$_2$ nanorods using a wet chemical method for photovoltaic and humidity sensing applications [89], while MnO$_2$ nanorods were fabricated using a hydrothermal synthesis method for supercapacitor applications [90]. The work of hydrothermally growth GaOOH nanorods in width of 200–500 nm presented a low-cost and large-scale production strategy to prepare nanorods for practical applications [91].

Many works have been reported for the greener synthesis of 1 D Cu(OH)$_2$ and CuO nanostructures [92–96]. However, only a very few reports have demonstrated cost-effective and efficient synthesis routes to fabricate ultrathin copper hydroxide/oxide nanowires. In previous reports, ultrathin $Cu(OH)_2$/CuO NWs were synthesized using either both weak and strong bases [97] (i.e., aqueous ammonia and NaOH/KOH solutions) or by the interaction between a copper complex and NaOH at the aqueous-organic interface [98]. Sundar et al. demonstrated a bio-surfactant assisted synthesis method to produce CuO nanowires, suggesting the crystal growth process follows the oriented attachment mechanism [99].

Very recently, our group fabricated scalable and reproducible ultrathin CuO nanowires from self-assembled ultrathin $Cu(OH)_2$ nanowires, using a facile, greener and surfactant-free sol–gel approach, as shown in **Figure 5** [100]. As depicted in **Figure 6**, we observed the nanocrystals self-assembly into a certain crystallographic orientation and after 45 min stirring it forms smoother surfaces of colloidal nanowires with a uniform diameter of 6 ± 2 nm from 1 hour to 4 hours stirring time intervals. The time-dependent X-ray diffractometer (XRD) analysis is supported to identify the crystal growth plane of nanowires along the [020] facet. Upon annealing the ultrathin $Cu(OH)_2$ nanowires on the Si substrate at 300°C for 1 hour, we fabricated high aspect ratio CuO nanowires over a large area, as shown in **Figure 7**. The respective XRD and XPS spectroscopies confirm the chemical composition of both $Cu(OH)_2$ and CuO nanowires and their purity. The calculated optical band gaps for $Cu(OH)_2$ and CuO nanowires are 1.51 and 1.10 eV, respectively.

Wang et al. fabricated uniform ultrafine $Cu(OH)_2$ and CuO nanowires using a simple wet chemical route for lithium-ion batteries [101]. Another study reported the aqueous phase synthesis of $Cu(OH)_2$ nanowires with diameters of about 10–20 nm [102]. They have initiated to study the OA crystal growth mechanism of $Cu(OH)_2$

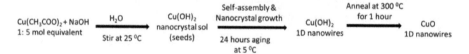

Figure 5.
Reaction scheme for the synthesis of ultrathin $Cu(OH)_2$ and CuO nanowires using sol–gel hydrolysis followed by directed self-assembly and annealing.

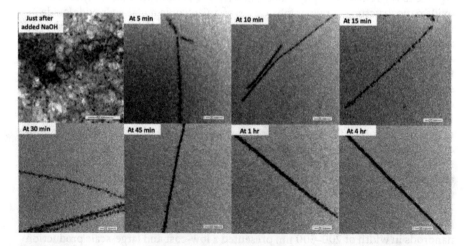

Figure 6.
Time-dependent TEM images at different stirring time intervals after the addition of NaOH during the synthesis of $Cu(OH)_2$ NWs (re-created from the original data).

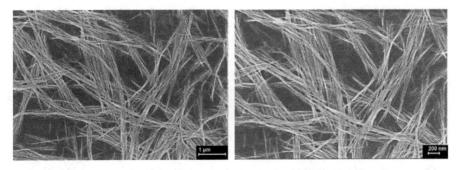

Figure 7.
SEM image of CuO nanowires fabricated on Si-substrate (re-created from the original data).

NWs, utilizing the power of high-resolution transmission electron microscopy (HRTEM) analysis. Very recently, Bhusari et al. reported the sol–gel approach to preparing $Cu(OH)_2$ NWs with a maximum diameter of 30 nm by varying the pH and temperature of the solution [103]. However, there is no experimental work reported for in detail investigations of their crystal growth mechanism, reproducibility, scalability and optimization of reaction parameters to get desired dimensions of nanowires for real applications in optoelectronics.

4. Summary and future prospective

One-dimensional ultrathin transition metal hydroxides/oxide nanowires are ideal building blocks for the miniaturization of next-generation high-performance integrated circuits (ICs), nanoelectronics and optoelectronic devices due to their quantum confinement and high aspect ratio. Although, we have witnessed the rapid progress and significant achievements of TMOS nanowire fabrication, still there are certain issues that need to be addressed.

1. Lack of better understanding of the guiding principles that control the size and shape of anisotropic nanowires.

It is crucial to develop more efficient, greener, scalable, and reproducible synthesis methods that can control the size of the nanowires. Aggregation of ultrathin nanowires is the major challenge of most available solution-based synthesis methods that are used to fabricate metal hydroxides/oxide nanowires. Fabrication of single ultrathin nanowires using solution-based strategies has not been studied for efficient and cost-effective devices. The controlled synthesis of ultrathin transition metal hydroxides/oxide nanowires is still in the preliminary stage and the guiding principles that control the size and shape are poorly understood and rarely explored. The existing crystal growth kinetic models have explained the crystal growth of nanostructures based on their nanoparticle diameter changes with time. These kinetic models cannot explain the crystal growth of one-dimensional nanostructures along certain specific directions. The attempts describing the kinetics of OA-based elongation of 1D nanostructures are still untapped and understudied. Therefore, exploration on crystal growth processes to generate effective novel kinetic models from experimental data will provide the prospects for the preparation of size and shape-controlled 1D nanostructures.

2. The insufficient yield and quality of NWs for device fabrication.

In general, each synthetic strategy has its advantages and disadvantages. Some synthesis gives high purity ultrathin nanowires, but the yield is poor. Many laboratory-scale synthesis routes and device fabrication are inappropriate for real commercial device fabrication due to this reason. It is challenging to develop strategies that have all the cost-effective features of commercialization. Therefore, more attention needs to pay to the different types of cost-effective fabrication methods of metal hydroxides/oxide nanowires that yield high quality and yield.

3. Lack of sufficient work for optimizing the fabrication process of NWs with desired electronic and optical properties for commercialization.

Optimization of synthesis routes to reduce the reaction time, better yield with desired dimensions of nanowires and their properties is essential to maximize the performance efficiency of NWs based devices. This is the main reason for the inappropriateness of laboratory-scale nanodevices for practical applications. The performance of optoelectronic devices can be improved by tuning the composition, dimensions, and crystal structure of TMOS NWs. These optimization studies are rarely reported for applications and this area can certainly strengthen by having strong collaborations between industries/entrepreneurs and scientists. When exploring the electronic properties of TMOS NWs, understanding their bandgap, electronic structure, electrical conductivity, dielectric constants and electron and hole mobilities are important to accelerate miniaturization and high-density integration of components of the next decade electronic devices. The factors such as light absorption, regeneration and recombination processes and charge carrier mobility of TMOS NWs are necessary to critically evaluate for future optical devices.

5. Conclusions

The controlled synthesis of anisotropic one-dimensional TMOS nanowires with versatile properties is most fascinating for different applications due to their superior performances based on quantum confinement effects and high aspect ratio. Their ultrathin TMOS nanowires have been attracted recently, however, many nanowires synthesis strategies are not sustainably produced. The fabrication of optoelectronic devices from ultrathin TMOS nanowires is clearly in its infancy. Currently, it is essential to conduct fundamental research on greener synthesis of ultrathin TMOS NWs, and the mechanisms to control the size and shape of 1D nanowires, as well as an evaluation of the optical and electrical properties of these NWs for industrial applications. Augmenting the OA crystal growth mechanism, we successfully fabricated scalable and reproducible sub 6 nm $Cu(OH)_2$ NWs and CuO NWs using a greener sol–gel hydrolysis followed by directed self-assembly and crystal growth of $Cu(OH)_2$ nanocrystals for potential electronic and optical applications. As TMOS NWs are excellent candidates for next-generation optoelectronic devices, their optical and electrical properties need to control by manipulating the fabrication conditions and fundamental understanding of controlling factors that can be optimally designed.

Acknowledgements

Authors acknowledge the Joint School of Nanoscience and Nanoengineering, a member of the South-eastern Nanotechnology Infrastructure Corridor (SENIC) and National Nanotechnology Coordinated Infrastructure (NNCI), supported by the NSF (Grant ECCS-1542174).

Conflict of interest

There is no conflict of interest to declare.

Author details

Gayani Pathiraja and Hemali Rathnayake*
Department of Nanoscience, Joint School of Nanoscience and Nanoengineering,
University of North Carolina at Greensboro, Greensboro, NC, USA

*Address all correspondence to: hprathna@uncg.edu

IntechOpen

References

[1] Gao X, Zhu H, Pan G, Ye S, Lan Y, Wu F, et al. Preparation and electrochemical characterization of anatase nanorods for lithium-inserting electrode material. The Journal of Physical Chemistry. B. 2004;**108**: 2868-2872. DOI: 10.1021/jp036821i

[2] Mirzaei A, Lee J-H, Majhi SM, Weber M, Bechelany M, Kim HW, et al. Resistive gas sensors based on metal-oxide nanowires. Journal of Applied Physics. 2019;**126**:241102. DOI: 10.1063/1.5118805

[3] Fang M, Xu WB, Han S, Cao P, Xu W, Zhu D, et al. Enhanced urea oxidization electrocatalysis on spinel cobalt oxide nanowires via on-site electrochemical defect engineering. Materials Chemistry Frontiers. 2021;**5**:3717-3724. DOI: 10.1039/D0QM01119C

[4] Zhang G, Xiao X, Li B, Gu P, Xue H, Pang H. Transition metal oxides with one-dimensional/one-dimensional-analogue nanostructures for advanced supercapacitors. Journal of Materials Chemistry A. 2017;**5**:8155-8186. DOI: 10.1039/C7TA02454A

[5] Li Y, Yang X-Y, Feng Y, Yuan Z-Y, Su B-L. One-dimensional metal oxide nanotubes, nanowires, nanoribbons, and nanorods: Synthesis, characterizations, properties and applications. Critical Reviews in Solid State and Materials Sciences. 2012;**37**:1-74. DOI: 10.1080/10408436.2011.606512

[6] Comini E. Metal oxides nanowires chemical/gas sensors: Recent advances. Materials Today Advances. 2020;**7**:100099. DOI: 10.1016/j.mtadv.2020.100099

[7] Zhou Z, Lan C, Wei R, Ho JC. Transparent metal-oxide nanowires and their applications in harsh electronics. Journal of Materials Chemistry C. 2019;**7**:202-217. DOI: 10.1039/C8TC04501A

[8] Zeng H, Zhang G, Nagashima K, Takahashi T, Hosomi T, Yanagida T. Metal–oxide nanowire molecular sensors and their promises. Chem. 2021;**9**:41. DOI: 10.3390/chemosensors9020041

[9] Yao Y, Sang D, Zou L, Wang Q, Liu CA. Review on the properties and applications of WO_3 nanostructure-based optical and electronic devices. Nanomaterials. 2021;**11**:2136. DOI: 10.3390/nano11082136

[10] Devan RS, Patil RA, Lin J-H, Ma Y-R. One-dimensional metal-oxide nanostructures: Recent developments in synthesis, characterization, and applications. Advanced Functional Materials. 2012;**22**:3326-3370. DOI: 10.1002/adfm.201201008

[11] Cademartiri L, Ozin GA. Ultrathin nanowires—A materials chemistry perspective. Advanced Materials. 2009;**21**:1013-1020. DOI: 10.1002/adma.200801836

[12] Decher G. Fuzzy nanoassemblies: Toward layered polymeric multicomposites. Science. 1997;**277**: 1232-1237

[13] Ramgir N, Datta N, Kaur M, Kailasaganapathi S, Debnath AK, Aswal DK, et al. Metal oxide nanowires for chemiresistive gas sensors: Issues, challenges and prospects. Colloids and Surfaces A: Physicochemical and Engineering Aspects. 2013;**439**:101-116

[14] Shen G, Chen P-C, Ryu K, Zhou C. Devices and chemical sensing applications of metal oxide nanowires. Journal of Materials Chemistry. 2009;**19**:828-839. DOI: 10.1039/B816543B

[15] Zhou S, Liu L, Lou S, Wang Y, Chen X, Yuan H, et al. Room-temperature ferromagnetism of diamagnetically-doped ZnO aligned

nanorods fabricated by vapor reaction. Applied Physics A: Materials Science & Processing. 2011;**102**:367-371. DOI: 10.1007/s00339-010-6011-7

[16] Wang Song J, Summers CJ, Ryou JH, Li P, Dupuis RD, Wang ZL. Density-controlled growth of aligned ZnO nanowires sharing a common contact: A simple, low-cost, and mask-free technique for large-scale applications. The Journal of Physical Chemistry. B. 2006;**110**:7720-7724. DOI: 10.1021/jp060346h

[17] Jimenez-Cadena G, Comini E, Ferroni M, Vomiero A, Sberveglieri G. Synthesis of different ZnO nanostructures by modified PVD process and potential use for dye-sensitized solar cells. Materials Chemistry and Physics. 2010;**124**:694-698. DOI: 10.1016/j.matchemphys.2010.07.035

[18] Thangala J, Vaddiraju S, Malhotra S, Chakrapani V, Sunkara MK. A hot-wire chemical vapor deposition (HWCVD) method for metal oxide and their alloy nanowire arrays. Thin Solid Films. 2009;**517**:3600-3605. DOI: 10.1016/j.tsf.2009.01.051

[19] Yang Q, Lu Z, Liu J, Lei X, Chang Z, Luo L, et al. Metal oxide and hydroxide nanoarrays: Hydrothermal synthesis and applications as supercapacitors and nanocatalysts. Progress in Natural Science: Materials International. 2013;**23**:351-366. DOI: 10.1016/j.pnsc.2013.06.015

[20] Filipič G, Cvelbar U. Copper oxide nanowires: A review of growth. Nanotechnology. 2012;**23**:194001. DOI: 10.1088/0957 4484/23/19/194001

[21] Qin Y, Li X, Wang F, Hu M. Solvothermally synthesized tungsten oxide nanowires/nanorods for NO$_2$ gas sensor applications. Journal of Alloys and Compounds. 2011;**509**:8401-8406. DOI: 10.1016/j.jallcom.2011.05.100

[22] Qurashi A, Tabet N, Faiz M, Yamzaki T. Ultra-fast microwave

synthesis of ZnO nanowires and their dynamic response toward hydrogen gas. Nanoscale Research Letters. 2009;**4**:948. DOI: 10.1007/s11671-009-9317-7

[23] Thiagarajan S, Sanmugam A, Vikraman D. Facile methodology of sol–gel synthesis for metal oxide nanostructures. In: Chandra U, editor. Recent Applications in Sol–Gel Synthesis. InTechOpen; 2017. Available from: https://www.intechopen.com/chapters/55242. DOI: 10.5772/intechopen.68708

[24] Sztaberek L, Mabey H, Beatrez W, Lore C, Santulli AC, Koenigsmann C. Sol–gel synthesis of ruthenium oxide nanowires to enhance methanol oxidation in supported platinum nanoparticle catalysts. ACS Omega. 2019;**4**:14226-14233. DOI: 10.1021/acsomega.9b01489

[25] Livage J, Henry M, Sanchez C. Sol–gel chemistry of transition metal oxides. Progress in Solid State Chemistry. 1988;**18**:259-341. DOI: 10.1016/0079-6786(88)90005-2

[26] Zha J, Roggendorf H. Sol–gel science, the physics and chemistry of sol–gel processing, Ed. by Brinker CJ, Scherer GW, Academic Press, Boston 1990, xiv, 908 pp., bound—ISBN 0-12-134970-5. Advanced Materials. 1990;**3**:522. DOI: 10.1002/adma.19910031025

[27] Sui R, Charpentier P. Synthesis of metal oxide nanostructures by direct sol–gel chemistry in supercritical fluids. Chemical Reviews. 2012;**112**:3057-3082. DOI: 10.1021/cr2000465

[28] Tseng TK, Lin YS, Chen YJ, Chu HA. Review of photocatalysts prepared by sol–gel method for VOCs removal. International Journal of Molecular Sciences. 2010;**11**:2336-2361. DOI: 10.3390/ijms11062336

[29] Foo KL, Hashim U, Muhammad K, Voon CH. Sol–gel synthesized zinc oxide

nanorods and their structural and optical investigation for optoelectronic application. Nanoscale Research Letters. 2014;**9**:429. DOI: 10.1186/1556-276X-9-429

[30] Ahn S-E, Ji HJ, Kim K, Kim GT, Bae CH, Park SM, et al. Origin of the slow photoresponse in an individual sol–gel synthesized ZnO nanowire. Applied Physics Letters. 2007;**90**: 153106. DOI: 10.1063/1.2721289

[31] Woo K, Lee HJ, Ahn J-P, Park YS. Sol–gel mediated synthesis of Fe_2O_3 nanorods. Advanced Materials. 2003;**15**:1761-1764. DOI: 10.1002/adma.200305561

[32] Dhanasekaran V, Soundaram N, Kim S-I, Chandramohan R, Mantha S, Saravanakumar S, et al. Optical, electrical and microstructural studies of monoclinic CuO nanostructures synthesized by a sol–gel route. New Journal of Chemistry. 2014;**38**:2327. DOI: 10.1039/c4nj00084f

[33] Mutin PH, Vioux A. Nonhydrolytic processing of oxide-based materials: Simple routes to control homogeneity, morphology, and nanostructure. Chemistry of Materials. 2009;**21**: 582-596. DOI: 10.1021/cm802348c

[34] Vioux A. Nonhydrolytic sol–gel routes to oxides. Chemistry of Materials. 1997;**9**:2292-2299. DOI: 10.1021/cm970322a

[35] Schubert U. Chemistry and fundamentals of the sol–gel process. In: The Sol–Gel Handbook. VCH-Wiley Verlag. Weinheim: GmbH; 2015. pp. 1-28

[36] Goodwin DW. Transition metal oxides. Crystal chemistry, phase transition and related aspects. NBS 49 by C. N. R. Rao and G. V. Subba Rao. Acta Crystallographica. Section B. 1975;**31**:2943. DOI: 10.1107/S0567740875009399

[37] Tran TH, Nguyen VT. Copper oxide nanomaterials prepared by solution methods, some properties, and potential applications: A brief review. International Scholarly Research Notices. 2014;**2014**:1-14. DOI: 10.1155/2014/856592

[38] Marcus P, Mansfeld FB. Analytical Methods in Corrosion Science and Engineering. CRC Press; 2005. Available from: https://www.intechopen.com/chapters/55242

[39] Hou H, Zhu Y, Hu Q. 3D Cu(OH) 2 hierarchical frameworks: Self-assembly, growth, and application for the removal of TSNAs. Journal of Nanomaterials. 2013;**2013**:1-8. DOI: 10.1155/2013/797082

[40] Costas A, Preda N, Florica C, Enculescu I. Metal oxide nanowires as building blocks for optoelectronic devices. In: Nanowires—Recent Progress. IntechOpen; 2020. DOI: 10.5772/intechopen.87902. Available from: https://www.intechopen.com/chapters/55242

[41] Uhlrich JJ, Olson DC, JWP H, Kuech TF. Surface chemistry and surface electronic properties of ZnO single crystals and nanorods. Journal of Vacuum Science and Technology A: Vacuum, Surfaces and Films. 2009;**27**:328-335. DOI: 10.1116/1.3085723

[42] Da Silva LF, Lopes OF, Catto AC, Avansi W, Bernardi MI, Li MS, et al. Hierarchical growth of ZnO nanorods over SnO 2 seed layer: Insights into electronic properties from photocatalytic activity. RSC Advances. 2016;**6**:2112-2118

[43] Krishnan S, Haseeb ASMA, Johan MR. One dimensional CuO nanocrystals synthesis by electrical explosion: A study on structural, optical and electronic properties. Journal of Alloys and Compounds. 2014;**586**:360-367. DOI: 10.1016/j.jallcom.2013.10.014

[44] Park S, Kim C-H, Lee W-J, Sung S, Yoon M-H. Sol–gel metal oxide dielectrics for all-solution-processed electronics. Materials Science & Engineering R: Reports. 2017;**114**:1-22. DOI: 10.1016/j.mser.2017.01.003

[45] Djurišić AB, Ng AMC, Chen XY. ZnO nanostructures for optoelectronics: Material properties and device applications. Progress in Quantum Electronics. 2010;**34**:191-259. DOI: 10.1016/j.pquantelec.2010.04.001

[46] Willander M. Zinc Oxide Nanostructures: Advances and Applications. Singapore: Pan Stanford Publishing; 2014

[47] Willander M, Zhao QX, Hu Q-H, Klason P, Kuzmin V, Al-Hilli SM, et al. Fundamentals and properties of zinc oxide nanostructures: Optical and sensing applications. Superlattices and Microstructures. 2008;**43**:352-361. DOI: 10.1016/j.spmi.2007.12.021

[48] Zhao Y, Liu J, Zhou Y, Zhang Z, Xu Y, Naramoto H, et al. Preparation of MoO$_3$ nanostructures and their optical properties. Journal of Physics. Condensed Matter. 2003;**15**:L547

[49] Kumari L, Li WZ, Xu JM, Leblanc RM, Wang DZ, Li Y, et al. Controlled hydrothermal synthesis of zirconium oxide nanostructures and their optical properties. Crystal Growth & Design. 2009;**9**:3874-3880. DOI: 10.1021/cg800711m

[50] Rajendran V, Anandan K. Different ionic surfactants assisted solvothermal synthesis of zero-, three and one-dimensional nickel oxide nanostructures and their optical properties. Materials Science in Semiconductor Processing. 2015;**38**:203-208. DOI: 10.1016/j.mssp.2015.03.058

[51] Gu F, Wang S, Cao H, Li C. Synthesis and optical properties of SnO$_2$ nanorods. Nanotechnology. 2008;**19**:095708

[52] Miao W, Minmin Z, Zuochen L. Optical and electric properties of aligned-growing Ta2O5 nanorods. Materials Transactions. 2008;**49**:0809160552

[53] Lugo-Ruelas M, Amézaga-Madrid P, Esquivel-Pereyra O, Antúnez-Flores W, Pizá-Ruiz P, Ornelas-Gutiérrez C, et al. Synthesis, microstructural characterization and optical properties of CuO nanorods and nanowires obtained by aerosol assisted CVD. Journal of Alloys and Compounds. 2015;**643**:S46-S50. DOI: 10.1016/j.jallcom.2014.11.119

[54] Penn RL. Imperfect oriented attachment: Dislocation generation in defect-free nanocrystals. Science. 1998;**281**:969-971. DOI: 10.1126/science.281.5379.969

[55] Penn RL, Banfield JF. Oriented attachment and growth, twinning, polytypism, and formation of metastable phases: Insights from nanocrystalline TiO 2. American Mineralogist. 1998;**83**:1077-1082. DOI: 10.2138/am-1998-9-1016

[56] Zhang H, Penn RL, Lin Z, Cölfen H. Nanocrystal growth via oriented attachment. CrystEngComm. 2014;**16**: 1407. DOI: 10.1039/c4ce90001d

[57] Xue X, Penn RL, Leite ER, Huang F, Lin Z. Crystal growth by oriented attachment: Kinetic models and control factors. CrystEngComm. 2014;**16**:1419. DOI: 10.1039/c3ce42129e

[58] He W, Wen K, Niu Y. Nanocrystals from Oriented-Attachment for Energy Applications (Springer Briefs in Energy). Cham: Springer International Publishing; 2018. DOI: 10.1007/978-3-319-72432-4

[59] Ning J, Men K, Xiao G, Zou B, Wang L, Dai Q, et al. Synthesis of narrow band gap SnTe nanocrystals: Nanoparticles and single crystal

nanowires via oriented attachment. CrystEngComm. 2010;**12**:4275. DOI: 10.1039/c004098n

[60] Ribeiro C, EJH L, Giraldi TR, Longo E, Varela JA, Leite ER. Study of synthesis variables in the nanocrystal growth behavior of tin oxide processed by controlled hydrolysis. The Journal of Physical Chemistry. B. 2004;**108**: 15612-15617. DOI: 10.1021/jp0473669

[61] Ribeiro C, EJH L, Longo E, Leite ER. A kinetic model to describe nanocrystal growth by the oriented attachment mechanism. ChemPhysChem. 2005;**6**:690-696. DOI: 10.1002/cphc.200400505

[62] Moldovan D, Yamakov V, Wolf D, Phillpot SR. Scaling behavior of grain-rotation-induced grain growth. Physical Review Letters. 2002;**89**:206101. DOI: 10.1103/PhysRevLett.89.206101

[63] Leite ER, Giraldi TR, Pontes FM, Longo E, Beltran A, Andres J. Crystal growth in colloidal tin oxide nanocrystals induced by coalescence at room temperature. Applied Physics Letters. 2003;**83**:1566-1568

[64] Li D, Nielsen MH, Lee JR, Frandsen C, Banfield JF, De Yoreo JJ. Direction-specific interactions control crystal growth by oriented attachment. Science. 2012;**336**:1014-1018

[65] Penn RL. Kinetics of oriented aggregation. The Journal of Physical Chemistry. B. 2004;**108**:12707-12712

[66] Zhang J, Lin Z, Lan Y, Ren G, Chen D, Huang F, et al. Multistep oriented attachment kinetics: Coarsening of ZnS nanoparticle in concentrated NaOH. Journal of the American Chemical Society. 2006;**128**: 12981-12987. DOI: 10.1021/ja062572a

[67] Kirchner HOK. Coarsening of grain-boundary precipitates.

Metallurgical Transactions. 1971;**2**: 2861-2864. DOI: 10.1007/BF02813264

[68] Joesten R, Fisher G. Kinetics of diffusion-controlled mineral growth in the Christmas Mountains (Texas) contact aureole. GSA Bulletin. 1988;**100**:714-732

[69] Zhang J, Huang F, Lin Z. Progress of nanocrystalline growth kinetics based on oriented attachment. Nanoscale. 2010;**2**:18-34. DOI: 10.1039/B9NR00047J

[70] Huang F, Zhang H, Banfield JF. Two-stage crystal-growth kinetics observed during hydrothermal coarsening of nanocrystalline ZnS. Nano Letters. 2003;**3**:373-378

[71] Wang F, Richards VN, Shields SP, Buhro WE. Kinetics and mechanisms of aggregative nanocrystal growth. Chemistry of Materials. 2014;**26**:5-21

[72] Huang F, Zhang H, Banfield JF. The role of oriented attachment crystal growth in hydrothermal coarsening of nanocrystalline ZnS. The Journal of Physical Chemistry. B. 2003;**107**: 10470-10475. DOI: 10.1021/jp035518e

[73] Wang Y, Zhang J, Yang Y, Huang F, Zheng J, Chen D, et al. NaOH concentration effect on the oriented attachment growth kinetics of ZnS. The Journal of Physical Chemistry. B. 2007;**111**:5290-5294. DOI: 10.1021/jp0688613

[74] Zhang J, Wang Y, Zheng J, Huang F, Chen D, Lan Y, et al. Oriented attachment kinetics for ligand capped nanocrystals: Coarsening of thiol-PbS nanoparticles. The Journal of Physical Chemistry. B. 2007;**111**:1449-1454. DOI: 10.1021/jp067040v

[75] EJH L, Ribeiro C, Longo E, Leite ER. Growth kinetics of tin oxide nanocrystals in colloidal suspensions under hydrothermal conditions.

Chemical Physics. 2006;**328**:229-235. DOI: 10.1016/j.chemphys.2006.06.032

[76] Zhuang Z, Zhang J, Huang F, Wang Y, Lin Z. Pure multistep oriented attachment growth kinetics of surfactant-free SnO_2 nanocrystals. Physical Chemistry Chemical Physics. 2009;**11**:8516. DOI: 10.1039/b907967j

[77] Gunning RD, O'Sullivan C, Ryan KM. A multi-rate kinetic model for spontaneous oriented attachment of CdS nanorods. Physical Chemistry Chemical Physics. 2010;**12**:12430-12435

[78] Zhan H, Yang X, Wang C, Liang C, Wu M. Multiple growth stages and their kinetic models of anatase nanoparticles under hydrothermal conditions. Journal of Physical Chemistry C. 2010;**114**:14461-14466. DOI: 10.1021/jp1062308

[79] Yin S, Huang F, Zhang J, Zheng J, Lin Z. The effects of particle concentration and surface charge on the oriented attachment growth kinetics of CdTe nanocrystals in H 2 O. Journal of Physical Chemistry C. 2011;**115**: 10357-10364. DOI: 10.1021/jp112173u

[80] Segets D, Hartig MAJ, Gradl J, Peukert WA. Population balance model of quantum dot formation: Oriented growth and ripening of ZnO. Chemical Engineering Science. 2012;**70**:4-13. DOI: 10.1016/j.ces.2011.04.043

[81] Xue X, Huang Y, Zhuang Z, Huang F, Lin Z. Temperature-sensitive growth kinetics and photoluminescence properties of CdS quantum dots. CrystEngComm. 2013;**15**:4963. DOI: 10.1039/c3ce40478a

[82] Huang Y, Zhuang Z, Xue X, Zheng J, Lin Z. Growth kinetics study revealing the role of the MPA capping ligand on adjusting the growth modes and PL properties of CdTe QDs. CrystEngComm. 2014;**16**:1547-1552. DOI: 10.1039/C3CE41684D

[83] Wolff A, Hetaba W, Wißbrock M, Löffler S, Mill N, Eckstädt K, et al. Oriented attachment explains cobalt ferrite nanoparticle growth in bioinspired syntheses. Beilstein Journal of Nanotechnology. 2014;**5**:210-218. DOI: 10.3762/bjnano.5.23

[84] Hazarika S, Mohanta D. Oriented attachment (OA) mediated characteristic growth of Gd2O3 nanorods from nanoparticle seeds. Journal of Rare Earths. 2016;**34**:158-165. DOI: 10.1016/S1002-0721(16)60009-1

[85] Chen Z, Han C, Wang F, Gao C, Liu P, Ding Y, et al. Precise control of water content on the growth kinetics of ZnO quantum dots. Journal of Crystal Growth. 2019;**511**:65-72. DOI: 10.1016/j.jcrysgro.2019.01.039

[86] Pacholski C, Kornowski A, Weller H. Self-assembly of ZnO: From nanodots to nanorods. Angewandte Chemie, International Edition. 2002;**41**:1188-1191

[87] Davis K, Yarbrough R, Froeschle M, White J, Rathnayake H. Band gap engineered zinc oxide nanostructures via a sol–gel synthesis of solvent driven shape-controlled crystal growth. RSC Advances. 2019;**9**:14638-14648. DOI: 10.1039/C9RA02091H

[88] Cao X, Wang N, Wang L. Ultrathin ZnO nanorods: Facile synthesis, characterization and optical properties. Nanotechnology. 2010;**21**:065603. DOI: 10.1088/0957-4484/21/6/065603

[89] Chaurasiya N, Kumar U, Sikarwar S, Yadav BC, Yadawa PK. Synthesis of TiO_2 nanorods using wet chemical method and their photovoltaic and humidity sensing applications. Sensors International. 2021;**2**:100095. DOI: 10.1016/j.sintl.2021.100095

[90] Jayachandran M, Rose A, Maiyalagan T, Poongodi N, Vijayakumar T. Effect of various aqueous electrolytes on the electrochemical

performance of α-MnO₂ nanorods as electrode materials for supercapacitor application. Electrochimica Acta. 2021;**366**:137412. DOI: 10.1016/j. electacta.2020.137412

[91] Liang H, Meng F, Lamb BK, Ding Q, Li L, Wang Z, et al. Solution growth of screw dislocation driven α-GaOOH nanorod arrays and their conversion to porous ZnGa 2 O 4 nanotubes. Chemistry of Materials. 2017;**29**:7278-7287. DOI: 10.1021/acs.chemmater.7b01930

[92] Xiang JY, Tu JP, Zhang L, Zhou Y, Wang XL, Shi SJ. Self-assembled synthesis of hierarchical nanostructured CuO with various morphologies and their application as anodes for lithium ion batteries. Journal of Power Sources. 2010;**195**:313-319. DOI: 10.1016/j. jpowsour.2009.07.022

[93] Lu C, Qi L, Yang J, Zhang D, Wu N, Ma J. Simple template-free solution route for the controlled synthesis of Cu(OH) 2 and CuO nanostructures. The Journal of Physical Chemistry. B. 2004;**108**:17825-17831. DOI: 10.1021/ jp046772p

[94] Wang W, Lan C, Li Y, Hong K, Wang GA. Simple wet chemical route for large-scale synthesis of Cu(OH)2 nanowires. Chemical Physics Letters. 2002;**366**:220-223. DOI: 10.1016/ S0009-2614(02)01571-3

[95] Wen X, Zhang W, Yang S, Dai ZR, Wang ZL. Solution phase synthesis of Cu(OH) 2 nanoribbons by coordination self-assembly using Cu 2 S nanowires as precursors. Nano Letters. 2002;**2**: 1397-1401. DOI: 10.1021/nl025848v

[96] Zhang P, Zhang L, Zhao G, Feng FA. Highly sensitive nonenzymatic glucose sensor based on CuO nanowires. Microchimica Acta. 2012;**176**:411-417. DOI: 10.1007/s00604-011-0733-x

[97] Du GH, Van Tendeloo G. Cu(OH)2 nanowires, CuO nanowires and CuO nanobelts. Chemical Physics Letters. 2004;**393**:64-69. DOI: 10.1016/j. cplett.2004.06.017

[98] Song X, Sun S, Zhang W, Yu H, Fan W. Synthesis of Cu(OH) 2 nanowires at aqueous–organic interfaces. The Journal of Physical Chemistry. B. 2004;**108**:5200-5205. DOI: 10.1021/jp036270w

[99] Sundar S, Venkatachalam G, Kwon S. Biosynthesis of copper oxide (CuO) nanowires and their use for the electrochemical sensing of dopamine. Nanomaterials. 2018;**8**:823. DOI: 10.3390/nano8100823

[100] Pathiraja G, Yarbrough R, Rathnayake H. Fabrication of ultrathin CuO nanowires augmenting oriented attachment crystal growth directed self-assembly of Cu(OH)₂ colloidal nanocrystals. Nanoscale Advances. 2020;**2**:2897-2906. DOI: 10.1039/ D0NA00308E

[101] Wang F, Tao W, Zhao M, Xu M, Yang S, Sun Z, et al. Controlled synthesis of uniform ultrafine CuO nanowires as anode material for lithium-ion batteries. Journal of Alloys and Compounds. 2011;**509**:9798-9803. DOI: 10.1016/j.jallcom.2011.07.109

[102] Xu H, Wang W, Zhu W, Zhou L, Ruan M. Hierarchical-oriented attachment: From one-dimensional Cu(OH) 2 nanowires to two-dimensional CuO nanoleaves. Crystal Growth & Design. 2007;**7**:2720-2724. DOI: 10.1021/cg060727k

[103] Bhusari R, Thomann J-S, Guillot J, Leturcq R. Morphology control of copper hydroxide based nanostructures in liquid phase synthesis. Journal of Crystal Growth. 2021;**570**:126225. DOI: 10.1016/j.jcrysgro.2021.126225

Chapter 8

One-Dimensional Metal Oxide Nanostructures for Chemical Sensors

Esther Hontañón and Stella Vallejos

Abstract

The fabrication of chemical sensors based on one-dimensional (1D) metal oxide semiconductor (MOS) nanostructures with tailored geometries has rapidly advanced in the last two decades. Chemical sensitive 1D MOS nanostructures are usually configured as resistors whose conduction is altered by a charge-transfer process or as field-effect transistors (FET) whose properties are controlled by applying appropriate potentials to the gate. This chapter reviews the state-of-the-art research on chemical sensors based on 1D MOS nanostructures of the resistive and FET types. The chapter begins with a survey of the MOS and their 1D nanostructures with the greatest potential for use in the next generation of chemical sensors, which will be of very small size, low-power consumption, low-cost, and superior sensing performance compared to present chemical sensors on the market. There follows a description of the 1D MOS nanostructures, including composite and hybrid structures, and their synthesis techniques. And subsequently a presentation of the architectures of the current resistive and FET sensors, and the methods to integrate the 1D MOS nanostructures into them on a large scale and in a cost-effective manner. The chapter concludes with an outlook of the challenges facing the chemical sensors based on 1D MOS nanostructures if their massive use in sensor networks becomes a reality.

Keywords: metal oxide semiconductors, one-dimensional nanostructures, nanowires, nanofibers, heteronanostructures, chemical vapor deposition, hydrothermal synthesis, electrospinning, chemiresistors, field-effect transistors, electrohydrodynamic printing

1. Introduction

The chemical sensor market requires high performance on all 4S parameters: sensitivity, selectivity, speed, and stability. Moreover, the prevalence of the internet of things (IoT) and wireless sensor networks, as the technology of choice for pervasive real-time monitoring, demands miniaturized sensors with very low power and low cost [1]. Over the past few decades, metal oxides semiconductors (MOS) have been intensively used in chemical sensors for a wide variety of applications. On one hand, MOS can be produced by low-cost wet-chemical synthesis routes from earth-abundant low-cost precursors and are, in general, non-toxic. On the other hand, MOS exhibit unique electronic, chemical and physical properties that are often sensitive to changes in their environment, making them suitable for chemical sensing [2–4]. Finally, the MOS sensors are relatively easy to be miniaturized and integrated into microfabrication processes. In this regard, the micro-electro-mechanical system (MEMS) technology has enabled the miniaturization of chemical sensors using MOS by allowing the implementation of heating and sensing elements by thin-film fabrication [5, 6]. Although, the MOS microsensors fabricated with MEMS technology are already on the market due to their balance in performance and cost, they still suffer from high power consumption and lack of selectivity. The former is connected with the high working temperatures (>200°C), which are necessary to activate sensing mechanisms such as redox reactions, and the latter refers to the non-specificity of MOS surfaces to gaseous analytes. Several strategies have been devised to reduce the power and control the response of the MOS sensors towards specific gases by structural and/or chemical modification of the bulk and/or the surface of the sensing material [7]. Also, a most widespread solution known as the electronic nose uses a broadly responsive array of sensors to generate complex multi-dimensional measurement data in combination with pattern recognition software to interpret the resulting signal pattern [8]. Another strategy relies on the operation of sensors in dynamic mode; this implies the active variation of control parameters such as the working temperature or bias voltage by the sensor electronics, allowing application-specific optimization of the sensor performance or the target application [9, 10]. Finally, UV light-activation has been exploited for the room-temperature operation of the MOS sensors with improved detection capabilities such as sensitivity, selectivity, or response/recovery time [11].

In this context, nanotechnology has emerged as a very promising route to overcome the current drawbacks of the MOS chemical sensors. Enormous research efforts have been devoted to designing and developing MOS nanomaterials for chemical sensing with the ultimate goal of achieving the sensitivity and selectivity levels required for real-world applications while operating the sensors at low temperature, ideally at room temperature [12]. All this, to pave the way for a new generation of chemical sensors consuming very low or zero power using nano-structured MOS layers mounted on cheap, abundant, and easy-to-process substrate materials such as polymers, paper, or fabrics. Then, the fabrication costs of the MOS sensors could be reduced by using state-of-the-art technologies to pattern the electrodes and the MOS nanostructures on top of the substrate, making it possible to produce chemical sensors on a large scale [13]. This would facilitate the introduction of chemical sensors based on MOS in growing markets such as smartphones and wearable devices.

Nanomaterials of all dimensionalities (0D, 1D, 2D, and 3D) and diversity of MOS such as TiO_2, ZnO, SnO_2, CuO, NiO and many others are being investigated for high-performance detection of gases, chemicals and biomolecules [14–20], mainly for applications in the fields of environmental monitoring, healthcare and health diagnostics [21, 22]. Nanomaterials serve as building blocks for the assembly of nanostructured layers with high surface area and porosity, leading to more sensitive and faster chemical sensors. Furthermore, novel synthesis techniques allow fine-tuning of the composition, morphology, and structure of the nanomaterials' surface, which, together with possible alterations in the electronic and chemical properties at the nanoscale, could contribute to enhancing the chemical affinity of the MOS nanostructures for specific species [23, 24]. Recently one-dimensional (1D) MOS nanostructures have gained increased attention for chemical sensing because they are more applicable to nanoelectronics and nanodevices due to their high-surface-area-to-volume ratio 1D morphology [25, 26]; this means that a significant fraction of the atoms is surface atoms that participate actively in surface reactions. In 1D nanostructures the width and thickness are confined to the nanoscale, while the length spans from the micrometric to the millimetric range, allowing 1D nanostructures to contact the microscopic and macroscopic world for many physical measurements.

This chapter surveys the latest achievements in the design and development of 1D MOS nanostructures for chemical sensing. The focus lies on conductometric sensors, specifically resistive and field-effect-transistor-based (FET) sensors, where MOS finds the broadest application. Also, this survey is limited to the 1D nanostructures that have demonstrated the greatest potential for use in conductometric sensors, namely nanowires and nanofibers. Other 1D nanostructures such as nanorods, nanotubes, nanobelts, nanoribbons, or nanoneedles have also been investigated, but to a lesser extent. The techniques for the synthesis of nanowires and nanofibers based on MOS are presented and discussed in terms of their complexity, the potential for integrating the 1D MOS nanostructures into sensor architectures and scalability. Finally, the chapter summarizes the challenges ahead and the prospects for progress in this field.

2. Sensitive materials

The heart of a chemical sensing device is the sensing element or sensor, which consists of two components: a receptor and a transducer. The receptor is the sensitive or active material that has an affinity for and interacts with a specific analyte, stimulating the transducer and causing a change in some physical or chemical property of the material that is ultimately converted into an electrical signal. The analyte may be chemical species (e.g., molecules, ions) or biological species (e.g., microorganisms, biomolecules) in the gas or liquid phase. The receptor function may rely on various principles: physical, where no chemical reaction takes place; chemical, in which a chemical reaction with the participation of the analyte gives rise to the analytical signal; and biochemical, when a biochemical process involving the analyte and a biological recognition element is the source of the analytical signal. The latter is called a biosensor, which is regarded as a subgroup within the group of chemical sensors. According to the operating principle of the transducer, chemical sensors may be classified as optical, electrochemical, electrical, gravimetric, magnetic or thermometric [27]. In electrically-transduced sensors, the signal arises from the changes in the properties of the sensitive material such as conductivity, work function, or permittivity. These changes are converted into variations in electrical parameters of the sensor such as capacitance, inductance or resistance and finally into changes in the device current or voltage. Electrically-based sensors are one of the most investigated chemical sensors due to their simplicity, portability, compatibility with standard electronics, non-line-of-sight detection, the capability of continuous monitoring and potential for wireless transmission.

2.1 Metal oxide semiconductors

MOS has been used as primarily sensitive materials in conductometric chemical sensors due to their outstanding physical and chemical properties. MOS can be produced by a large number of cost-effective synthetic methods, have shown to be active to detect chemical analytes, and their energy band alignment has proved to be suitable to immobilize biomolecules (e.g., enzymes, antibodies, DNA) [2, 18]. MOS are classified according to their conductivity as n-type and p-type, in which the charge carriers are electrons and holes, respectively. N-type MOS (e.g., SnO_2, ZnO, TiO_2, In_2O_3, WO_3, V_2O_5) are the most representative materials for sensing gases and bioanalytes. Some of them are already commercially used, as it is the case in gas sensors, due to their higher sensitivity compared to p-type MOS (e.g., NiO, CuO, Co_3O_4, Cr_2O_3, Mn_3O_4) or other MOS that present both n- and

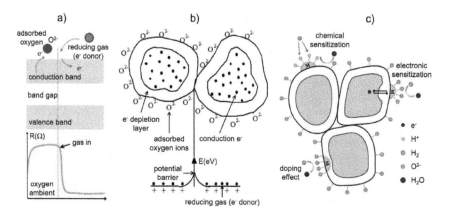

Figure 1.
Schematic illustration of the sensing mechanisms in an n-type metal oxide semiconductor: (a) Interaction with ambient oxygen and with a reducing gas and variation of the sensor resistance; adapted with permission from [32], Copyright 2012 Authors, licensee IntechOpen. (b) Electron depletion layer at the grain surface and intergrain barrier potential; adapted with permission from [33], Copyright 2012 Authors. (c) Hydrogen detection mechanisms in a metal oxide semiconductor functionalized with metal nanoparticles; adapted with permission from [51], Copyright 2018 Elsevier.

p-type behavior (e.g., Fe_2O_3, HgO_2) [28, 29]. Also, n-type MOS are thermally stable and have the possibility to work at lower partial pressure, in contrast to p-type MOS which are less stable due to their tendency to exchange lattice oxygen easily with air. Another practical reason for the use of n-type MOS in conductometric sensors is related to the preferred direction for resistance change during detection of reducing gases (the vast majority of gaseous analytes), which simplifies the peripheral electronics for measurements and improves the reproducibility of the output signal. Nonetheless, p-type MOS should not be underestimated as sensitive materials, as significant improvement in sensor performance can be achieved by incorporating p-type MOS with commonly used n-type MOS [30]. As an example, p-type MOS have been used as good catalysts to promote selective oxidation of various volatile organic compounds. Moreover, the distinctive oxygen adsorption of p-type MOS may be used to design high-performance gas sensors that show low humidity dependence and rapid recovery speed [31].

Figure 1a and **b** illustrate the sensing mechanisms in an n-type MOS [32, 33]. The molecules of the ambient oxygen are adsorbed and then ionized by capturing electrons from the conduction band. Thus, an electron depletion layer of a given width, known as Debye length, forms at the surface of the MOS grains and a potential barrier develops at the boundaries of adjacent grains, caused by the negative surface charge due to the adsorbed oxygen ions. As a result, the density of free electrons is reduced and the electron conduction is hindered, increasing the sensor resistance. In the presence of a reducing gas (electron donor), the gas molecules react with the adsorbed oxygen ions, which are released to the ambient, while the trapped electrons return to the conduction band. Then, the density of free electrons increases and the intergrain barrier potential is reduced, resulting in a decrease in the sensor resistance. Many optimization strategies aim at modulating or triggering variations in the electron depletion region or barrier potential in a controlled manner to improve the receptor function in chemical sensors based on MOS.

2.2 Functionalized metal oxide semiconductors

Since the efficiency of the chemical receptor material is surface-dependent, previous research on MOS sensors proved that MOS with sizes within the Debye length, typically of the order of 2–100 nm, are attractive in chemical sensing as they provide higher surface-area-to-volume-ratio, as compared to bulk materials. Also, it was found that nanostructures (e.g., 1D MOS) can provide specific crystal facets and electronic properties to the surface that enhance the performance of the receptor [34]. Further, the functionalization or modification of MOS nanostructures by loading, doping, surface-decoration or hybridization with second-phase constituents (e.g., noble metals, other MOS, carbon-based materials) showed other ways to enhance and extend the capabilities of the receptor [35, 36]. Similarly, in biochemical sensors, the functionalization of the surface by immobilizing biomolecules that act as biological recognition elements of specific analytes (e.g., glucose, urea, cancer cells, viruses) is mandatory to sensitize the MOS surface [2, 18, 37].

The rationale for these improvements or further sensitization is generally connected with an increase in the density of active sites (e.g., defects, oxygen vacancies) or charge carriers (doping effect). Also, it is related to the catalytic activation (chemical sensitization) and the formation of interfaces (electronic sensitization), either metal-semiconductor or semiconductor-semiconductor interfaces. The latter is known as heterojunctions as they involve two dissimilar semiconductors, whilst the combination of multiple heterojunctions together in a system is called a heterostructure. The materials chosen for these interfaces dictate the principles of sensing; for example, in gas sensing, adsorption, reaction, and electronic behavior. On one side, the intimate electrical contact at the interface between the two components equilibrates the Fermi level across the interface to the same energy, usually resulting in charge transfer and further extending the region of charge depletion/accumulation. On the other side, the mix of the two components leads to synergistic behavior. This means that each component serves a different purpose that is complementary to the other, so that the synergistic effect of the two-component system is greater than the effect of each element. The synergistic effect is possible due to three common features: geometric effects, electronic effects, and chemical effects. These are the basis for the improved sensing properties such as enhanced sensitivity and recovery speed or reduced operating temperature of the MOS heteronanostructures [38–42]. These approaches are also used to improve further the selectivity, particularly towards gas analytes. The usual combination includes wide bandgap MOS of n-type as the host material and second-phase constituents commonly chosen from noble metals (e.g., Pt, Pd, Ag, Au) [43–45] or other transition metal oxides (e.g., p-type: NiO, CuO, Cr_2O_3, Fe_2O_3, Co_3O_4, Mn_3O_4; n-type: SnO_2, ZnO, In_2O_3, WO_3, TiO_2) [46, 47]. Recent combinations also make use of carbon-based materials (e.g., carbon nanotubes, graphene) [48, 49] and two-dimensional (2D) inorganic materials (e.g., transition metal dichalcogenides TMDC, transition metal carbides and nitrides MXenes, phosphorene) [50].

Generally, the modifiers or additives (i.e., metals, MOS, rGO) enhance the sensing properties of the host MOS by surface- (chemical sensitization) and/or interface- (electronic sensitization) dependent effects. As an example, **Figure 1c)** illustrates the synergistic mechanisms, both chemical and electronic, enabling the detection of hydrogen by a MOS functionalized with metal nanoparticles [51]. The chemical sensitization includes spill-over (i.e., the enrichment of the MOS surface with reactive species through catalysis), whereas the electronic sensitization involves Fermi level control (i.e., changes in the chemical state of the additive with active species). To date, there is no clear evidence of whether any of these mechanisms are superior. Nevertheless, the literature indicates the need to combine both mechanisms to induce a better performance of MOS. The possibility of tuning the sensor performance via chemical and electronic sensitization depends strongly on the characteristics (size, shape, distribution, composition, oxidation states) of the MOS and additives forming the heteronanostructure and the formation of an intimate electric contact at the interface of the components. The optimal combination of components significantly improves the selectivity and sensitivity of MOS sensors to specific gases, especially for sensors operated at room temperature.

In summary, MOS nanomaterials modified with second-phase materials and controlled interfaces are essential for sensing chemical species more efficiently. Among several nanoscale materials, 1D structures have been shown to be most relevant to chemical sensing. They are projected as potential structures for molecular-level sensing, particularly when integrated as single elements. Below are discussed the most common synthetic methods to achieve 1D MOS nanostructures and their integration with appropriate transducers for their application as chemical sensors.

3. Synthesis of 1D metal oxide nanostructures

Among a host of 1D nanostructures, nanowires and nanofibers bring the greatest potential for use in the next generation of chemical sensors based on MOS [52–60]. Both nanomaterials have a high specific surface area and a large surface-area-to-volume ratio. However, their aspect ratio, crystallinity and surface properties may differ markedly because of the different synthesis techniques and conditions used, which in turn result in different chemical sensing performances. The techniques for the synthesis of 1D MOS structures can be classified in two general approaches: top-down and bottom-up technologies. Top-down technologies are subtractive technologies that rely on microfabrication methods, mainly lithography and etching processes, to reduce the lateral dimensions of a MOS film to nanometer size. Usually, the 1D nanostructures produced by this approach are amorphous or polycrystalline structures. Top-down technologies are well developed in the semiconductor industry, but need expensive equipment limiting their broad application in the academic sector. Bottom-up technologies, in contrast, are additive technologies that consist of the assembly of molecular building blocks that lead to the formation of nanostructures. Bottom-up technologies are generally enabled by vapor- or solution-based techniques and are considered cost-effective solutions for large-scale production of 1D nanostructures [24]. The nanomaterials produced by this approach may have monocrystalline or polycrystalline characteristics depending upon the specific processing steps of the vapor- or solution-based techniques employed.

Different routes are commonly used to synthesize 1D MOS nanostructures [61–63]. Whist chemical vapor deposition (CVD) is one of the most representative vapor-phase methods for forming 1D nanostructures; hydrothermal synthesis and electrospinning-assisted synthesis are the most representative amongst the liquid-phase methods. However, CVD and hydrothermal synthesis share common mechanisms for 1D structure formation based on nucleation and growth processes, whereas 1D structures formed by electrospinning rely on the generation of guiding polymer-based fibers using an electrostatic field to shape the 1D structure. The following sub-sections discuss separately the synthesis of 1D MOS nanostructures based on nucleation and growth processes and those mediated by electrospinning, herein called nanowires and nanofibers, respectively. A sub-section dedicated to the functionalization of these 1D nanostructures with second-phase materials is also included in this section.

3.1 Synthesis of nanowires based on nucleation and growth processes

The group of synthetic methods employed to form chemical-sensitive 1D structures usually involves CVD and hydrothermal processes. Generally, CVD refers to methods based on chemical reactions of gaseous reactants in an activated environment. In CVD synthesis, chemical precursors in the vapor phase are delivered to a reactor, where external energy (heat, light, or plasma) is added to the system to initiate the deposition reactions. These reactions can be homogeneous or heterogeneous; the first occurs in the gas phase whilst the second occurs between gas-phase species and a solid substrate (usually involving an initial gas phase reaction with the formation of reactive intermediate species). Homogeneous reactions form typically non-adherent powders and by-products whereas heterogeneous reactions lead to nucleation and growth processes on solid substrate surfaces that result in the formation of a solid material [64]. Tuning the reaction conditions by adjusting pressures, precursors, and type of activation, amongst other CVD-type related parameters, can promote the formation of either planar films or 1D structures such as nanowires. In contrast, hydrothermal synthesis refers to chemical reaction methods run in a sealed and heated aqueous solution at appropriate temperatures (100–1000 °C) and pressures (1–100 MPa); the name solvothermal is commonly used when organic solvents are involved in the solution. In hydrothermal/solvothermal synthesis, the properties of the synthesized structures are controlled by adjusting parameters such as the solution pH, the chemical species concentration, and the oxidation-reduction potential, apart from the temperature and pressure. The nucleation of the structures can also be adjusted by adding inorganic additives. Thus, hydrothermal/solvothermal synthesis can generate large good quality crystals while keeping control over the material composition [65]. Both CVD and hydrothermal/solvothermal methods have been suitable to deliver MOS nanoparticles and nanostructures including nanowires of ZnO [66–68], SnO$_2$ [68, 69], or WO$_3$ [70, 71], to cite a few examples. The formation of nanowires by these methods generally responds to two mechanisms: one assisted by adding catalyst seeds and the other without catalysts.

3.1.1 Mechanisms involving catalyst

This growth mechanism is characterized by the use of molten metal catalysts over the substrate that can rapidly adsorb a fluid (in gaseous, liquid, or supercritical phase) containing the MOS precursor to supersaturation levels. According to the type of fluid, this mechanism is known as vapor-liquid-solid (VLS), solution-liquid-solid (SLS) or supercritical-fluid-liquid-solid (SFLS) [24, 72, 73]. The adsorption of the fluid can be molecular or dissociative followed by adatoms diffusion on the catalyst, the substrate, or the nanowire sidewalls. Diffusion across the substrate and the sidewalls needs to be rapid to avoid nucleation events. In this regard, the nanowires can grow from the top or from the bottom of the catalyst as single nanowires, as depicted in **Figure 2a–c**, respectively [74]. In many cases, there is a one-to-one correspondence between the catalyst particle size and the nanowire, although this is not a rule. Multiple nanowires can also grow, without

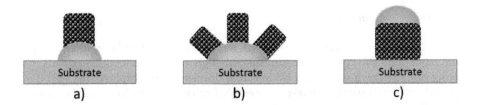

Figure 2.
Catalytic growth of nanowires from (a and b) the top and (c) the bottom of the catalyst particle. Nanowires in (a) and (c) may have a one-to-one correspondence with the catalyst particle size, whereas wires in (b) do not have direct correspondence with the catalyst particle size; reprinted from.

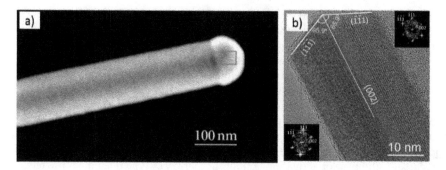

Figure 3.
(a) High-resolution SEM image of a ZnO nanowire grown by VLS mechanism using binary allow of Cu/Au as a catalyst; reprinted with permission from [75], Copyright 2012 Authors. (b) HRTEM image of CuO nanowires grown by non-catalytic mechanism assisted by twin boundary defects (the insets show fast Fourier transform images from the areas of the crystals indicated by red squares); reprinted with permission from [78], Copyright 2014 American Chemical Society.

direct correspondence with the catalyst size, but with other structural factors such as the curvature and lattice matching at the catalyst-nanowire interface. VLS- and SFLS-related methods have a wide selection of catalysts (Au is the most common) and deliver high-quality nanowires with wide synthetic tunability. However, they usually require high temperature (>400 °C) and/or high pressure, and specialized equipment. SLS-related methods, in contrast, have the advantage of requiring low temperature (200–350 °C), although this fact restricts to a certain degree the choice of catalyst to low melting point catalyst such as Ga, In, or Sn. These mechanisms have brought up discussions of whether the catalyst particles reach a liquid-phase or stay in the solid phase. Nevertheless, and since any of these possible ways is ruled out, currently one can also find literature reports that state the growth of nanowires by vapor-solid-solid (VSS), solution-solid-solid (SSS), or supercritical-fluid-solid-solid (DFSS) mechanisms, with the results suggesting that the dynamics of nanowire growth is not affected by the phase of the catalyst particle [72]. An example of ZnO nanowires grown by VLS using binary allow of Cu/Au as the catalyst is displayed in **Figure 3a)** [75]. Cu/Au catalysts have better adhesion properties than pure Au catalysts providing advantages to pattern vertically aligned nanowires as demonstrated for ZnO nanowires.

Catalytic growth of nanowires brings a fine control over the wire geometry, specially diameters and lengths. However, the nanowires yielded by catalyst-based routes incorporate catalyst atoms (impurities) into their structure, influencing the nanowires' physical and chemical properties and possible intended applications. Therefore, non-catalytic alternative routes (without using catalyst particles) are also being used and explored to grow nanowires. The next sub-section deals with this type of mechanism.

3.1.2 Mechanisms without catalysts

Also known as the catalyst-free growth mechanism. It is usually represented by the vapor-solid (VS) growth mechanism, although it also includes other growth processes, such as those assisted by defects or droplets. In the vapor-solid (VS) growth, mass transport is achieved preferably from the vapor phase. The nanowire crystallization originates from the direct vapor condensation, without needing the assistance of defects. In this process, the initially condensed molecules form seed crystals that serve as nucleation sites. Once an atomic layer is nucleated, the subsequent atomic layers grow at a faster rate than the wire edge, whose edge is consumed by mass transport to the newly formed layer [63, 76]. Previous, *in-situ* TEM observations of catalyst-free VS grown tungsten oxide nanowires showed that the wires' edges grow approximately 20 times slower than the newly formed atomic layers [77]. Non-catalytic growth of nanowires assisted by defects, in contrast, relies on line defects that act as nucleation centers. In this type of growth, mass transport can be provided from a vapor or by adatoms diffusion along with the growing structure. Previous reports connect the final 1D morphological structure with the defect types. For instance, defects such as screw dislocations showed to lead to nanowires with cylindrical shape [76], whereas planar defects (twin boundaries) have proved to serve as points for preferential nucleation (reducing the nucleation energy barrier) on the nanowire tip and lead to prism-like 1D morphologies [78], as shown in **Figure 3b**).

The group of non-catalyst mechanisms can also involve growth processes assisted by liquid native droplets (form by the native metal of a MOS, e.g., Cu for CuO, or Zn for ZnO). Due to the consumption or crystallization of the metal droplet during reactions and nanowire growth, these metal droplets are not considered as catalysts, despite the nanowire growth process following similar principles to those of catalyst growth via the VLS mechanism. Hence, the nanowires yielded utilizing native metal droplets do not display the droplets at the top/bottom of the wire structure or introduce impurities as in VLS [76]. Previous experiments corroborated this growth mechanism by *in-situ* TEM analysis of Al_2O_3 nanowires grown from Al liquid particles. The studies revealed a layer-by-layer growth at the Al liquid droplet and Al_2O_3 nanowire interface, promoted by the surrounding oxygen and Al_2O gases, following the so-called oscillatory mass transport, which is also characteristic in the growth of the nanowires by VLS and VS [79].

3.2 Polymer-assisted synthesis of nanofibers

Metal oxide nanofibers can be synthesized by chemical bottom-up routes like the ones described in the previous sub-section, as well as by top-down routes either mechanical or spinning methods [80–82]. Among the latter, electrospinning is the most widely used method for the production of metal oxide nanofibers, mainly due to its high simplicity and ease of use, low-cost setup, ability to mass-production of continuous fibers, and flexibility in controlling the diameter, morphology, structure, and alignment of the fibers [82–84]. The typical setup for electrospinning is depicted in **Figure 4a)** [85]. This is rather simple and operates at ambient conditions, although the mechanism of electrospinning is complex [86, 87]. An

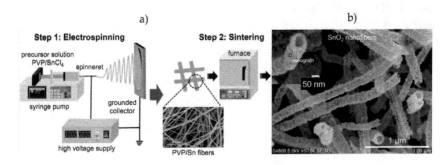

Figure 4.
(a) Schematic illustration of the synthesis of SnO_2 nanofibers by (step 1) electrospinning of the precursor solution followed by (step 2) high-temperature sintering of the as-spun fibers; adapted with permission from [85], Copyright 2018 Authors, licensee MDPI. (b) SEM micrograph of the hollow SnO_2 nanofibers resulting from (a); reprinted with permission from [97], Copyright 2013 eXpress Polymer Letters.

electrospinning system consists of three major components: a high voltage supply, a spinneret (e.g., glass capillary tube, metallic needle, pipette tip) and a grounded collector. The high voltage source injects the charge of a given polarity into a polymer solution, which is fed (e.g., syringe pump) at a constant rate to the spinneret. An electric field is thus established between the spinneret and the collector and when it reaches a critical value, the liquid reaching the spinneret tip forms a cone (Taylor cone) that emits a liquid jet through its apex. This charged liquid jet is stable only at the tip of the cone and undergoes an unstable and rapid stretching and whipping process downstream of the spinneret, which leads to the formation of a long and thin thread. As the liquid jet is continuously elongated and the solvent is evaporated, its diameter can be reduced from hundreds of micrometers down to the sub-micrometer scale. The electrospun polymer fibers are deposited randomly on the plate due to the attraction of the collector placed in front of the spinneret.

The diameter, morphology, and structure of the electrospun fibers are key factors that need to be controlled for practical applications [88, 89]. They depend on a multitude of parameters related to the solution, setup and electrospinning process [90–94]. Target fibers can be obtained by properly choosing the polymer and the solvent and their concentration in the solution, thereby adjusting the electrospinning-relevant solution properties such as viscosity, surface tension, and electrical conductivity. The setup orientation (e.g., vertical, horizontal), spinneret type (e.g., needleless, single-needle, coaxial-needle) and nozzle diameter, solution feed rate, applied voltage, and spinneret-to-collector distance tremendously influence the fiber features. These features are also affected by the evaporation conditions such as ambient temperature and humidity.

Nanofibers of the MOS used for chemical sensing can be produced by electrospinning of a solution containing a polymer (e.g., polyvinylpyrrolidone PVP, polyvinylalcohol PVA, polyvinylacetate PVAc, polyethylene oxide PEO) and an inorganic precursor (e.g., acetates, nitrates, chlorides) in a solvent (e.g., deionized water DIW, ethanol EtOH, isopropyl alcohol IPA, dimethyl formamide DMF) followed by a sintering step, also known as annealing or calcination, at high temperature [84, 95, 96]. After obtaining the electrospun polymer/metal composite fibers, the polymer is eliminated by the sintering process. The evaporation of residual solvent and water from the fibers occurs in the first place, at low temperature. As the temperature rises, the polymer decomposes gradually and fiber shrinkage takes place. Simultaneously the inorganic precursor undergoes a complex transformation including outwards radial diffusion, nucleation, condensation, crystallization and oxidation processes to finally form metal oxide nanograins aligned along the preceding as-spun fibers [84]. A simplified schematic of the process is shown in **Figure 4a)** [85], while **Figure 4b)** displays the SnO_2 nanofibers obtained by sintering the as-spun fibers [97]. The nanofibers have diameters of less than 100 nm that are rather uniform along the entire length of the nanofibers, which are in the order of hundreds of micrometers. They are hollow nanofibers and have an ultra-thin porous granular wall consisting of SnO_2 nanograins.

Sintering parameters such as the heating rate, sintering temperature and time, cooling rate, and atmosphere largely influence the diameter, structural morphology, crystallinity and grain size of the MOS nanofibers [93], which in turn determine their sensing performance [98–100]. Hollow nanofibers have about twice the active surface area of solid nanofibers of the same diameter, which increases the sensor response, as gases can interact with both the outer and inner surface of the nanofibers [101]. While MOS nanowires are single-crystalline with no noticeable grains, electrospun MOS nanofibers are composed of polycrystalline grains. Higher sintering temperatures or longer sintering times result in better crystallinity, which in turn leads to higher sensor response. Also, the grain size increases with the increasing sintering temperature and time, but the sensor response decreases with the increasing grain size due to the lower surface area. Hence, it is necessary to optimize the size and crystallinity of the grains in the MOS nanofibers simultaneously to attain superior sensing behavior [102–107]. The diameter and wall thickness of the hollow MOS nanofibers increase with the increasing concentration of polymer and inorganic precursors in the solution, respectively [108]. Reducing both the diameter and wall thickness also improves the sensing properties of the electrospun MOS nanofibers [98]. A decrease in the diameter of the nanofiber contributes to decreasing both the grain size and the time of gas diffusion into the nanofiber, resulting in enhanced sensor response and reduced response time. Nevertheless, there is a limit in the process of enhancing sensor performance by decreasing the grain size, since an excessive decrease of grain size (<10 nm) leads to a loss of structural stability and, as a consequence, to changes in both surface and catalytic properties of the material. Moreover, the porosity of the wall determines the accessibility of the gases to the inner surface of the hollow fibers, and several strategies have been adopted to increase the porosity of the electrospun MOS nanofibers. One strategy is to remove the polymer from the as-spun polymer/metal fibers by exposing the fibers to RF oxygen plasma before sintering [109, 110], as can be seen in **Figure 5**. This technique allows tuning the porous structure of the hollow MOS nanofibers by controlling the etching power and time. Finally, in a recent work 1D hierarchical structures of ZnO have been prepared by growing ZnO nanowires over electrospun polyacrylonitrile (PAN) fibers loaded with Au nanoparticles, as seeds for catalytic VLS growth [111]. **Figure 6** shows the resulting ZnO cactus-like structures, whose surface area is far larger than that of electrospun ZnO nanofibers of the same diameter.

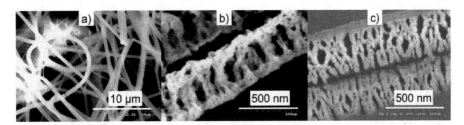

Figure 5.
SEM micrographs of electrospun PVA/Sn fibers (a) without post-treatment and (b) after exposure to RF oxygen plasma, (c) highly porous hollow SnO₂ nanofibers obtained by sintering of the PVA/Sn fibers in (b). Reprinted with permission from [110], Copyright 2009 Elsevier.

Figure 6.
(a) SEM micrograph of a ZnO nano-cactus like structure obtained by VLS growth of ZnO nanowires over electrospun fibers of polyacrylonitrile (PAN) loaded with Au nanoparticles. (b) Higher magnification SEM micrograph of the dashed area in a), showing Au nanoparticles at the tip of ZnO nanowires (dotted circles) and EDX spectrum of the ZnO structure (inset). Reprinted with permission from [111], Copyright 2021 Elsevier.

3.3 Functionalization of 1D metal oxide nanostructures

Functionalization or modification with second-phase constituents (modifiers or additives) can strongly change the physical and chemical properties of the 1D MOS nanostructures. Tailoring the properties (e.g., shape, size, load, dispersion) of the additives is a very challenging task because not all methods allow for the control of their properties and homogeneous dispersion over the 1D MOS nanostructures. The functionalization or modification of 1D MOS nanostructures with second-phase materials may be of the decorative-type, only at surface level, when incorporating low amounts of the additives at the surface. It can also involve single mixtures, when mixing 1D nanostructures and the additives randomly, or doping, when additive atoms incorporate in the intrinsic material structure [25, 112]. Generally, the methods to functionalize or modify 1D MOS nanostructures fall into two categories: one-step processes in which the additive materials are incorporated simultaneously during the 1D nanostructure formation or multiple-step processes in which the additives are incorporated over the pre-synthetized 1D MOS nanostructures [113]. In both cases, the incorporation may involve a precursor for the targeted additive or pre-formed particles, as shown in **Figure 7**. Routes 1 and 2 (representing one-step processes) may be enabled by sol-gel, hydrothermal synthesis, CVD, and electrospinning. Routes 3 and 4 (representing multiple-step processes) rely on techniques, such as dip- or spin-coating, to introduce pre-formed additives or a broad type of techniques that can incorporate the additive from a precursor or as ions. For instance, high energy ion implantation, immersion in solutions containing the additive precursor followed by a photocatalytic reduction or heat treatment, sputtering, hydrothermal synthesis, CVD including atomic layer deposition (ALD), amongst others.

1D heteronanostructures have gained great attention due to their hybrid properties, which may induce synergies between the host material (MOS) and the guest material (modifier or additive), resulting in improved and/or new attributes for the chemical detection [114–116]. In the first place, p-type MOS of transition metals (e.g., CuO, NiO, Fe_2O_3) and noble metals (e.g., PdO, Ag_2O) have been applied to develop ultrasensitive chemical sensors by tuning the electrical properties of the n-type MOS by forming n-p heterojunctions. Also, n-n heterojunctions may lead to enhanced sensing performance of the nanostructures composed of two n-type MOS with different work functions (e.g., SnO_2-ZnO, WO_3-SnO_2, TiO_2-SnO_2, SnO_2-In_2O_3), as compared to single-phase MOS nanostructures. In addition, the combination of MOS and graphene may significantly improve the performance of MOS gas sensors

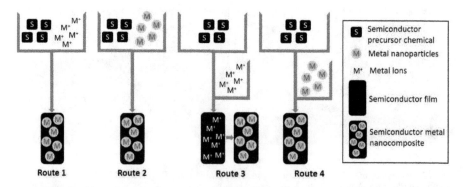

Figure 7.
Routes to the functionalization or modification of 1D MOS nanostructures with second-phase materials (modifiers or additives); adapted with permission from [113], Copyright 2006 American Chemical Society.

at room temperature [48, 117]. Graphene is a family of two-dimensional (2D) nanomaterials (e.g., pristine graphene PG, graphene oxide GO, reduced graphene oxide
rGO, graphene quantum dots GQD, graphene nanoplatelets GnP) that are obtained
from natural graphite or synthesized chemically from organic compounds, and
have different morphology, physical, chemical, and electronic properties. Among
them, rGO is the best choice for gas sensing applications because it has oxygen
functional groups, defects and vacancies on its surface, which favor the adsorption
of gases. Moreover, rGO behaves as a p-type semiconductor and is stable at high
temperature [118, 119]. When added to 1D MOS nanostructures, rGO nanosheets
can increase the overall sensing surface and adsorption sites, and form n-p or p-p
type heterojunctions with the MOS grains, thereby modulating the resistance of
the MOS gas sensors. Also, noble metals (e.g., Au, Ag, Pt, Pd) and transition metals
(e.g., Ni, Cu, Co) act as effective modifiers or additives, particularly due to their
catalytic properties.

The functionalization or modification of 1D MOS nanostructures based on
nucleation and growth processes usually rely on two-step processes, in which the
second-phase material is incorporated in a subsequent step after the synthesis
of the 1D nanostructure; routes 3 and 4 in **Figure 7**. Hence, the methods for the
functionalization step are varied. They may include, for instance, sputtering as in
recent reports that showed the incorporation of DC pulsed sputtered Au nanoparticles over the surface of hydrothermally synthesized ZnO nanowires, as displayed
in **Figure 8a**) [120]. In this method, the size and density of the Au nanoparticles
decrease as the sputtering pressure increases (e.g., from 5 to 20 mTorr) due to the
dependency of the mean free path and rate of gas phase collisions on the process
pressure. The routes to functionalize 1D nanostructures in a second-step process
also involve a broad variety of CVD methods. Among them, for instance, aerosol-
assisted (AA) CVD has demonstrated to be useful to incorporate both metals and
MOS nanoparticles over 1D nanostructures. This method allowed for the incorporation of dispersed nanoparticles based on n-type or p-type MOS from metals such as
Pd [121], Ni, Co, or Ir [122]. It also allowed for the formation of core-shell 1D nanostructures based on WO_3 nanowires covered by a Ce_2O_3 thin film [123]. Similarly,
flame-assisted CVD has shown to functionalize 1D nanostructures including SnO_2
nanowires with Au and Pd nanoparticles [124]. This method has also been used
to modify the surface of SnO_2 nanowires with an amorphous carbon layer [125].
Other methods for functionalization in a second-step may also combine the merits
of several techniques, including sol-gel, dip-coating, and flame spray pyrolysis,
as is the case of the sol-flame method. This method incorporates the second phase

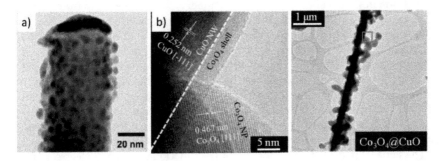

Figure 8.
*(a) TEM micrograph of a ZnO nanowire functionalized with sputtered Au nanoparticles; reprinted with
permission from [120], Copyright 2021 Elsevier. (b) Functionalization of CuO nanowires by sol-flame:
HRTEM (left) and TEM (right) micrographs of a CuO nanowire functionalized with Co_3O_4 nanoparticles;
reprinted with permission from [126], Copyright 2013 American Chemical Society.*

Figure 9.
(a) SEM micrograph of WO₃ nanowires functionalized with noble metal nanoparticles (Au, Pt) in a single-step process by AACVD. HRTEM micrographs of WO₃ nanowires functionalized with nanoparticles of (b) Au and (c) Pt, with insets showing the size distribution and lattice fringes of the nanoparticles; reprinted from [128].

material by dip-coating the nanowires with a sol-gel precursor solution and anneal-ing them over a flame for a few seconds. During the flame treatment, the metal salt precursor decomposes chemically to the final metal or MOS and nucleates locally over the nanowire. An example of this process used to functionalize CuO nanowires with Co_3O_4 is displayed in **Figure 8b**) [126].

The functionalization or modification of 1D MOS nanostructures based on nucleation and growth processes can also be achieved by one-step processes. However, their use is less common, despite the advantage of reducing processing steps. Examples of 1D nanostructures functionalized by a one-step process include those achieved by the AACVD of two metal precursors from one-pot simultane-ously [127]. In this method, the precursor leading the formation of nanowires is in a higher concentration than the precursor for the modifier. **Figure 9** displays exam-ples of the WO_3 nanowires functionalized with Au and Pt nanoparticles obtained by this method [128]. The functionalization of nanowires with MOS from metals such as Fe [129] and Cu [31] was also achieved by this method.

There is much less published work on the functionalization or modification of electrospun MOS nanofibers for application in chemical sensors than on nanowires. Most works chose to incorporate the additives or their precursors into the solution containing the polymer and the inorganic precursor of the MOS (i.e., routes 1 and 2 in **Figure 7**). Thus, for example, composite nanofibers are prepared by dissolv-ing inorganic precursors of the involved MOS in suitable solvents and mixing the solutions with the polymer solution, usually by magnetic stirring. Then, the inorganic precursors are distributed uniformly in the polymer by electrospinning and the metals are oxidized upon sintering of the electrospun polymer/metal fibers. Intimate contact between the MOS nanocrystals is achieved in the composite nanofibers [130–133]. Another method uses the coaxial-electrospinning configura-tion [84, 134], for which polymer solutions are prepared with each of the inorganic precursors separately. The solution with the precursor of the main metal oxide leaves the spinneret through a central circular nozzle, while the solution with the precursor of the second metal oxide exits the spinneret through an annular nozzle that surrounds and is concentric to the circular nozzle, as depicted in **Figure 10a**) [135]. After sintering of the electrospun fibers, composite nanofibers with a core-shell structure are obtained, in which the two metal oxides occupy distinct zones with a well-defined interface [136–138], as can be seen in **Figure 10b**).

The same strategy has been adopted for the functionalization of electrospun MOS nanofibers with additives such as graphene (rGO) and metals. In this case, the rGO flakes and metal nanoparticles are dispersed or their precursors are dissolved in a liquid (e.g., DIW, EtOH, IPA), usually by ultrasonic agitation. The colloidal dis-persions or solutions so obtained are added to the solution with the polymer and the

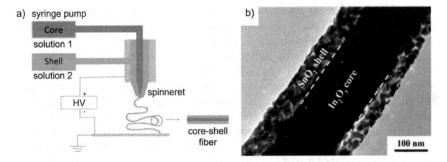

Figure 10.
(a) Layout of a typical spinneret used for coaxial electrospinning; adapted with permission from [135], Copyright 2019 Authors, licensee MDPI. (b) TEM micrograph of an In_2O_3-SnO_2 core-shell nanofiber obtained by sintering of coaxially electrospun PVP/In-PVP/Sn fibers; reprinted with permission from [138], Copyright 2016 American Chemical Society.

inorganic precursor of the host MOS and, then, magnetically stirred until a homogeneous electrospinnable solution is achieved. As an example, **Figure 11** shows TEM images of an electrospun nanofiber of SnO_2 loaded with rGO [139]. Double-shell hollow nanofibers are usually obtained, with the rGO nanosheets on top of the MOS nanograins [139–143]. Generally, it has been found that the rGO-loaded MOS nanofibers are more sensitive to specific gases and that the optimal operating temperature (i.e., the temperature at which the sensor response reaches a maximum) is lower than that of the pure MOS nanofibers. This improved sensing behavior is attributed to the formation of local n-p or p-p (MOS-rGO) heterojunctions.

Some authors have attempted to further improve the gas sensing capabilities of electrospun single-phase MOS nanofibers [144], MOS composite nanofibers [145, 146], and rGO-loaded MOS nanofibers [147] by functionalization of the nanofibers with metal nanoparticles. For this purpose, they chose also the routes 1 and 2 in **Figure 7**, leading to the dispersion of the metal nanoparticles, formed *in-situ* or pre-formed, in the solution with the polymer, inorganic precursors, and eventually rGO. Sensors based on hybrid nanofibers composed of ZnO, rGO and nanoparticles of Au or Pd prepared by electrospinning showed enhanced sensitivity towards reducing gases and volatile organic compounds, for the pure and rGO-loaded ZnO nanofibers [147].

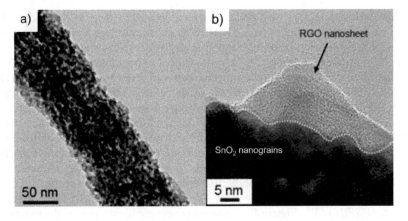

Figure 11.
(a) Low-magnification and (b) high-magnification TEM micrographs of an rGO-loaded SnO_2 nanofiber obtained by sintering of an electrospun PVA/Sn/rGO fiber; reprinted with permission from [139], Copyright 2015 American Chemical Society.

It was proven therefore the synergistic combination of the catalytic effects of the noble metal nanoparticles and the hybrid sensing mechanism, which combines the effects of radial resistance modulation, intergrain (ZnO/ZnO) modulation and local n-p (ZnO-rGO) heterojunctions. Moreover, it has been observed that there is an optimal value of the load of rGO (<1 wt%) and metal nanoparticles (1–4 wt%) in the nanofibers, above which the gas sensing performance of the hybrid nanofibers does not show further improvement or starts degrading. This result is attributed to the agglomeration of rGO nanosheets and metal nanoparticles in the polymer, resulting in the formation of agglomerates on the surface of the nanofibers and hence in an increased density of p-p (rGO-rGO) heterojunctions and metal-metal contacts, in detriment of n-p or p-p (MOS-rGO) heterojunctions and metal-MOS contacts. The electrospinning of polymer solutions containing nanomaterials is very challenging as to achieve an even distribution of the nanomaterials in the polymer fibers, since the large specific energy of the solution promotes the agglomeration of the nanomaterials [148, 149]. To overcome this problem, the nanomaterials can be loaded onto the surface of the fibers either after electrospinning or after sintering the as-spun fibers, i.e., routes 3 and 4 in **Figure 7**. This approach however adds complexity and costs as it introduces an additional process step (e.g., sputtering, ALD) for which dedicated equipment is required [150, 151].

4. Integration of 1D metal oxide nanostructures into chemical sensors

The chemical sensing capabilities of the MOS have been exploited primarily in electrically-transduced sensors of the resistive and field-effect transistor (FET) types [55]. Chemiresistive sensors or chemiresistors and FET sensors are the most investigated and exploited sensing configurations owing to their simplicity, ease of fabrication and operation, and feasibility of miniaturization. The most widely used architectures of chemiresistors and FET sensors are presented in this section. There follows a discussion on the methods and techniques that allow 1D MOS nanostructures to be integrated as sensitive material into such architectures, paying attention to aspects such as sensor reproducibility and potential for scaling up sensor fabrication.

4.1 Sensor architectures

Chemiresistive sensors are bipolar devices addressed to measure the electrical resistance of a semiconductor (e.g., 1D MOS nanostructure) as the sensing material bridging two electrodes or interdigitated electrodes (IDE) supported by an insulating substrate. Typically, FET sensors consist of a semiconductor (e.g., 1D MOS nanostructure) as the conducting channel connected by the source and drain electrodes. This semiconductor is placed on the top of an insulated gate electrode so that its conductance can be regulated by varying the bias voltage of the gate electrode. This classical architecture for electronic metal-oxide field-effect transistors (MOSFET) is usually similar for FET sensors addressed to gas-phase analytes. In contrast, the architecture of FET sensors for liquid phase analytes differs from the traditional MOSFET, since the gate electrode (or reference electrode) is immersed into the liquid analyte with the conducting channel being sensitive to the ions of the analyte. Therefore, this architecture is known as an ion-sensitive field-effect transistor (ISFET) [37]. **Figure 12** displays a schematic illustration of the two types of sensor architectures (resistive and FET) targeted in this section.

Nowadays, the fabrication of these transducer platforms exploits micro/nano-fabrication technologies, usually based on silicon as substrate and

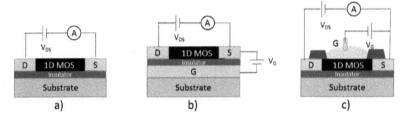

Figure 12.
Schematic illustration of a resistive (a) and FET transducing platform for gas (b) and liquid (c) phase analytes. NW: nanowires, D: drain electrode, S: source electrode, and G: gate electrode.

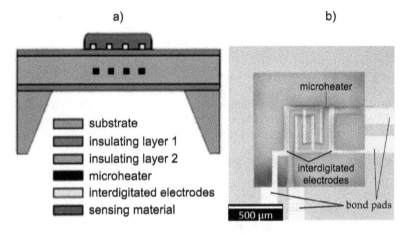

Figure 13.
(a) Layout of a typical microresistor (lateral view); adapted with permission from [6], Copyright 2018 Authors, licensee MDPI. (b) Optical image of a microhotplate and its indicated components (top view).

micro-electro-mechanical systems (MEMS) technology. This facilitates their integration as arrays in monolithic microchips as well as the incorporation of microheaters with low thermal losses (the so-called microhotplates) [5, 6, 74]. **Figure 13a**) displays the typical architecture of a microresistor consisting of a microhotplate and a sensing material on its top, and **Figure 13b**) shows an image of a microhotplate. This consists of a thin layer (a few micrometers thick) of a dielectric material, also called a membrane, supported by a silicon substrate at its periphery. The microheater is embedded within the membrane and insulated from the interdigitated electrodes patterned on top of the membrane. The use of silicon and MEMS technologies allows for the incorporation of integrated circuits along with the driving and signal conditioning circuitry or other smart features (e.g., wireless communication) to build electronic noses with potentially low-cost production [152, 153]. However, recently other substrate materials (e.g., polymers) and technologies (e.g., printing) are being explored and optimized to provide also integrated elements driven by the use of optimized active 1D MOS nanostructures that can operate at room temperature or close to it [13, 154].

4.2 Assembly of 1D metal oxide nanostructures on transducer platforms

To enable practical use of 1D MOS nanostructures, these structures must bring the interdigitated electrodes (chemiresistor) or the source and drain electrodes (FET) into contact. This allows the electrical current to flow and the resistance (or conductance) changes to be monitored. Such connection can be attained either by a single 1D nanostructure or by multiple 1D nanostructures [52, 54, 98]. Due to the requirement of precise alignment between the 1D nanostructure and the patterned electrodes, the fabrication process of the individual 1D nanostructure device is rather complicated, time-consuming and expensive. Therefore, to simplify the fabrication process and electrical signal measurement, the multiple 1D nanostructure devices become the most widely accepted configuration for practical applications.

4.2.1 Coating methods

The assembly of multiple 1D nanostructures for chemical sensors usually involves the use of a two-step process [52], following any of the routes sketched in **Figure 14a**). In the first route (top-electrode architecture), either 1D nanostructures are synthesized directly on a blank substrate or pre-synthetized 1D nanostructures are transferred on the substrate. Then, the electrodes are deposited by sputtering on the substrate with the 1D nanostructures on their top with the help of a mask. Conversely, in the second route (bottom-electrode architecture), the electrodes are deposited firstly on the blank substrate and the 1D nanostructures are either synthesized or transferred on the substrate with the patterned electrodes on its top. Technologically, bottom-electrode architectures are preferred for chemical sensors, as they may facilitate the direct integration of 1D nanostructures. This type of architecture also prevents the introduction of contaminants into the sensitive materials as the processing steps for the definition of the electrode are performed before the integration of the sensitive material.

The synthesis of 1D MOS nanostructures directly onto the sensor substrate [155], either a blank substrate or a substrate with patterned electrodes, is the preferred choice, since it reduces the fabrication time and costs of the sensors. Also, it reduces the incorporation of contaminants by avoiding the use of transfer media (often liquids) that tend to degrade the surface properties of the sensitive structures. On the other hand, the direct integration of 1D MOS nanostructures demands substrate materials such as silicon (Si), alumina (Al_2O_3), quartz, or high-temperature-resistant polymers (e.g., polyimide PI) [156] capable to withstand the thermal steps required for the synthesis of the MOS nanostructures and the

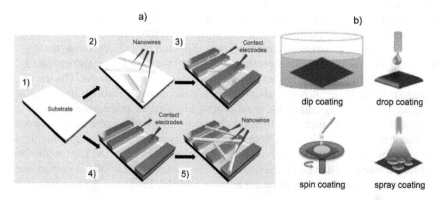

Figure 14.
(a) Schematic diagram of the architectures (top- or bottom-electrodes) used in multiple 1D nanostructures based chemical sensors: (1) blank substrate, (2) nanowires transferred onto the blank substrate, (3) electrodes patterned onto the substrate with nanowires, (4) electrodes patterned onto the blank substrate, (5) nanowires transferred onto the substrate with patterned electrodes; adapted with permission from [52], Copyright 2021 Authors, licensee MDPI. (b) Schematic illustration of wet-coating methods for the transfer of nanomaterials from a liquid dispersion to a substrate; adapted from [158].

high operating temperatures of the MOS chemical sensors (>200 °C). However, since the most advanced functional nanomaterials for chemical sensing based on MOS suggest that the next generation of MOS chemical sensors could work at room temperature, the need for heating elements could be omitted in the future, opening the possibility to use more abundant substrate materials that are cheaper and easier to process compared to those above. In this respect, steps are already being taken towards the synthesis of 1D MOS nanostructures on tiny substrates of flexible stretchable soft polymers (e.g., polyethylene terephthalate PET, polytetrafluoroethylene PTFE, polyaniline PANI, polyimide PI), textiles, or paper [157], which are of major interest for use in wearable sensor devices for emerging applications (e.g., healthcare). To achieve this, the processing temperatures are required to be below the glass transition temperature of polymers (<400 °C PI) or the thermal degradation temperature of the substrates (<100 °C textiles, paper). This limits the application of some synthetic procedures as those based on high-temperature chemical vapor deposition or electrospinning, as they demand temperatures of at least 400°C for either precursor decomposition or sintering steps. Then, the transfer of the pre-synthesized 1D nanostructures on flexible substrates seems to be the only feasible alternative to date.

The transfer of the pre-formed 1D MOS nanostructures to the sensor substrate may be a complex, time-consuming and expensive approach. In transfer methods, firstly, the 1D nanostructures need to be detached from the substrate on which they were synthesized and subsequently be dispersed in a liquid, usually a volatile organic solvent. There is a diversity of wet-coating techniques to transfer nanomaterials in suspension in a liquid to a substrate. The most common ones are illustrated in **Figure 14b**) [158]. Basically, the choices are either immersing the sensor substrate in the dispersion (dip-coating) or dosing the dispersion in form of droplets (drop coating, spin coating, spray coating) that settle on the substrate. The liquid evaporates from the in-flight droplets and/or the substrate, which can be heated at a constant temperature during or at the end of the transfer process.

The two approaches described previously result in multiple 1D MOS nanostructures forming a mat- or web-like deposit; this is a stacked network of 1D MOS nanostructures lying randomly in all directions of space. The deposits often display a bi-modal pore size distribution with relatively large pores (sub-micron to a few microns) side by side with nanosized pores (e.g., electrospun nanofibers), facilitating effective gas transport into the sensing layer. In addition, there are many cross-points between the 1D nanostructures in the network, which may favor the current percolation and the whole conductivity of the film. Nevertheless, several weaknesses have been identified in these mat-type sensitive layers that bring into question their usefulness for chemical sensors.

Even if the above techniques allow for a high degree of control of the parameters relevant to the synthesis or transfer of the 1D nanostructures, properties such as the size, thickness, porosity, or nano/microstructure vary greatly from one deposit to another, which lead to differences in the chemical sensing performance between sensors. To overcome this problem, wire structures connected in parallel are the ideal architecture to achieve a well-defined conduction channel that is easy to modulate by the interactions of the analyte and surface. In such a structure the grain boundaries or nanowire-nanowire interfaces are removed and thus the sensitivity of the system depends only on the nanowire surface due to an efficient transfer of charge with a lower probability of recombination. Usual approaches to achieve such integration of 1D nanostructures involve alignment methods based on the Langmuir-Blodgett technique or electric and magnetic field-assisted orientation techniques [159] such as dielectrophoresis [160]. For instance, a recent report based on the dielectrophoresis method proposed the use of a nanoelectrode array

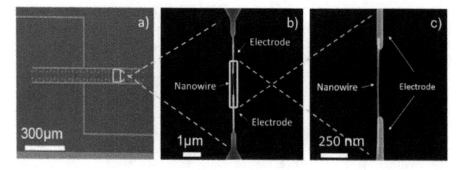

Figure 15.
(a) View of a nanoelectrode array containing several single-nanowires connected in parallel. (b), (c) Detailed views of a pair of nanoelectrodes with a nanowire interconnected across them; reprinted from [161].

system for the integration of various single-nanowires connected in parallel [161]. This system allowed for the selective integration of single-nanowires connected across various faced electrodes using the dielectrophoresis method, as can be observed in **Figure 15**.

Several methods have also been developed to orient and align electrospun fibers, either mechanically (e.g., rotating drum, rotating sharp disk) [84, 162–164] or by the action of electric fields [165–167]. In the latter, gaps of an insulating material (e.g., air, quartz, polystyrene) are introduced on the surface of a conductive substrate, and patterned electrodes (e.g., interdigitated electrodes) can also be used to obtain oriented nanofibers. When an insulating gap is introduced into the collector, it changes the structure of the external electric field. As a result, the directions of the electrostatic forces acting on a fiber that is sitting across the gap will be altered. In addition, once the charged fiber has moved into the vicinity of the electrodes, charges on the fiber will induce opposite charges on the surface of the electrodes. These opposite charges will further attract the fiber to the electrodes. These two types of electrostatic forces simultaneously pull the fiber towards the edge of the two electrodes and acting on different portions of a fiber will eventually lead to its uniaxial alignment through the gap, as can be seen in **Figure 16** [167].

It has been observed that the electrospun MOS nanofibers have a weak interfiber interaction and poor adherence to the electrodes and the substrate, resulting in high contact resistance and inferior mechanical properties. Thus, various treatments have been applied to the as-spun polymer/inorganic precursor fibers before sintering. These are traditional treatments such as irradiation with UV-light [168] or hot-pressing [169, 170], and a novel nanoscale welding technique [171]. The latter is simple and easy to apply, does not require specific equipment, preserves the interconnected structure of nanofibers, and is also the most efficient in terms of enhancement of the interfiber connections and interfacial adhesion. All this makes it the optimal treatment for application in chemical sensors.

Usually, conventional wet-coating techniques do not allow for highly-localized highly-dispersed coating of very small areas. Ideally, the sensing material should cover only the zone spanned by the electrodes, this is the active surface of the sensor whose area in the case of the microsensors is typically less than 0.2 mm^2. In practice, however, it is necessary to use masks to keep the region around the electrodes clean, particularly the bond pads of the electrodes and microheater. In this regard, the electrohydrodynamic jet printing technique [172–174], also called e-jet printing, is a precise way to transfer nanomaterials in a liquid on a substrate with greater resolution and repeatability than standard wet-coating and printing techniques (e.g., screen printing, aerosol-jet printing, ink-jet printing). In e-jet printing,

Figure 16.
(a) Layout of the setup to electrospin fibers as uniaxially aligned arrays, where the collector consists of two conductive strips separated by a void gap. SEM micrographs of uniaxially aligned electrospun nanofibers of (b) TiO$_2$ and (c) Sb-doped SnO$_2$; the insets show enlarged SEM micrographs of the nanofibers. Adapted with permission from [167], Copyright 2003 American Chemical Society.

droplets of a liquid suspension, also called an ink, are pulled out from a nozzle by the action of an external electric field. Charges accumulate at the liquid surface and the coulombic repulsion causes the liquid meniscus at the nozzle tip to deform into a conical shape (Taylor cone). When the electric field exceeds a critical value, the stress from the surface charge repulsion at the cone apex exceeds the surface tension and a small droplet is emitted towards the substrate. The key to high-resolution droplet printing is to use electric potentials below those required for droplet atomization (electrospray) and small nozzles (<100 μm). Then, the deposited droplets can be as small as hundreds of nanometers.

Highly integrated arrays of MEMS gas sensors have been prepared by e-jet printing [175, 176], as depicted in **Figure 17**. Firstly, long electrospun MOS nanofibers were fragmented into smaller pieces by ultrasonication and dispersed in a liquid. The electrodes were then coated with the fragments of the MOS nanofibers by using an e-jet printing system including a pulse-modulated voltage supply and a moving stage to control the droplet size and droplet position, respectively. The nozzle used was less than 100 μm in inner diameter and the dot pattern size of the MOS nanofibers was in the order of 40–60 μm. It is worth mentioning that e-jet printing was accomplished without nozzle clogging and that all the microsensors survived the process, in contrast to the standard wet-coating techniques, where the failure of the microsensor is frequently observed owing to the impact of big droplets, typically of tens of micrometers, causing the fragile dielectric membrane to crack. Furthermore, using a smaller nozzle, adjusting the applied voltage and pulse width, and tightly controlling the X-Y moving platform a continuous jetting regime was

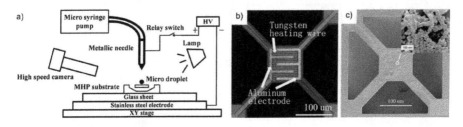

Figure 17.
(a) Layout of a typical setup for e-jet printing mounted on an X-Y positioning stage, (b) Optical microscope micrograph of the microhotplate (MHP), (c) SEM micrograph of the MHP coated with fragments of Pd-loaded SnO$_2$ nanofibers (inset) by using the e-jet printing setup in (a). Adapted with permission from [175], Copyright 2018 Authors.

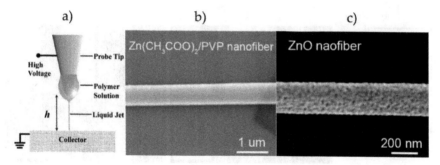

Figure 18.
(a) Sketch of a typical electrohydrodynamic direct-writing (EDW) setup; reprinted with permission from [186], Copyright 2006 American Chemical Society. SEM images of (b) PVP/Zn fiber obtained by EDW over a silicon substrate and (c) ZnO nanofiber obtained by sintering of the PVP/Zn fiber in (b); reprinted with permission from [189], Copyright 2013 Elsevier.

achieved, enabling to print line patterns less than 40μm in width with a maintained geometry and homogeneous distribution of the fragments of MOS nanofibers [176].

4.2.2 Patterning methods

Conventional methods based on micro- and nano-fabrication top-down processes that include steps of deposition, lithography, and etching may be used to pattern 1D MOS nanostructures and assist the growth of 1D MOS nanostructures in selected areas of a device. These processes are usually employed to define the area, density, and size of 1D nanostructures by patterning catalyst seeds that assist the growth of 1D nanostructures via VLS growth mechanisms. Other approaches also include forming templates to assist the directional growth of 1D nanostructures with subsequent template etching to free the 1D nanostructures. Hence, these processes generally lead to vertical aligned 1D nanostructures that are bridged with top and bottom contact as resistors or as FETs, for instance, by adding a vertical surrounded-gate [177, 178]. These approaches are not discussed in this chapter and more details can be found in the following literature [55, 158, 179, 180].

The electrospinning technique is more suited for the formation of nanofiber mats on areas that largely exceed the active area of the chemical sensors. Generally, the size of the area covered with the electrospun fibers decreases with the decreasing nozzle-to-substrate distance, which is accompanied by a reduction of the applied potential to preserve the electric field and prevent the onset of electrical discharges. The near-field electrospinning (NFES) method, also known as the electrohydrodynamic direct-writing (EDW) method, exploits this behavior [181–186]. The working principle of EDW is similar to that of the traditional electrospinning but with some distinctive features, namely the small nozzle diameter (<50 μm), the low applied voltage (<1 kV), the short distance between the nozzle and the substrate (0.5–5 mm), and the use of a moving nozzle and/or substrate, which facilitates position-tunable alignment and on-demand patterning of fibers on the substrate. EDW uses the stability region of the liquid jet and supplies discrete droplets of the polymer solution, in the same way as a dip pen does. This technique has been proven for direct-writing individual MOS nanofibers on silicon substrate for use in both chemiresistive [187, 188] and FET [189] sensor devices. As an example, **Figure 18** shows images of a ZnO nanofiber obtained by sintering a PVP/Zn nanofiber written by EDW on a substrate.

5. Summary and outlook

This chapter provides a general overview of the status and recent advances in developing chemical sensors based on 1D MOS nanostructures. The contents focus on the most frequent strategies and methods used to achieve higher sensing performances to cope with present and future applications, which ultimately drive innovation in chemical sensors. The applications usually impose requirements in terms of capabilities and efficiency of chemical detection, miniaturization, fabrication costs, power consumption, sensor stability, and lifetime. The combination of these needs poses difficult challenges to a diversity of enabling technologies encompassing materials synthesis, nanotechnology, nano- and micro-fabrication, and printing technologies, amongst others. In this context, the survey covers the state-of-the-art and advances of the main synthetic methods to produce 1D sensitive materials used in chemical sensors, particularly, nanowires and nanofibers based on MOS. It also tackles the effect of incorporating second-phase materials to bring improved and/or new chemical sensing attributes to single-phase 1D MOS nanostructures and the routes to form this type of heteronanostructures. Finally, it discusses the most common chemical sensing architectures to enable the response of 1D nanostructures via resistive or FET measurements, as well as the techniques used to assemble the nanostructures onto these sensor platforms (of resistive and FET type).

The vast reports and prospects in chemical sensors indicate that the trends of these devices are directed towards further miniaturization and operation at the minimum power consumption. The last will be achieved mainly by reducing the operating temperature (at which sensitive MOS materials are stimulated) to levels close to room temperature. Improvements on traditional functional parameters such as the sensitivity, stability, speed of response, and selectivity to application-dependent target analytes are also expected. All these improvements may ultimately come true by gaining a better understanding of the synergistic sensing effects in 1D heteronanostructures composed of MOS, metals and/or 2D nanomaterials such as graphene, TMDC, or MXenes. So that these combinations can be tailored more precisely shortening the try and error steps currently employed when tuning sensitive heteronanostructures. Certainly, this knowledge needs to be developed in parallel to robust and reproducible routes to synthesize 1D MOS heteronanostructures. The new synthetic routes must be optimized to enable the high dispersion, homogeneous distribution, and maximal interfacial area between the MOS and the second-phase constituents, with the minimum penalty in terms of sustainability, cost-effectiveness, and scalability. One-step routes that allow to grow, form, or assemble 1D MOS nanostructures directly onto the transducer platforms are pursued in the first place, followed by the techniques that allow the transfer of pre-formed 1D MOS nanostructures onto the transducer platforms with good control over the nanostructure orientation, alignment, and electrical contacts. When applied to the fabrication of chemical sensors, both types of techniques need to ensure the localized integration of 1D MOS nanostructures over the active area of the sensor device. This means that direct and transfer methods for the integration of 1D nanostructures must be adapted to operate efficiently either, in small sensitive areas, in agreement with the materials used as platforms, for instance, silicon-based MEMS or temperature-sensitive substrates (polymers, textiles, etc.). Therefore, the integration methods must also ensure control in the sub-millimeter range, optimal connectivity and conductivity between the nanostructures bridging the interdigitated electrodes, and strong adhesion of the nanostructures to the transducer platform.

Traditional synthetic methods based on nucleation and growth processes, such as hydrothermal synthesis and chemical vapor deposition, must be exploited to allow one-step integration of 1D MOS nanostructures with transducer platforms. These techniques are appropriate for direct integration, particularly with silicon-based platforms. However, their potential is not fully exploited as most of the works on chemical sensors fabrication turn to a removal and re-deposition approach for integrating 1D MOS nanostructures synthetized by these techniques.

The electrohydrodynamic techniques emerge as a promising alternative for integrating 1D MOS nanostructures onto chemiresistive and FET platforms due to its high-precision for printing areas of less than 0.1 mm^2 (e-jet printing) or its high-resolution to write directly (on electrodes and substrate patterns) lines of less than 100 nm in width. The direct printing of 1D MOS nanostructures may improve the reproducibility of chemical sensors. The application of electrohydrodynamic techniques to manufacture chemical sensors using MOS is still at an early stage. Much effort needs to be done to optimize the composition and properties of the dispersion and polymer solutions (inks), nozzle geometry and dimensions, and process conditions, before these techniques, become a rapid and cost-effective tool for large-scale fabrication of chemical sensors based on MOS. However, their availability in the short term could push forward the application of 1D nanostructures and emerging flexible substrates for chemical sensors.

Acknowledgements

The programme Interreg V Sudoe of the EU (Grant SEO2/P1/E569, NanoSen-AQM) is acknowledged for funding this publication. SV acknowledges the support of MCIN/AEI/10.13039/501100011033, via Grant PID2019-107697RB-C42 (ERDF A way of making Europe).

Author details

Esther Hontañón[1]* and Stella Vallejos[2]

1 Institute of Physical and Information Technologies, Spanish National Research Council, Madrid, Spain

2 Institute of Microelectronics of Barcelona, Spanish National Research Council, Cerdanyola del Vallès, Barcelona, Spain

*Address all correspondence to: esther.hontanon@csic.es

IntechOpen

References

[1] Gomes JB, Rodrigues JJPC, Rabêlo RAL, Kumar N, Kozlov S. IoT-enabled gas sensors: Technologies, applications and opportunities. Journal of Sensors and Actuators Networks. 2019;**8**:29. DOI: 10.3390/jsan8040057

[2] Wang H, Ma J, Zhang J, Feng Y, Vijjapu MT, Yuvaraja S, et al. Gas sensing materials roadmap. Journal of Physics: Condensed Matter. 2021;**33**:303001. DOI: 10.1088/1361-648X/abf477

[3] Saruhan B, Fomekong RL, Nahirniak S. Review: Influences of semiconductor metal oxide properties on gas sensing characteristics. Frontiers in Sensors. 2021;**2**:657931. DOI: 10.3389/fsens.2021.657931

[4] Serban I, Enesca A. Metal oxides-based semiconductors for biosensors applications. Frontiers in Chemistry. 2020;**8**:354. DOI: 10.3389/fchem.2020.00354

[5] Briand D, Courbat J. Micromachined semiconductor gas sensors. In: Jaaniso R, Tan OK, editors. Semiconductor Gas Sensors. Woodhead Publishing Series in Electronic and Optical Materials. 2nd ed. Amsterdam: Elsevier; 2020. pp. 413-464. DOI: 10.1016/B978-0-08-102559-8.00013-6

[6] Liu H, Zhang L, Li KHH, Tan OK. Microhotplates for metal oxide semiconductor gas sensor applications-Towards the CMOS-MEMS monolithic approach. Micromachines. 2018;**9**:557. DOI: 10.3390/mi9110557

[7] Korotcenkov G. Gas response control through structural and chemical modification of metal oxide films: State of the art and approaches. Sensors and Actuators B. 2005;**107**:209-232. DOI: 10.1016/j.snb.2004.10.006

[8] Park SY, Kim Y, Kim T, Eom TH, Kim SY, Jang HW. Chemoresistive materials for electronic nose: Progress, perspectives, and challenges. InfoMat. 2019;**1**(7):289-316. DOI: 10.1002/inf2.12029

[9] Schütze A, Sauerwald T. Dynamic operation of semiconductor sensors. In: Jaaniso R, Tan OK, editors. Semiconductor Gas Sensors. Woodhead Publishing Series in Electronic and Optical Materials. 2nd ed. Amsterdam: Elsevier; 2020. pp. 385-412. DOI: 10.1016/B978-0-08-102559-8.00012-4

[10] Palacio F, Fonollosa J, Burgués J, Gomez JM, Marco S. Pulsed-temperature metal oxide gas sensors for microwatt power consumption. IEEE Access. 2020;**8**:70938-70946. DOI: 10.1109/ACCESS.2020.2987066

[11] Kumar R, Liu X, Kumar M. Room-temperature gas sensors under photoactivation: From metal oxides to 2D materials. Nano-Micro Letters. 2020;**12**:164. DOI: 10.1007/s40820-020-00503-4

[12] Majhi SM, Mirzaei A, Kim HW, Kim SS, Kim TW. Recent advances in energy-saving chemiresistive gas sensors: A review. Nano Energy. 2021;**79**:105369. DOI: 10.1016/j.nanoen.2020.105369

[13] Dai J, Ogbeide O, Macadam N, Sun Q, Yu W, Li Y, et al. Printed gas sensors. Chemical Society Reviews. 2020;**49**:1756-1789. DOI: 10.1039/c9cs00459a

[14] Fazio E, Spadaro S, Corsaro C, Neri G, Leonardi SG, Neri F, et al. Metal-oxide based nanomaterials: Synthesis, characterization and their applications in electrical and electrochemical sensors. Sensors. 2021;**21**:2494. DOI: 10.3390/s21072494

[15] Naresh V, Lee N. A review on biosensors and recent development of

nanostructured materials-enabled biosensors. Sensors. 2021;**21**:1109. DOI: 10.3390/s21041109

[16] Wang Z, Zhu L, Sun S, Wang J, Yan W. One-dimensional nanomaterials in resistive gas sensor: From material design to application. Chemosensors. 2021;**9**:198. DOI: 10.3390/chemosensors9080198

[17] Malik R, Towner VK, Mishra YK, Lin L. Functional gas sensing nanomaterials: A panoramic view. Applied. Physics Reviews. 2020;7:021301. DOI: 10.1063/1.5123479

[18] Nunes D, Pimentel A, Goncalves A, Pereira S, Branquinho R, Barquinha P, et al. Metal oxide nanostructures for sensor applications. Semiconductor Science and Technology. 2019;**34**: 043001. DOI: 10.1088/1361-6641/ab011e

[19] Qadir A, Le TK, Malik M, Min-Dianey KAA, Saeed I, Yu Y, et al. Representative 2D-material-based nanocomposites and their emerging applications: A review. RCS Advances. 2021;**11**:23860-23880. DOI: 10.10139/d1ra03425a

[20] Pham VP, Nguyen MT, Park JW, Kwak SS, Nguyen DHT, Mun MK, et al. Chlorine-trapped CVD bilayer graphene for resistive pressure sensor with high detection limit and high sensitivity. 2D Materials. 2017;**4**(2):025049. DOI: 10.1088/2053-1583/aa6390

[21] Chen X, Leishman M, Bagnall D, Nasiiri N. Nanostructured gas sensors: From air quality and environmental monitoring to healthcare and medical applications. Nanomaterials. 2021;**11**: 1927. DOI: 10.3390/nano11081927

[22] Kumar DK, Reddy KR, Sadhu V, Shetti NP, Reddy CV, Chouhan RS, et al. Metal oxide-based nanosensors for health care and environmental applications. In: Kanchi S, Sharma D, editors. Nanomaterials in Diagnostic Tools and Devices. 1st ed. Amsterdam: Elsevier; 2020. pp. 113-129. DOI: 10.1016/C2018-0-02240-9

[23] Xue S, Cao S, Huang Z, Yang D, Zhang G. Improving gas-sensing performance based on MOS nanomaterials: A review. Materials. 2021;**14**:4263. DOI: 10.3390/ma14154263

[24] Sun Z, Liao T, Kou L. Strategies for designing metal oxide nanostructures. Science China Materials. 2017;**60**:1. DOI: 10.1007/s40843-016-5117-0

[25] Yang B, Myung NV, Tran T-T. 1D metal oxide semiconductor materials for chemiresistive gas sensors: A review. Advanced Electronic Materials. 2021;7:2100271. DOI: 10.1002/aelm.202100271

[26] Kaur N, Singh M, Comini E. One-dimensional nanostructured chemoresistive sensors. Langmuir. 2020;**36**:6326-6344. DOI: 10.1021/acs.langmuir.0c00701

[27] Hulanicki A, Glab S, Ingman F. Chemical sensors definitions and classification. Pure and Applied Chemistry. 1991;**63**(9):1247-1250. DOI: 10.1351/pac199163091247

[28] Deng Y. Sensing mechanism and evaluation criteria of semiconducting metal oxides gas sensors. In: Semiconducting Metal Oxides for Gas Sensing. 1st ed. Singapore: Springer Nature; 2019. pp. 23-51. DOI: 10.1007/978-981-13-5853-1

[29] Korotcenkov G. Metal oxides for solid-state gas sensors: What determines our choice? Materials Science and Engineering B. 2007;**139**:1-23. DOI: 10.1016/j.mceb.2007.01.044

[30] Kim H-J, Lee J-H. Highly sensitive and selective gas sensors using p-type oxide semiconductors: Overview. Sensors and Actuators B. 2014;**192**:607-627. DOI: 10.1016/j.snb.2013.11.005

[31] Annanouch FE, Haddi Z, Vallejos S, Umek P, Guttmann P, Bittencourt C, et al. Aerosol-assisted CVD-grown WO_3 nanoneedles decorated with copper oxide nanoparticles for the selective and humidity-resilient detection of H_2S. ACS Applied Materials & Interfaces. 2015;**7**:6842-6851. DOI: 10.1021/acsami.5b00411

[32] Choopun S, Hongsith N, Wongrat E. Metal oxide nanowires for gas sensors. In: Peng X, editor. Nanowires—Recent Advances. London: IntechOpen; 2012. pp. 3-23. DOI: 10.5772/54385

[33] Vergara A, Llobet E. Sensor selection and chemo-sensory optimization: Toward an adaptable chemo-sensory system. Frontiers in Neuroengineering. 2012;**4**:19. DOI: 10.3389/fneng.2011.00019

[34] Gurlo A. Nanosensors: Towards morphological control of gas sensing activity. SnO_2, In_2O_3, ZnO and WO_3 case studies. Nanoscale. 2011;**3**:154-165. DOI: 10.1039/C0NR00560F

[35] Li Z, Li H, Wu Z, Wang M, Luo J, Torun H, et al. Advances in designs and mechanisms of semiconducting metal oxide nanostructures for high-precision gas sensors operated at room temperature. Materials Horizon. 2019;**6**: 470-506. DOI: 10.1039/c8mh01365a

[36] Yang S, Lei G, Xu H, Lan Z, Wang Z, Gu H. Metal oxide based heterojunctions for gas sensors: A review. Nanomaterials. 2021;**11**:1026. DOI: 10.3390/nano11041026

[37] Sadighbayan D, Hasanzadeh M, Ghafar-Zadeh E. Biosensing based on field-effect transistors (FET): Recent progress and challenges. Trends in Analytical Chemistry. 2020;**133**:116067. DOI: 10.1016/j.trac.2020.116067

[38] Joshi N, Baunger ML, Shimiziu FM, Riul A, (Jr), Oliveira O N (Jr). Insights into nano-heterostructured materials for gas sensing. A review. Multifunctional Materials. 2021;**4**(3): 032002. DOI: 10.1088/2399-7532/ac1732

[39] Walker JM, Akbar SA, Morris PA. Synergistic effects in gas sensing semiconducting oxide nano-heterostructures: A review. Sensors and Actuators B. 2019;**286**:624-640. DOI: 10.1016/j.sintl.2019.01.049

[40] Deng Y. Semiconducting metal oxides: Composition and sensing performance. In: Semiconducting Metal Oxides for Gas Sensing. 1st ed. Singapore: Springer Nature; 2019. pp. 77-103. DOI: 10.1007/978-981-13-5853-1

[41] Zappa D, Galstyan V, Kaur N, Arachchige HMMM, Sisman O, Comini E. Metal oxide-based heterostructures for gas sensors—A review. Analytical Chimica Acta. 2018;**1039**:1-23. DOI: 10.1016/j.aca.2018.09.020

[42] Miller DR, Akbar SA, Morris PA. Nanoscale metal oxide-based heterojunctions for gas sensing: A review. Sensors and Actuators B. 2014;**204**:250-272. DOI: 10.1016/j.snb.2014.07.074

[43] Tobaldi DM, Leonardi SG, Movlaee K, Lajaunie L, Seabra MP, Arenal R, et al. Hybrid noble metals/metal-oxide bifunctional nano-heterostructure displaying outperforming gas-sensing and photocromic performances. ACS Omega. 2018;**3**:9846-9859. DOI: 10.1021/acsomega.8b01508

[44] Müller SA, Degler D, Feldmann C, Türk M, Moos R, Fink K, et al. Exploiting synergies in catalysis and gas sensing using noble metal-loaded oxide composites. ChemCatChem. 2018;**10**: 864-880. DOI: 10.1002/cctc.201701545

[45] Rai P, Majhi SM, Yu Y-T, Lee J-H. Noble metal@metal oxide semiconductor core@shell nano-architectures as a new

platform for gas sensors applications. RCS Advances. 2015;5:76229-76248. DOI: 10.1039/c5ra14322e

[46] Jian Y, Hu W, Zhao Z, Cheng P, Haick H, Yao M, et al. Gas sensors based on chemi-resistive hybrid functional nanomaterials. Nano-Micro Letters. 2020;12:71. DOI: 10.1007/s40820-020-0407-5

[47] Sowmya B, John A, Panda PK. A review on metal-oxide based p-n and n-n heterostructured nano-materials for gas sensing applications. Sensors International. 2021;2:100085. DOI: 10.1016/j.sintl.2021.100085

[48] Sun D, Luo Y, Debliquy M, Zhang C. Graphene-enhanced metal oxide gas sensors at room temperature: A review. Beilstein Journal of Nanotechnology. 2018;9:2832-2844. DOI: 10.3762/bjnano.9.264

[49] Aroutiounian VM. Metal oxide gas sensors decorated with carbon nanotubes. Lithuanian Journal of Physics. 2015;55(4):319-329. DOI: 10.3952/physics.v55i4.3230

[50] Bhati VS, Kumar M, Banerjee R. Gas sensing performance of 2D nanomaterials/metal oxide nanocomposites: A review. Journal of Materials Chemistry C. 2021;9:8776-8808. DOI: 10.1039/d1tc01857d

[51] Li Z, Yao Z, Haidry AA, Plecenik T, Xie L, Sun L, et al. Resistive-type hydrogen gas sensor based on TiO_2: A review. International Journal of Hydrogen Energy. 2018;43:21114-21132. DOI: 10.1016/j.ijhydene.2018.09.051

[52] Zeng H, Zhang G, Nagashima K, Takahashi T, Hosomi T, Yanagida T. Metal-oxide nanowire molecular sensors and their promises. Chemosensors. 2021;9:41. DOI: 10.3390/chemosensors9020041

[53] Wang F, Zu H. Application of one-dimensional metal oxide semiconductor in field effect transistor. In: Zhou Y, editor. Semiconducting Metal Oxide Thin Film Transistors. 1st ed. Bristol: IOP Publishing; 2021. pp. 6-1-6-25. DOI: 10.1088/978-0-7503-2556-1ch6

[54] Korotcenkov G. Current trends in nanomaterials for metal oxide-based conductometric gas sensors: Advantages and limitations. Part 1: 1D and 2D nanostructures. Nanomaterials. 2020;10:1392. DOI: 10.3390/nano10071392

[55] Wang Y, Duan L, Deng Z, Liao J. Electrically transduced gas sensors based on semiconducting metal oxide nanowires. Sensors. 2020;20:6781. DOI: 10.3390/s20236781

[56] Aliheidari N, Aliahmad N, Agarwal M, Dalir H. Electrospun nanofibers for label-free sensor applications. Sensors. 2019;19:3587. DOI: 10.3390/s19163587

[57] Comini E. Metal oxide nanowire chemical sensors: Innovation and quality of life. Materials Today. 2016;10(10):559-567. DOI: 10.1016/j.mattod.2016.05.016

[58] Comini E, Baratto C, Faglia G, Ferroni M, Ponzoni A, Zappa D, et al. Metal oxide nanowire chemical and biochemical sensors. Journal of Materials Research. 2013;28(21):2911-2931. DOI: 10.1557/jmr.2013.304

[59] Ding B, Wang M, Wang X, Yu J, Sun G. Electrospun nanomaterials for ultrasensitive sensors. Materials Today. 2010;13(11):16-27. DOI: 10.1016/S1369-7021(10)70200-5

[60] Comini E, Sberveglieri G. Metal oxide nanowires as chemical sensors. Materials Today. 2010;13(7-8):36-44. DOI: 10.1016/S1369-7021(10)70126-7

[61] Lockman Z, editor.1-Dimensional Metal Oxide Nanostructures: Growth,

Properties and Devices. 1st ed. Zhang S, editor. Advances in Materials Science and Engineering. Boca Raton: CRC Press Taylor & Francis Group; 2020. p. 347. DOI: 10.1201/9781351266727

[62] Devan RS, Patil RA, Lin J-H, Ma Y-R. One-dimensional metal-oxide nanostructures: Recent developments in synthesis, characterization, and applications. Advanced Functional Materials. 2012;**22**:3326-3370. DOI: 10.1002/adfm.201201008

[63] Comini E, Baratto C, Faglia G, Ferroni M, Vomiero A, Sberveglieri G. Quasi-one dimensional metal-oxide semiconductors: Preparation, characterization and application as chemical sensors. Progress in Materials Science. 2009;**54**:1-67. DOI: 10.1016/j. pmatsci.2008.06.003

[64] Vallejos S, Di Maggio F, Shujah T, Blackman C. Chemical vapour deposition of gas sensitive metal oxides. Chemosensors. 2016;**4**:4. DOI: 10.3390/ chemosensors4010004

[65] Feng S-H, Li G-H. Hydrothermal and solvothermal synthesis. In: Xu R, Xu Y, editors. Modern Inorganic Synthetic Chemistry. 2nd ed. Amsterdam: Elsevier; 2017. pp. 73-104. DOI: 10.1016/ B978-0-444-63591-4.00004-5

[66] Alshehri NA, Lewis AR, Pleydell-Pearce C, Maffeis TGG. Investigation of the growth parameters of hydrothermal ZnO nanowires for scale up applications. Journal of the Saudi Chemical Society. 2018;**22**(5):538-545. DOI: 10.1016/j.jscs.2017.09.004

[67] Chang P-C, Fan Z, Wang D, Tseng W-Y, Chiou W-A, Hong J, et al. ZnO nanowires synthesized by vapor trapping CVD method. Chemistry of Materials. 2004;**16**:5133-5137. DOI: 10.1021/cm049182c

[68] Tonezzer M, Armellini C, Toniutti L. Sensing performance of electronic noses: A comparison between ZnO and SnO$_2$ nanowires. Nanomaterials. 2021;**11**:2773. DOI: 10.3390/nano11112773

[69] Zhou JX, Yin Z. Raman spectroscopy and photoluminescence study of single-crystalline SnO$_2$ nanowires. Solid State Communications. 2006;**138**(5):242-246. DOI: 10.1016/j. ssc.2006.03.007

[70] Nekita S, Nagashima K, Zhang G, Wang Q, Kanai M, Takahashi T, et al. Face-selective crystal growth of hydrothermal tungsten oxide nanowires for sensing volatile molecules. ACS Applied Nano Materials. 2020;**3**(10): 10252-10260. DOI: 10.1021/ acsanm.0c02194

[71] Kaur N, Zappa D, Poli N, Comini E. Integration of VLS-grown WO$_3$ nanowires into sensing devices for the detection of H$_2$S and O$_3$. ACS Omega. 2019;**4**(15):16336-16343. DOI: 10.1021/ acsomega.9b01792

[72] Wang F, Dong A, Buhro WE. Solution-liquid-solid synthesis, properties, and applications of one-dimensional colloidal semiconductor nanorods and nanowires. Chemical Reviews. 2016;**116**:10888-10933. DOI: 10.1021/acs.chemrev.5b00701

[73] Kolasinski KW. Catalytic growth of nanowires: Vapor-liquid-solid, vapor-solid-solid, solution-liquid-solid and solid-liquid-solid growth. Current Opinion on Solid State and Materials Science. 2006;**10**:182-191. DOI: 10.1016/ j.cossms.2007.03.002

[74] Vallejos S, Gràcia I, Chmela O, Figueras E, Hubálek J, Cané C. Chemoresistive micromachined gas sensors based on functionalized metal oxide nanowires: Performance and reliability. Sensors and Actuators B. 2016;**235**:525-534. DOI: 10.1016/j. snb.2016.05.102

[75] Qi H, Glaser ER, Caldwell JD, Prokes SM. Growth of vertically aligned ZnO nanowire arrays using bilayered metal catalysts. Journal of Nanomaterials. 2012;**2012**:260687. DOI: 10.1155/2012/260687

[76] Rackauskas S, Nasibulin AG. Nanowire growth without catalysts: Applications and mechanisms at the atomic scale. ACS Applied Nano Materials. 2020;**3**(8):7314-7324. DOI: 10.1021/acsanm.0c01179

[77] Zhang Z, Wang Y, Li H, Yuan W, Zhang X, Sun C, et al. Atomic-scale observation of vapor-solid nanowire growth via oscillatory mass transport. ACS Nano. 2016;**10**(1):763-769. DOI: 10.1021/acsnano.5b05851

[78] Rackauskas S, Jiang H, Wagner JB, Shandakov SD, Hansen TW, Kauppinen EI, et al. In situ study of noncatalytic metal oxide nanowire growth. Nano Letters. 2014;**14**(10): 5810-5813. DOI: 10.1021/nl502687s

[79] Oh SH, Chisholm MF, Kauffman Y, Kaplan WD, Luo W, Rühle M, et al. Oscillatory mass transport in vapor-liquid-solid growth of sapphire nanowires. Science. 2010;**330**(6003): 489-493. DOI: 10.1126/science.1190596

[80] Barhoum A, Pal K, Rahier H, Uludag H, Kim IS. Nanofibers as new-generation materials: From spinning and nano-spinning fabrication techniques to emerging applications. Applied Materials Today. 2019;**17**:1-35. DOI: 10.1016/j.apmt.2019.06.015

[81] Gugulothu D, Barhoum A, Nerella R, Ajmer R, Bechelany M. Fabrication of nanofibers: Electrospinning and non-electrospinning techniques. In: Barhoum A, Bechelany M, Makhlouf ASH, editors. Handbook of Nanofibers. Part I: Fundamental Aspects, Experimental Setup, Synthesis, Properties and Characterization. 1st ed. Switzerland: Springer Nature; 2019. pp. 45-77. DOI: 10.1007/978-3-319-53655-2_6

[82] Kailasa S, Reddy MSB, Maurya MR, Rani BG, Rao KV, Sadasivuni KK. Electrospun nanofibers: Materials, synthesis parameters, and their role in sensing applications. Macromolecular Materials and Engineering. 2021;**306**: 2100410. DOI: 10.1002/mame.202100410

[83] Alghoraibi I, Alomari S. Different methods for nanofiber design and fabrication. In: Barhoum A, Bechelany M, Makhlouf ASH, editors. Handbook of Nanofibers. Part I: Fundamental Aspects, Experimental Setup, Synthesis, Properties and Characterization. 1st ed. Switzerland: Springer Nature; 2019. pp. 79-124. DOI: 10.1007/978-3-319-53655-2_11

[84] Homaeigohar S, Davoudpour Y, Habibi Y, Elbahri M. The electrospun ceramic hollow nanofibers. Nanomaterials. 2017;7:383. DOI: 10.3390/nano7110383

[85] Chen Y, Lu W, Guo Y, Zhu Y, Lu H. Synthesis, characterization and photocatalytic activity of nanocrystalline first transition-metal (Ti, Mn, Co, Ni and Zn) oxide nanofibers by electrospinning. Applied Sciences. 2019;**9**:8. DOI: 10.3390/app9010008

[86] Bagchi S, Brar R, Singh B, Ghanshyam C. Instability controlled synthesis of tin oxide nanofibers and their gas sensing properties. Journal of Electrostatics. 2015;**78**:68-78. DOI: 10.1016/j.elstat.2015.11.001

[87] Li Z, Wang C. Introduction of electrospinning. In: One-Dimensional Nanostructures. Electrospinning Technique and Unique Nanofibers. Springer Briefs in Materials. Berlin: Springer; 2013. pp. 1-13. DOI: 10.1007/978-3-642-36427-3_1

[88] Yang X, Wang J, Guo H, Liu L, Xu W, Duan G. Structural design toward functional materials by electrospinning: A review. e-Polymers. 2020;**20**:682-712. DOI: 10.1515/epoly-2020-0068

[89] Wang C, Wang J, Zeng L, Qiao Z, Liu X, Liu H, et al. Fabrication of electrospun polymer nanofibers with diverse morphologies. Molecules. 2019;**24**:834. DOI: 10.3390/molecules 24050834

[90] Xue J, Wu T, Dai Y, Xia Y. Electrospinning and electrospun nanofibers: Methods, materials and applications. Chemical Reviews. 2019;**119**:5298-5415. DOI: 10.1021/acs. chemrev.8b00593

[91] Al-Hazeem NZA. Nanofibers and electrospinning method. In: Zykas GZ, Mitropoulos AC, editors. Novel Nanomaterials: Synthesis and Applications. London: IntechOpen; 2018. pp. 191-210. DOI: 10.5772/ intechopen.72060

[92] Williams GR, Raimi-Abraham BT, Luo CJ. Electrospinning fundamentals. In: Nanofibers in Drug-Delivery. London: UCL Press; 2018. pp. 24-59. DOI: 10.2307/j.ctv550dd1.6

[93] Li Z, Wang C. Effects of working parameters on electrospinning. In: One-Dimensional Nanostructures. Electrospinning Technique and Unique Nanofibers. Springer Briefs in Materials. Berlin: Springer; 2013. pp. 15-28. DOI: 10.1007/978-3-642-36427-3_2

[94] Bhardwaj N, Kundu SC. Electrospinning: A fascinating fiber fabrication technique. Biotechnology Advances. 2010;**28**:325-347. DOI: 10.1016/ j.biotechadv.2010.01.004

[95] Korotcenkov G. Electrospun metal oxide nanofibers and their conductometric gas sensor application. Part 1: Nanofibers and features of their forming. Nanomaterials. 2021;**11**:1544. DOI: 10.3390/nano11061544

[96] Esfahani H, Jose R, Ramakrishna S. Electrospun ceramic nanofiber mats today: Synthesis, properties, and applications. Materials. 2017;**10**:1238. DOI: 10.3390/ma10111238

[97] Xia X, Dong XJ, Wei QF, Cai YB, Lu KY. Formation mechanism of porous hollow SnO$_2$ nanofibers prepared by one-step electrospinning. eXPRESS Polymer Letters. 2012;**6**(2):169-176. DOI: 10.3144/expresspolymlett.2012.18

[98] Korotcenkov G. Electrospun metal oxide nanofibers and their conductometric gas sensor application. Part 2: Gas sensors and their advantages and limitations. Nanomaterials. 2021;**11**:1555. DOI: 10.3390/ nano11061555

[99] Deng Y. Semiconducting metal oxides: Morphology and sensing performance. In: Semiconducting Metal Oxides for Gas Sensing. Singapore: Springer Nature; 2019. pp. 53-75. DOI: 10.1007/978-981-13-5853-1

[100] Wang C, Yin L, Zhang L, Xiang D, Gao R. Metal oxide gas sensors: Sensitivity and influencing factors. Sensors. 2010;**10**:2088-2106. DOI: 10.3390/s100302088

[101] Katoch A, Abideen ZU, Kim J-H, Kim SS. Influence of hollowness variation on the gas-sensing properties of ZnO hollow nanofibers. Sensors and Actuators B. 2016;**232**:98-704. DOI: 10.1016/j.snb.2016.04.013

[102] Katoch A, Choi S-W, Kim SS. Nanograins in electrospun oxide nanofibers. Metals and Materials International. 2015;**21**(2):213-221. DOI: 10.1007/s12540-015-4319-8

[103] Katoch A, Sun G-J, Choi S-W, Byun J-H, Kim SS. Competitive influence of grain size and crystallinity on gas sensing performances of ZnO nanofibers. Sensors and Actuators B. 2013;**185**:411-416. DOI: 10.1016/j. snb.2013.05.030

[104] Wal RL, Hunter GW, Xu JC, Kulis MJ, Berger GM, Ticich TM. Metal-oxide nanostructure and gas-sensing performance. Sensors and

Actuators B. 2009;**138**:113-119.
DOI: 10.1016/j.snb.2009.02.020

[105] Comini E. Metal-oxide nano-crystals for gas sensing. Analytical Chimica Acta. 2006;**568**:28-40.
DOI: 10.1016/j.aca.2005.10.069

[106] Rothschild A, Komem Y. On the relationship between the grain size and gas-sensitivity of chemo-resistive metal-oxide gas sensors with nanosized grains. Journal of Electroceramics. 2004;**13**:97-301. DOI: 10.1007/s10832-004-5178-8

[107] Rothschild A, Komem Y. The effect of grain size on the sensitivity of nanocrystalline metal-oxide gas sensors. Journal of Applied Physics. 2004;**95**: 6374-6380. DOI: 10.1063/1.1728314

[108] Liao Y, Fukuda T, Wang S. Electrospun metal oxide nanofibers and their energy applications. In: Rahman MM, Asiri AM, editors. Nanofiber Research—Reaching New Heights. London: IntechOpen; 2016. pp. 169-190. DOI: 10.5772/63414

[109] Du H, Yang W, Sun Y, Yu N, Wang J. Oxygen-plasma-assisted enhanced acetone-sensing properties of ZnO nanofibers by electrospinning. ACS Applied Materials & Interfaces. 2020;**12**: 23084-23093. DOI: 10.1021/acsami. 0c03498

[110] Zhang Y, Li J, An G, He X. Highly porous SnO_2 fibers by electrospinning and oxygen plasma etching and its ethanol sensing properties. Sensors and Actuators B. 2010;**144**:43-48.
DOI: 10.1016/j.snb.2009.10.012s

[111] Islam M, Srivastava AK, Basavaraja BM, Sharma A. "Nano-on-micro" approach enables synthesis of ZnO nano-cactus for gas sensing applications. Sensors International. 2021;**2**:100084. DOI: 10.10161/j. sint.2021.100084

[112] Tomic M, Setka M, Vojkukva L, Vallejos S. VOCs sensing by metal oxides, conductive polymers, and carbon-based materials. Nanomaterials. 2021;**11**:552. DOI: 10.3390/nano11020552

[113] Palgrave RG, Parkin IP. Aerosol assisted chemical vapor deposition using nanoparticle precursors: A route to nanocomposite thin films. Journal of the American Chemical Society. 2006; **128**(5):1587-1597. DOI: 10.1021/ ja055563v

[114] Andre RS, Mattoso LHC, Correa DS. Electrospun ceramic nanofibers and hybrid-nanofiber composites for gas sensing. ACS Applied Nano Materials. 2019;**2**:4026-4042. DOI: 10.1021/acsanm.9b01176

[115] Imran M, Motta N, Shafiei M. Electrospun one-dimensional nanostructures: A new horizon for gas sensing materials. Beilstein Journal of Nanotechnology. 2018;**9**:2128-2170. DOI: 10.3762/bjnano.9.202

[116] Abideen ZU, Kim J-H, Lee J-H, Kim J-Y, Mirzaei A, Kim HW, et al. Electrospun metal oxide composite nanofibers gas sensors: A review. Journal of the Korean Ceramic Society. 2017;**54**(5):366-379. DOI: 10.4191/ kcers.2017.54.5.12

[117] Xia Y, Li R, Chen R, Wang J, Xiang L. 3D architecture graphene/ metal oxide hybrids for gas sensors: A review. Sensors. 2018;**18**:1456. DOI: 10.3390/s18051456

[118] Mahji SM, Mirzaei A, Kim HW, Lim SS. Reduced graphene oxide (rGO)-loaded metal-oxide nanofiber gas sensors: An overview. Sensors. 2021;**21**:1352. DOI: 10.3390/s211041352

[119] Gao Y, Zhang B, Cheng M, Zhao L, Lu G. P2GS.9—Gas sensors based on metal oxides decorated by reduced graphene oxide with enhanced gas sensing properties. In: Proceedings of the

17th International Meeting on Chemical Sensors (IMCS 2018); 15-19 July 2018; Vienna. Berlin: AMA Association for Sensors and Measurement; 2018. pp. 786-787. DOI: 10.5162/IMCS2018/P2GS.9

[120] Tuscharoen S, Hicheeranun W, Chananonnawathorn C, Horprathum M, Kaewkhao J. Effect of sputtered pressure on Au nanoparticles formation decorated ZnO nanowire arrays. Materials Today: Proceedings. 2021;43:2624-2628. DOI: 10.1016/j.matpr.2020.04.626

[121] Annanouch FE, Haddi Z, Ling M, Di Maggio F, Vallejos S, Vilic T, et al. Aerosol-assisted CVD-grown PdO nanoparticle-decorated tungsten oxide nanoneedles extremely sensitive and selective to hydrogen. ACS Applied Materials & Interfaces. 2016;8(16): 10413-10421. DOI: 10.1021/acsami.6b00773

[122] Navarrete E, Llobet E. Synthesis of p-n heterojunctions via aerosol assisted chemical vapor deposition to enhance the gas sensing properties of tungsten trioxide nanowires: A mini-review. Journal of Nanoscience and Nanotechnology. 2021;21(4):2462-2471. DOI: 10.1166/jnn.2021.19105

[123] Tomic M, Setka M, Chmela O, Gràcia I, Figueras E, Cané C, et al. Cerium oxide-tungsten oxide core-shell nanowire-based microsensors sensitive to acetone. Biosensors. 2018;8(4): 116. DOI: 10.3390/bios8040116

[124] Choi MS, Mirzaei A, Na HG, Kim S, Kim DE, Lee KH, et al. Facile and fast decoration of SnO_2 nanowires with Pd embedded SnO_{2-x} nanoparticles for selective NO_2 gas sensing. Sensors and Actuators B. 2021;340:129984. DOI: 10.1016/j.snb.2021.129984

[125] SnO_2 nanowires decorated by insulating amorphous carbon layers for improved room-temperature NO_2 sensing. Sensors and Actuators B.

2021;326:128801. DOI: 10.1016/j.snb.2020.128801

[126] Feng Y, Cho IS, Rao PM, Cai L, Zheng X. Sol-flame synthesis: A general strategy to decorate nanowires with metal oxide/noble metal nanoparticles. Nano Letters. 2013;13(3):855-860. DOI: 10.1021/nl300060b

[127] Vallejos S, Stoycheva T, Umek P, Navio C, Snyders R, Bittencourt C, et al. Au nanoparticle-functionalized WO_3 nanoneedles and their application in high sensitivity gas sensor devices. Chemical Communications. 2011;47: 565-567. DOI: 10.1039/C0CC02398A

[128] Vallejos S, Umek P, Stoycheva T, Annanouch F, Llobet E, Correig X, et al. Single-step deposition of Au- and Pt- nanoparticle-functionalized tungsten oxide nanoneedles synthesized via aerosol-assisted CVD, and used for fabrication of selective gas microsensors arrays. Advanced Functional Materials. 2013;23(10):1313-1322. DOI: 10.1002/adfm.201201871

[129] Vallejos S, Gràcia I, Figueras E, Cané C. Nanoscale heterostructures based on $Fe_2O_3@WO_{3-x}$ nanoneedles and their direct integration into flexible transducing platforms for toluene sensing. ACS Applied Materials & Interfaces. 2015;7(33):18638-18649. DOI: 10.1021/acsami.5b05081

[130] Shao H, Huang M, Fu H, Wang S, Wang L, Lu J, et al. Hollow WO_3/SnO_2 hetero-nanofibers. Controlled synthesis and high efficiency of acetone vapor detection. Frontiers in Chemistry. 2019;7:785. DOI: 10.3389/fchem.2019.00785

[131] Lu Z, Zhou Q, Wang C, Wei Z, Xu L, Gui Y. Electrospun $ZnO-SnO_2$ composite nanofibers and enhanced sensing properties to SF_6 decomposition byproduct H_2S. Frontiers in Chemistry. 2018;6:540. DOI: 10.3389/fchem.2018.00540

[132] Lee J-H, Kim J-Y, Mirzaei A, Kim HW, Kim SS. Significant enhancement of hydrogen-sensing properties of ZnO nanofibers through NiO loading. Nanomaterials. 2018;**8**:902. DOI: 10.3390/nano8110902

[133] Yan C, Lu H, Gao J, Zhu G, Yin F, Yang Z, et al. Synthesis of porous NiO-In$_2$O$_3$ composite nanofibers by electrospinning and their highly enhanced gas sensing properties. Journal of Alloys and Compounds. 2017;**699**:567-574. DOI: 10.1016/j.jallcom.2016.12.307

[134] Yousefzadeh M, Ghasemkhah F. Design of porous, core-shell, and hollow nanofibers. In: Barhoum A, Bechelany M, Makhlouf ASH, editors. Handbook of Nanofibers. Part I: Fundamental Aspects, Experimental Setup, Synthesis, Properties and Characterization. 1st ed. Switzerland: Springer Nature; 2019. pp. 157-214. DOI: 10.1007/978-3-319-53655-2_9

[135] Vijayan P, Al-Maadeed M. Self-repairing composites for corrosion protection: A Review on recent strategies and evaluation methods. Materials. 2019;**12**:2754. DOI: 10.3390/ma12172754

[136] Li F, Gao X, Wang R, Zhang T. Design of WO$_3$-SnO$_2$ core-shell nanofibers and their enhanced gas sensing performance based on different work function. Applied Surface Science. 2018;**442**:30-37. DOI: 10.1016/j.apsusc.2018.02.122

[137] Li F, Gao X, Wang R, Zhang T, Lu G. Study on TiO$_2$-SnO$_2$ core-shell heterostructure nanofibers with different work function and its application in gas sensor. Sensors and Actuators B. 2017;**248**:812-819. DOI: 10.1016/j.snb.2016.12.009

[138] Li F, Gao X, Wang R, Zhang T, Lu G, Barsan N. Design of core-shell heterostructure nanofibers with different work function and their sensing properties to trimethylamine.

ACS Applied Materials & Interfaces. 2016;**8**:19799-19806. DOI: 10.1021/acsami.6b04063

[139] Lee J-H, Katoch A, Choi S-W, Kun J-H, Kim HW, Kim SS. Extraordinary improvement of gas-sensing performances in SnO$_2$ nanofibers due to creation of local p-n heterojunctions by loading reduced graphene oxide nanosheets. ACS Applied Materials & Interfaces. 2015;**7**:3101-3109. DOI: 10.1021/am5071656

[140] Dang TK, Son NT, Lanh NT, Phuoc PH, Viet NN, Thong LV, et al. Extraordinary H$_2$S sensing performance of ZnO/rGO external and internal heterojunctions. Journal of Alloys and Compounds. 2021;**879**:160457. DOI: 10.1016/j.jallcom.2021.160457

[141] Salehi T, Taherizadeh A, Bahrami A, Allafchian A, Ghafarinia. Toward a highly functional hybrid ZnO nanofiber-rGO gas sensor. Advanced Engineering Materials. 2020;**22**: 202000005. DOI: 10.1002(adem. 202000005

[142] Masa S, Robés D, Hontañón E, Lozano J, Eqtesadi S, Narros A. Graphene-tin oxide composite nanofibers for low temperature detection of NO$_2$ and O$_3$. Sensors & Transducers. 2020;**246**(7):71-78

[143] Wang D, Tang M, Sun J. Direct fabrication of reduced graphene oxide@ SnO$_2$ hollow nanofibers by single-capillary electrospinning as fast NO$_2$ gas sensor. Journal of Nanomaterials. 2019;**2019**:1929540. DOI: 10.1155/2019/1929540

[144] Li W-T, Zhang X-D, Guo X. Electrospun Ni-doped SnO$_2$ nanofiber array for selective sensing of NO$_2$. Sensors and Actuators B. 2017;**244**: 509-521. DOI: 10.1016/j.snb.2017.01.022

[145] Park S, Bulemo PM, Koo W-T, Ko J, Kim I-D. Chemiresistive acetylene

sensor fabricated from Ga-doped ZnO nanofibers functionalized with Pt catalysts. Sensors and Actuators B. 2021;**343**:130137. DOI: 10.1016/j. snb.2021.130137

[146] Lalwani SK, Beniwal A, Sunny. Enhancing room temperature ethanol sensing using electrospun Ag-doped SnO_2-ZnO nanofibers. Journal of Materials Science: Materials in Electronics. 2020;**31**:17212-17224. DOI: 10.1007/s10854-020-04276-9

[147] Abideen ZU, Kim J-H, Mirzaei A, Kim HW, Kim SS. Sensing behavior to ppm-level gases and synergistic sensing mechanism in metal-functionalized rGO-loaded ZnO nanofibers. Sensors and Actuators B. 2018;**255**:1884-1896. DOI: 10.1016/j.snb.2017.08.210

[148] Al-Dhahebi AM, Gopinath SCB, Saheed MSM. Graphene impregnated electrospun nanofiber sensing materials: A comprehensive overview on bridging laboratory set-up to industry. Nano Convergence. 2020;**2**:27. DOI: 10.1186/ s4080-020-00237-4

[149] Chen C, Tang Y, Vlahovic B, Yan F. Electrospun polymer nanofibers decorated with noble metal nanoparticles for chemical sensing. Nanoscale Research Letters. 2017;**12**:451. DOI: 10.1186/s11671-017-2216-4

[150] Yang T, Zhan L, Huan CZ. Recent insights into functionalized electrospun nanofibrous films for chemo-/bio-sensors. Trends in Analytical Chemistry. 2020;**124**:115813. DOI: 10.1016/j. trac.2020.115813

[151] Kim J-H, Abideen ZU, Zheng Y, Kim SS. Improvement of toluene-sensing performance of SnO_2 nanofibers by Pt functionalization. Sensors. 2016;**16**:1857. DOI: 10.3390/s16111857

[152] Kim S, Brady J, Al-Badani F, Yu S, Hart J, Jung S, et al. Nanoengineering approaches toward artificial nose.

Frontiers in Chemistry. 2021;**9**:629329. DOI: 10.3389/fchem.2021.629329

[153] Hierlemann A. CMOS-based chemical sensors. In: Brand O, Fedder GK, editors. Advanced Micro and Nanosystems. Vol. 2. CMOS-MEMS. 1st ed. Weinheim: Wiley-VCH; 2005. p. 335-390. DOI: 10.1002/ 9783527616718.ch7

[154] Gao W, Ota H, Kiriya D, Takei K, Javey A. Flexible electronics toward wearable sensing. Accounts of Chemical Research. 2019;**52**:523-533. DOI: 10.1021/acs.accounts.8b00500

[155] Hung CM, Thanh DTT, Van Hieu N. On-chip growth of semiconductor metal oxide nanowires for gas sensors. A review. Journal of Science: Advanced Materials and Devices. 2017;**2**:263-285. DOI: 10.10161/j. jsadm.2017.07.009

[156] Yang D, Cho I, Kim D, Lim MA, Li Z, Ok JG, et al. Gas sensor by direct growth and functionalization of metal oxide/metal sulfide core-shell nanowires onto flexible substrates. ACS Applied Materials & Interfaces. 2019;**11**:24298-24307. DOI: 10.1021/ acsami.9b06951

[157] XiaoQi Z, HuanYu C. Flexible and stretchable metal oxide gas sensors for healthcare. Science China Technological Sciences. 2019;**62**:209-223. DOI: 10.1007/ s11431-018-9397-5

[158] Habibi M, Ahmadian-Yazdi M-R, Eslamian M. Optimization of spray coating for the fabrication of sequentially deposited planar perovskite solar cells. Journal of Photonics for Energy. 2017;**7**(2):022003. DOI: 10.1117/ 1.JPE.7.022003

[159] Li Y, Delaunay J-J. Progress toward nanowire device assembly technology. In: Prete P, editor. Nanowires. London: IntechOpen; 2010. 25 pp. DOI: 10.5772/ 39521

[160] Freer EM, Grachev O, Duan X, Martín S, Stumbo DP. High-yield self-limiting single-nanowire assembly with dielectrophoresis. Nature Nanotechnology. 2010;**5**:525-530. DOI: 10.1038/NNANO.2010.206

[161] Chmela O, Sadílek J, Domènech-Gil G, Samà J, Somer J, Mohan R, et al. Selectively arranged single-wire based nanosensor array systems for gas monitoring. Nanoscale. 2018;**10**:9087-9096. DOI: 10.1039/c8nr01588k

[162] Yuan H, Zhou Q, Zhang Y. Improving fiber alignment during electrospinning. In: Afshari M, editor. Electrospun Nanofibers. Woodhead Publishing Series in Textiles. Amsterdam: Elsevier; 2017. pp. 125-147. DOI: 10.1016/B978-0-08-100907-9.00006-4

[163] Prabhakaran MP, Vatankhah E, Ramakrishna S. Electrospun aligned PHBV/collagen nanofibers as substrates for nerve tissue engineering. Biotechnology and Bioengineering. 2013;**110**(10):2775-2784. DOI: 10.1002/bit.24937

[164] Munir MW, Ali U. Classification of electrospinning methods. In: Ghamsari MS, Dhara S, editors. Nanorods and Nanocomposites. London: IntechOpen; 2020. 19 pp. DOI: 10.5772/intechopen.88654

[165] Li D, Ouyang G, McCann JT, Xia Y. Collecting electrospun nanofibers with patterned electrodes. Nano Letters. 2005;5(5):913-916. DOI: 10.1021/nl0504235

[166] Li D, Wang Y, Xia Y. Electrospinning nanofibers as uniaxially aligned arrays and layer-by-layer stacked films. Advanced Materials. 2004;**18**(4):361-366. DOI: 10.1002/adma.200306226

[167] Li D, Wang Y, Xia Y. Electrospinning of polymeric and ceramic nanofibers as uniaxially aligned arrays. Nano Letters. 2003;**3**(8):1167-1171. DOI: 10.1021/nl0344256

[168] Meng Y, Liu G, Liu A, Guo Z, Sun W, Shan F. Photochemical activation of electrospun In_2O_3 nanofibers for high performance electronic devices. ACS Applied Materials & Interfaces. 2017;**9**:10805-10812. DOI: 10.1021/acsami.6b15916

[169] Kim I-D, Jeon E-K, Choi S-H, Choi D-K, Tuller HL. Electrospun SnO_2 nanofiber mats with thermo-compression step for gas sensing applications. Journal of Electroceramics. 2010;**25**:159-167. DOI: 10.1007/s10832-010-9607-6

[170] Yang D-J, Kamienchik I, Youn DY, Rothschild A, Kim I-D. Ultrasensitive and highly selective gas sensor based on electrospun SnO_2 nanofibers modified by Pd loading. Advanced Functional Materials. 2010;**20**:4258-4264. DOI: 10.1002/adfm.201001251

[171] Cui Y, Meng Y, Wang Z, Wang C, Liu G, Martins R, et al. High performance electronic devices based on nanofibers via a crosslinking welding process. Nanoscale. 2018;**10**:19427-19434. DOI: 10.1039/c8nr05420g

[172] Park J-U, Hardy M, Kang SJ, Barton K, Adair K, Mukhopadhyay DK, et al. High-resolution electrohydro-dynamic jet printing. Nature Materials. 2007;**6**:782-789. DOI: 10.1038/nmat1974

[173] Zou W, Yu H, Zhou P, Liu L. Tip-assisted electrohydrodynamic jet printing for high resolution microdroplet deposition. Materials and Design. 2019;**185**:107609. DOI: 10.10106/j.matdes.2019.107609

[174] Liang Y, Yong J, Yu Y, Nirmalathas A, Ganesan K, Evans R, et al. Direct electrohydrodynamic patterning of high-performance all metal oxide thin-film electronics. ACS Nano.

2019;**13**:13957-13964. DOI: 10.1021/
acsnano.9b05715

[175] Wu H, Yu J, Cao R, Yang Y, Tang Z.
Electrohydrodynamic inkjet printing of
Pd loaded SnO_2 nanofibers on a CMOS
micro hotplate for low power H_2
detection. AIP Advances. 2018;**8**:055307.
DOI: 10.1063/1.5029283

[176] Kang K, Yang D, Park J, Kim S,
Cho I, Yang H-H, et al. Micropatterning
of metal oxide nanofibers by
electrohydrodynamic (EHD) printing
towards highly integrated and
multiplexed gas sensor applications.
Sensors and Actuators B. 2017;**250**:574-
583. DOI: 10.1016/j.snb.2017.04.194

[177] Schmidt V, Riel H, Senz S, Karg S,
Riess W, Gösele U. Realization of a
silicon nanowire vertical surround-
gate field-effect transistor. Small.
2006;**2**(1):85-88. DOI: 10.1002/smll.
200500181

[178] Ng HT, Han J, Yamada T, Nguyen P,
Chen YP, Meyyappan M. Single crystal
nanowire vertical surround-gate
field-effect transistor. Nano Letters.
2004;**4**(7):1247-1252. DOI: 10.1021/
nl049461z

[179] Fan HJ, Werner P, Zacharia M.
Semiconductor nanowires: From
self-organization to patterned growth.
Small. 2006;**2**(6):700-717. DOI: 10.1002/
smll.200500495

[180] Zhang B, Gao P-X. Metal oxide
nanoarrays for chemical sensing: A
review of fabrication methods, sensing
modes, and their inter-correlations.
Frontiers in Materials. 2019;**6**:65.
DOI: 10.3389/fmats.2019.00055

[181] Mkhize N, Murugappan K,
Castell MR, Bhaskaran H.
Electrohydrodynamic jet printed
conducting polymer for enhanced
chemiresistive gas sensors. Journal of
Materials Chemistry C. 2021;**9**:4591-
4596. DOI: 10.1039/d0tc05719c

[182] Zhang Z, He H, Fu W, Ji D,
Ramakrishna S. Electro-hydrodynamic
direct-writing technology toward
patterned ultra-thin fibers: Advances,
materials and applications. Nano Today.
2020;**35**:100942. DOI: 10.1016/j.
nantod.2020.100942

[183] Xu W, Zhang S, Xu W. Recent
progress on electrohydrodynamic
nanowire writing. Science China
Materials. 2019;**62**(11):1709-1726.
DOI: 10.1007/s40483-019-9583-5

[184] He X-X, Zheng J, Yu G-F, You M-H,
Yu M, Ning X, et al. Near-field
electrospinning: Progress and
applications. The Journal of Physical
Chemistry. 2017;**121**:8663-8678.
DOI: 10.1021/acs.jpcc.6b12783

[185] Huang Y, Bu N, Duan Y, Pan Y,
Liu H, Yin Z, et al. Electrohydrodynamic
direct-writing. Nanoscale. 2013;**5**:12007-
12017. DOI: 10.1039/c3nr04329k

[186] Sun D, Chang C, Sha L, Lin L.
Near-field electrospinning. Nano Letters.
2006;**6**(4):839-842. DOI: 10.1021/
nl0602701

[187] Lim K, Jo Y-M, Yoon J-W, Lee J-H.
Metal oxide patterns of one-dimensional
nanofibers: On-demand, direct-write
fabrication, and application as a novel
platform for gas detection. Journal of
Materials Chemistry A. 2019;**7**:24919-
24928. DOI: 10.1039/c9ta09708b

[188] Ruggieri F, Di Camillo D, Lozzi L,
Santucci S, De Marcellis A, Ferri G, et al.
Preparation of nitrogen doped TiO_2
nanofibers by near field electrospinning
(NFES) technique for NO_2 sensing.
Sensors and Actuators B. 2013;**179**:
107-113. DOI: 10.1016/j.snb.2012.10.094

[189] Wang X, Zheng G, He G, Wei J,
Liu H, Lin Y, et al. Electrohydrodynamic
direct-writing ZnO nanofibers for
device applications. Materials Letters.
2013;**109**:58-61. DOI: 10.1016/j.
matlet.2013.07.051

Energy Storage Properties of Topochemically Synthesized Blue TiO₂ Nanostructures in Aqueous and Organic Electrolyte

Parthiban Pazhamalai, Karthikeyan Krishnamoorthy and Sang-Jae Kim

Abstract

This book chapter discusses the topochemical synthesis of blue titanium oxide (b-TiO₂) and their application as electrode material for supercapacitor devices in aqueous and organic electrolytes. The formation mechanism of b-TiO₂ via topochemical synthesis and their characterization using X-ray diffraction, UV–visible, photoluminescence, electron spin resonance spectroscopy, laser Raman spectrum, X-ray photoelectron spectroscopy, and morphological studies (FESEM and HR-TEM) are discussed in detail. The supercapacitive properties of b-TiO₂ electrode were studied using both aqueous (Na₂SO₄) and organic (TEABF₄) electrolytes. The b-TiO₂ based symmetric-type supercapacitor (SC) device using TEABF₄ works over a wide voltage window (3 V) and delivered a high specific capacitance (3.58 mF cm⁻²), possess high energy density (3.22 μWh cm⁻²) and power density (8.06 mW cm⁻²) with excellent cyclic stability over 10,000 cycles. Collectively, this chapter highlighted the use of b-TiO₂ sheets as an advanced electrode for 3.0 V supercapacitors.

Keywords: blue TiO₂, topochemical synthesis, electron-spin resonance, photoluminescence, supercapacitors

1. Introduction

The invent of graphene and graphene-like two dimensional materials has created great interest in the exploration of other two-dimensional materials family which includes hexagonal boron nitride, graphitic carbon nitride, transition metal dichalcogenides (TMDCs), layered metal-oxides/double hydroxides, MXenes (transition metal carbides/nitrides), metal–organic frameworks (MOFs), covalent-organic frameworks (COFs), polymers, black phosphorus (BP), siloxene/silicenes, and metallenes [1–7]. In this scenario, the preparation of the 2D materials is of great interest for the development of the new range of materials [8–10]. The preparation methodologies/strategies can be categorized into top-down technique and bottom-up technique [8, 11]. In general, the top-down technique mainly depends on the breakage of the bulk materials to the micro- and nano-scale, whereas in the bottom-up technique, the growth of the materials is from atomic level to the macro-scale structure [12–14]. Amongst the preparation of the materials, bottom-up techniques such as molecular beam epitaxy, chemical vapor deposition and so on produced high purity and efficient layered materials with desired crystal structure [15–17]. This technique involves in building the nano materials from atomic/molecular scale using heterogeneous/homogeneous chemical reactions, which tend to produce thermodynamically stable materials where the morphology/structure is controlled by both reaction kinetics and thermodynamics [8]. In addition, the building of nanomaterials with desired morphology/structure and composition in a single step is very challenging due to their inherent characteristics of the materials [8, 18]. Moreover, the synthesis of two-dimensional layered materials (metal nitrides/carbides) are hindered due to their non-layered crystal structure. And also, the preparation methods employed in the bottom-up approach is of high cost and the utilization of the prepared materials also requires highly sophisticated instrumentation for characterizing the same. In this regard, top-down technique is of great interest as the cost of production of the materials is less compared to the former with high amount of yield. Some of the top-down approaches for the production of the 2D materials are photolithographic technique, exfoliation (wet and solid phase), mechanical ball milling (wet and solid phase), chemical etching, and topochemical reaction [8, 11]. Amongst the other methods for the top-down technique, topo-chemical reaction methods is the promising strategy for the preparation of the 2D layered materials [8].

The topochemical route can be classified depends on the reaction methodologies such as adding, extracting or substituting elements to/from the source materials to form the new materials with the retention of the structure/morphology of source material [11]. This type of preparation technique adopts the "corner-overtaking" route, which circumvents the hindrances of direct synthesis through chemical reaction to multi-level steps synthesis [11]. In this aspect, the topochemical preparation route is utilized to synthesize the high value added 2D materials which are difficult to process. Compared to that of the direct synthesis/bottom up approach, top down method has higher advantage to prepare the 2D materials with controlled morphology, composition and material structure. The topochemical reaction route can be performed through various methodologies which includes the (i) selective etching of the elements [19, 20]; (ii) electrochemical methods [21, 22] (iii) high temperature treatment reaction [23, 24] and (iv) liquid phase reaction [25, 26]. In the selective etching methods, the 2D layered materials can be easily obtained by from the bulk counterpart and also it is possible to engineer the interface and surface of the prepared materials. However, the chemical used as etchant is toxic which also affect the quality of the raw materials for the preparation of materials. In this case, hydrofluoric acid is used as the etchant for the preparation of MXenes from the MAX phase element, similarly hydrochloric acid is used for the preparation of siloxene from calcium silicide [27–30]. In case of the electrochemical methods for the preparation, mild and controllable reaction is carried out with the application of electric field to promote the fast reaction [21, 22]. The limitation in this process is that it requires complex instrumentation and the mass production is very limited in the electrochemical method. In the high temperature treatment reaction, the preparation is associated with the controllable atmosphere induced chemical reaction and high temperature condition is employed for the preparation with the limitation of heterogeneity of gas/solid interface reaction, explosion risk at higher temperature and toxic gases as a byproduct of the reaction [23, 24]. Amongst the other techniques in the topochemical reaction, liquid phase exfoliation as the great advantages of high mass yield, wide range of controllable reaction parameters such as solvents, rich experimental/theoretical foundation and produce uniform and good crystalline materials. Hence the liquid phase topochemical methods attains more merits compared to the other methods.

In this book chapter, we have focused on preparing the 2D sheet like blue titanium oxide via liquid phase topochemical reaction on titanium boride. The TiO_2 prepared technique has electron-rich property which shows enormous property in the field of electrochemical energy storage. This electron-rich TiO_2 will be refereed as blue/black TiO_2 which inherits exceptional physical and chemical properties to that of other forms TiO_2 due to their disordered surface structure and Ti^{3+}/oxygen vacancies which leads to high conductivity, magnetic properties, and better chemical properties; yet to be explored for the energy storage application. Herein, the preparation, characterization and the electrochemical properties of blue titanium oxide with sheet-like nanostructures is investigated in detail in this chapter.

2. Preparation of blue titanium oxide (TiO_2) nanosheets

The blue TiO_2 nanosheets was prepared using simple hydrothermal assisted topochemical process as reported in the literature [31, 32]. Initially an appropriate amount of titanium boride was dissolved in 25 mL of hydrofluoric acid and transferred to Teflon lined autoclave for hydrothermal process at 180°C for 12 h. Upon completion of the hydrothermal process, the blue color precipitate is washed with double distilled water/ethanol and dried at 80°C for 12 h to obtain the blue TiO_2 nanosheets.

3. Physicochemical characterization of the blue TiO_2 nanosheets

The blue TiO_2 (blue titanium oxide) nanosheets were prepared using layered TiB_2 and HF as starting materials via hydrothermal process in acidic medium [31]. The mechanism for the formation of layer like blue titanium oxide from the precursor TiB_2 is due to the topochemical reaction as seen in the preparation of siloxenes and MXenes under acidic medium [19, 33]. The TiB_2 is a layered material with titanium and boride arranged in alternating layers. In this case the TiB2 reacts with the HF, which topochemically dissolves boron with the release of hydrogen gas. This generated hydrogen gas creates the oxygen vacancies via H_2 reduction of Ti^{4+} on the surface [31, 34]. The digital micrograph of the precursor and the final product is depicted in the **Figure 1(A)**, which shows the change in from gray TiB_2 to blue colored TiO_2. The X-ray diffraction pattern of the precursor TiB_2 and the prepared TiO_2 were provided in **Figure 1(B)**, which confirms the complete transformation of TiB_2 to TiO_2 and matched with the anatase TiO_2 (JCPDS No: 021-1272) [35]. The implication of the bule colored TiO_2 can be corelated with the vacancy of oxygen, disordered surface and deficiency of oxygen which can be confirmed using the spectroscopic analysis such as electron spin resonance spectroscopy (ESR), UV–visible (UV–vis) and photoluminescence (PL) respectively. The X-band electron spin resonance spectrum of the blue titanium oxide was obtained under the temperature of 77 K with the frequency of 9364 GHz and provided in **Figure 1(C)**. The presence of Ti^{3+} centers in the g- region (ranging from 1.87 to 1.99) which is associated to the resonation of Ti^{3+} species [36]. In addition, the presence of a number of partially overlapping signals in the ESR spectrum is matched well to the different types of Ti^{3+} centers present in the blue titanium oxide. The obtained ESR results is in agreement with the previous finding on the partially reduced TiO_2 [31]. **Figure 1(D)** shows the comparative UV–vis spectra of commercial TiO_2 and blue titanium oxide and the peak observed at 354 nm corresponds to the intrinsic absorption edge of TiO_2 which is red-shifted for the blue titanium oxide to that of the commercial TiO_2 confirming the increase in electron concentration on the surface of blue titanium

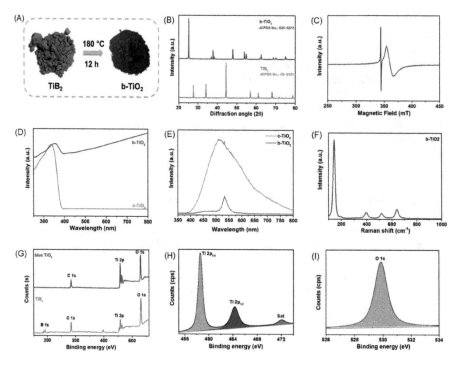

Figure 1.
(A) Digital micrograph of titanium boride and blue titanium oxide. (B) X-ray diffraction pattern of the titanium boride and blue titanium oxide. (C) Electron spin resonance spectrum of blue titanium oxide. (D) UV–visible spectrum of commercial anatase titanium oxide and blue titanium oxide. (E) Photoluminescence spectrum of commercial anatase TiO_2 and blue titanium oxide. (F) Raman spectrum of blue titanium oxide. (G) XPS survey spectrum of TiB_2 and blue titanium oxide. XPS core level spectrum of (H) Ti 2p and (I) O 1 s state in blue titanium oxide.

oxide. Furthermore, the blue titanium oxide shows increased absorbance compared to the commercial TiO_2 over 400–800 nm which is due to the absorbance of low-energy-photon and/or thermal-excitations of trapped electrons in localized states of defects [37–40]. **Figure 1(E)** shows the comparative PL spectra of commercial and prepared TiO_2. The PL spectra depicts the presence of peak over 510–540 nm for the blue titanium oxide compared to that of the commercial TiO_2 which shows a broad band over the region of 400–800 nm [41, 42]. The blue titanium oxide shows higher luminescence intensity to that of the commercial TiO_2 suggesting the low electron–hole recombination rate of blue titanium oxide [43, 44]. These studies confirming the higher electronic conductive states of blue titanium oxide compared to the commercial TiO_2. The laser Raman spectrum of the blue TiO_2 is presented in **Figure 1(F)** which discloses the occurrence of a major vibration band at 148.25 cm^{-1} corresponds to the E_g mode of anatase TiO_2 [45]. In addition to the major vibrational band, the occurrence of the three minor vibrational bands occurred at 396, 517, and 640 cm^{-1} which correspond to the B_{1g}, A_{1g}, and E_g vibrational modes of anatase TiO_2, respectively [38, 46].

The comparative X-ray photoelectron survey spectrum of TiB_2 and blue titanium oxide is provided in **Figure 1(G)**, which shows the presence of Ti 2p, B 1 s, C 1 s and O 1 s states [47]. The existence of oxygen and carbon peaks in the spectrum is due to atmospheric interaction with the sample. From the comparative XPS spectra of TiB_2 and blue titanium oxide, exhibits the complete removal of the boron from TiB_2 which is evident from the XPS spectrum. This study proves the compete dissolution of the boron during the hydrothermal reaction [47]. The XPS spectrum of Ti and O

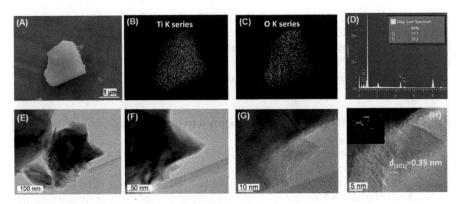

Figure 2.
FESEM and HR-TEM analysis of blue titanium oxide nanostructures.

oxidation states of elements present in the blue titanium oxide nanostructures were provided in **Figure 1(H and I)**. The XPS core level spectrum of Ti 2p state reveals the existence of the $2p_{3/2}$ and $2p_{1/2}$ at 458, and 464 eV respectively. In addition, the observance of small-satellite peak at 472.2 eV corresponding to the defect of oxygen in the blue titanium oxide [38, 48]. Overall, the core-level spectrum of the titanium highlights the Ti^{4+} oxidation state with littleTi^{3+} state due to oxygen deficiency in blue titanium oxide [49]. **Figure 1(I)** represents the O 1 s states of blue titanium oxide at 530 eV respectively [38, 48]. The calculation of the O/Ti atomic ratio of blue titanium oxide nanostructures is found to be 2.47 from XPS analysis which is in agreement with the literature [50].

To analyze the surface morphology of blue titanium oxide nanostructures field emission scanning electron micrograph (FE-SEM) with energy dispersive spectra (EDS) and high-resolution transmission electron microscope (HR-TEM) were performed. The FE-SEM of blue titanium oxide is provided in **Figure 2(A)**; reveals the presences of nano-sheets/plate-like blue titanium oxide as result of topochemical reaction. The elemental-mapping of Ti & O in the blue titanium oxide is shown in **Figure 2(B and C)** and EDS spectrum (**Figure 2(D)**) indicated the homogeneous distributions of Ti and O throughout the blue titanium oxide sheet/plate. With an O/Ti atomic ratio of about 3.92 present in the blue titanium oxide nanostructures [50, 51]. The HR-TEM micrographs of blue titanium oxide obtained under various levels of magnifications are shown in **Figure 2(E–H)**. The HR-TEM micrograph of blue titanium oxide (**Figure 2(E–F)** revealed the presence of sheet-like structures with the lateral size ranging from 200 nm in length and 250 nm in breadth. The magnified portion of **Figure 2(F)** is provided in **Figure 2(G)** reveals the presence of thinner blue titanium oxide sheets with few-layers. The lattice fringes (**Figure 2(H)**) of blue titanium oxide sheets exhibits d = 0.35 nm (an interplanar-spacing) corresponds to (101) plane and SAED pattern (inset of **Figure 2(H)**) reveals the clear diffraction spots signifying the crystallinity-nature of blue titanium oxide [37].

4. Electrochemical characterization of blue titanium oxide electrode

4.1 Preparation of electrode for electrochemical characterization

The working electrode was prepared by mixing the active material (blue titanium oxide) with carbon black and PVDF in an appropriate ratio of 80:15:5 using NMP as solvent. The slurry was spin coated on stainless steel substrate for the three electrode and two electrode characterizations. For the fabrication of symmetric supercapacitor, CR2032 coin cell configuration is utilized and the TiO_2-coated stainless-steel substrates with an area of 1.86 cm^2 as electrodes separated by a Celgard membrane and 1 M TEABF$_4$ in acetonitrile as the electrolyte. All the electrochemical measurements were performed using an Autolab PGSTAT302N electrochemical workstation.

4.2 Half-cell characterization of blue titanium oxide electrode

The electrochemical properties of the blue titanium oxide electrodes in aqueous electrolyte (half-cell) were inspected using cyclic voltammetry (CV), electrochemical impedance spectroscopy (EIS), and galvanostatic charge–discharge (CD) analysis, respectively. The half-cell electrochemical characterization of the blue titanium oxide electrode was measured in 1 M sodium sulfate electrolyte and provided in **Figure 3**. The cyclic voltammogram provided in **Figure 3(A)** exhibits the existence of the quasi-rectangular behavior suggesting the pseudocapacitive nature of charge-storage in the blue titanium oxide electrode [52, 53]. The indication of the pseudocapacitance nature in blue titanium oxide due to ion-intercalation/de-intercalation phenomenon in addition to the electric double layer/surface capacitance, we quantify the contribution on the overall capacitance using Dunn's method [54]. The plot ofslope of log (i) versus log (ν) obtained from the power law is provided in **Figure 3(B)** [55, 56]:

$$i = a \nu^b \tag{1}$$

where, "ν", "a" and "b" are scan rate (mV s^{-1}), adjustable parameters, respectively. The value of "b" is calculated from the slope of log i versus log ν. From the obtained b-value, the nature of the capacitance is evaluated as when b value is 0.5, reveals the diffusion mediated ion-intercalation/de-intercalation whereas the b is 1 represents the surface capacitive nature in the electrode. The contribution of the capacitance related to the surface and diffusion process can be quantified using the relation [54]:

$$I(V) = k_1 \nu + k_2 \nu^{1/2} \tag{2}$$

Were "$k_1 \nu$" and "$k_2 \nu^{1/2}$" related to the contributions from surface- and diffusion-mediated intercalation/de-intercalation process, respectively. The slope and intercept of the plot between $I(V)/v^{1/2}$ and $v^{1/2}$, provides the value of k_1 and k_2 to determine the capacitance contributions [54]. The overall charge stored and their contribution related to the diffusion and surface is calculated using the CV curve in **Figure 3(C)** and the contribution plot is provided in **Figure 3(D)**. From the **Figure 3(D)**, the diffusion capacitance is increasing from 42.10 to 77.29% when the scan rate is decreased from 50 to 5 mV s^{-1}. The diffusion capacitance is higher at the low scan rate which is due to the fact that at high scan rates, the electrolyte ions faces time constraints which limits the diffusion of electrolyte ions whereas low scan rates provide sufficient time for the electrolyte ions to diffuse into the interior surface of the blue titanium oxide electrode [57]. The charge discharge profile shows sloppy symmetric type curves which is evident from **Figure 3(E)** and delivered a high gravimetric specific capacitance of 19 F g-1 and areal capacitance of 19 mF cm^{-2} obtained at a constant current of 0.5 mA as seen in **Figure 3(F)**. The excellent electrochemical performance of the blue titanium oxide in the neutral electrolyte is due to the reason such as sheet-like structure, enhanced electrical conductivity with more oxygen vacant sites in blue titanium oxide.

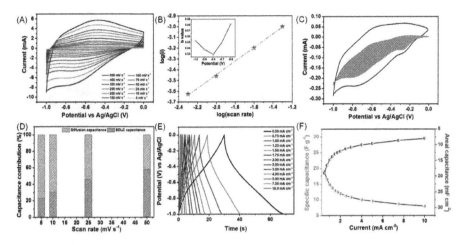

Figure 3.
Three electrode characterization of the blue titanium oxide electrode (A) cyclic voltammetry, (B) plot of log (I) vs. scan rate. The inset in (B) shows the plot of b-values vs. potential. (C) CV profile of blue titanium oxide electrode with the shaded part revealing the capacitive current. (D) Capacitance contribution plot. (E) Galvanostatic charge discharge analysis and the (F) plot of specific capacitance of the blue titanium oxide electrode.

4.3 Full-cell characterization of blue titanium oxide symmetric supercapacitor

In the view point of the practical application, a symmetric supercapacitor is fabricated using blue titanium oxide as electrode and organic liquid (TEABF$_4$) is used as electrolyte. The electrochemical performance of the fabricated blue titanium oxide SSC was tested for the electrochemical stability of the operating potential window by measuring the CV profiles over the increasing voltage from 0 to 3 3.0 V at a scan rate of 100 mV s^{-1} as shown in **Figure 4**, which shows the distorted rectangular shaped curves and also suggested that the blue titanium oxide SSC device can operate at a window of 3.0 V without any sign of evolution. The cyclic voltammogram of the fabricated SSC displayed typical rectangular shaped curves for all the CV scans at different rates ranging from 5 to 1000 mV s^{-1} (**Figure 5(A and B)**) suggesting the pseudocapacitive nature of charge-storage in blue titanium oxide via intercalation/de-intercalation phenomenon [58, 59]. The plot of the scan rate verses specific capacitance is provided in **Figure 5(C)**, depicts a high capacitance of 6.67 F g^{-1} measured at a scan rate of 5 mV s^{-1}. The EIS analysis of the blue titanium oxide SSC is provided in **Figure 5(D–F)** and represented in the form of Nyquist and Bode plots [60]. The high- and mid- frequency region can be used to determine the solution resistance (R$_s$) or equivalent series resistance (ESR) whereas the low- frequency region directly related to the frequency dependent ion diffusion kinetics of electrolyte ions to the electrode [19]. The values of R$_s$ and R$_{ct}$ from the Nyquist plot are determined to be 1.60 and 15.4 Ω, respectively. The presence of the Warburg element in the model circuit related to the frequency dependent ion diffusion kinetics of electrolyte ions to the electroactive surface as evident from **Figure 5(D)** [61]. **Figure 5(E)** shows the Bode phase angle plot of blue titanium oxide SSCs which tails the phase angle at the low- frequency region (0.01 Hz) is about 64.94°, thus, suggesting the pseudo-capacitive nature of the blue titanium oxide SSCs [33]. **Figure 5(F)** presents the dependence of specific capacitance as a function of applied frequency which is about 3.24 F g^{-1} for the blue titanium oxide SSCs, and the capacitance decreases with respect to an increase in frequency [62].

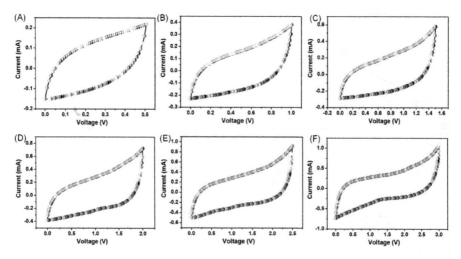

Figure 4.
CV profile of blue titanium oxide symmetric supercapacitor measured with increasing in voltage window.

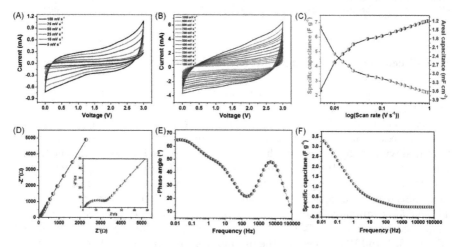

Figure 5.
Full cell characterization of the blue titanium oxide SSC. (A and B) CV profiles and (C) plot of specific capacitance of blue titanium oxide SSC. (D) Nyquist plot and the enlarged portion is provided in the inset; (E) bode phase angle plot and (F) effect of frequency on capacitance blue titanium oxide SSC.

The charge–discharge profiles of the fabricated SSC measured using various constant currents (0.5 to 10 mA) is provided in **Figure 6(A)**. The profiles indicate the existence of the quasi symmetric profiles suggesting the pseudocapacitive behavior of the blue titanium oxide SSCs [63] and the dependance of current on the capacitance is given in **Figure 6(B)**. A high specific capacitance of the 4.80 F g^{-1} is obtained from the GCD profile measured at a constant current of 0.5 mA. The rate capability of the blue titanium oxide SSC device is provided in **Figure 6(C)** which indicating the better capacitance retention when switched from low- to high- current suggesting superior rate capability [19]. The energy and power density plot is represented in the form of Ragone plot (**Figure 6(D)**) which holds a high-energy-density of 6.0 Wh kg^{-1} with a power of about 750 W kg^{-1}, respectively. The long tern cyclic life of the blue titanium oxide SSC depicts the capacitance retention of about 90.2% after 10,000 cycles (**Figure 6(E)**), highlighting the better cycle

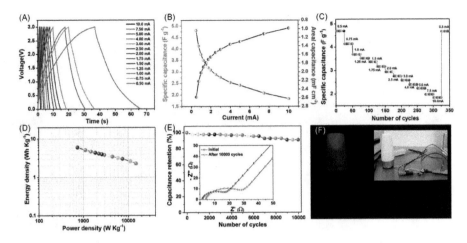

Figure 6.
Full cell characterization of the blue titanium oxide SSC. (A) CD profiles; (B) plot of specific capacitance; (C) rate capability; (D) Ragone plot; (E) cycling stability; the inset in (E) shows the Nyquist plot before and after 10000 cycles of blue titanium oxide SSC device. (F) Real time application of the blue titanium oxide SSC device.

life [25]. The impedance analysis after stability shows the increase in the ESR and charge transfer resistance such as 1.60 to 4.17 Ω and 15.4 to 24.4 Ω. **Figure 6(F)** shows the practical applications of the blue titanium oxide SSC glowing a blue lamp over 10 seconds after charged upto 3 V using a constant current of 0.5 mA.

5. Conclusion

In this book chapter we have demonstrated the preparation of the blue titanium oxide via topochemical reduction of titanium boride in hydrofluoric acid via hydrothermal process. In this we have discussed the formation of 2D layered titanium oxide with oxygen deficiency and the utilization in the energy storage sector. The electrochemical energy storage properties of the blue titanium oxide is tested via symmetric configuration type supercapacitor operated over a wide potential range of 3.0 V with good electrochemical stability and higher rate. The prepared blue titanium oxide shows higher output than those of other 2D electrode materials-based supercapacitors. From overall studies, the prepared materials shows better electrochemical performance which paves the way as a promising electrode material for next-generation smart energy storage sectors.

Acknowledgements

This research was supported by the Basic Science Research Program through the National Research Foundation of Korea (NRF) grant funded by the Korea government (MSIT) (2019R1A2C3009747, 2020R1A2C2007366, 2021R1I1A1A01049635 & 2021R1A4A2000934).

Conflict of interest

"The authors declare no conflict of interest."

Author details

Parthiban Pazhamalai[1], Karthikeyan Krishnamoorthy[1] and Sang-Jae Kim[1,2,3*]

1 Faculty of Applied Energy System, Nanomaterials and System Laboratory, Major of Mechatronics Engineering, Jeju National University, Jeju, Republic of Korea

2 Nanomaterials and System Laboratory, Major of Mechanical System Engineering, College of Engineering, Jeju National University, Jeju, Republic of Korea

3 Research Institute of Energy New Industry, Jeju National University, Jeju, Republic of Korea

*Address all correspondence to: kimsangj@jejunu.ac.kr

IntechOpen

References

[1] Randviir EP, Brownson DAC, Banks CE. A decade of graphene research: Production, applications and outlook. Materials Today. 2014;**17**:426-432. DOI: 10.1016/j.mattod.2014.06.001

[2] Shi Z, Cao R, Khan K, Tareen AK, Liu X, Liang W, et al. Two-dimensional tellurium: Progress, challenges, and prospects. Nano-Micro Letters. 2020;**12**:99. DOI: 10.1007/s40820-020-00427-z

[3] Zeng M, Xiao Y, Liu J, Yang K, Fu L. Exploring two-dimensional materials toward the next-generation circuits: From monomer design to assembly control. Chemical Reviews. 2018;**118**:6236-6296. DOI: 10.1021/acs.chemrev.7b00633

[4] Khan K, Tareen AK, Aslam M, Wang R, Zhang Y, Mahmood A, et al. Recent developments in emerging two-dimensional materials and their applications. Journal of Materials Chemistry C. 2020;**8**:387-440. DOI: 10.1039/C9TC04187G

[5] Forouzandeh P, Pillai SC. Two-dimensional (2D) electrode materials for supercapacitors. Materials Today: Proceedings. 2021;**41**:498-505. DOI: 10.1016/j.matpr.2020.05.233

[6] Pham VP. Direct growth of graphene on flexible substrates toward flexible electronics: A promising perspective. In: Flex. Electron. InTech; 2018. DOI: 10.5772/intechopen.73171

[7] Pham VP, Yeom GY. Recent advances in doping of molybdenum disulfide: Industrial applications and future prospects. Advanced Materials. 2016;**28**:9024-9059. DOI: 10.1002/adma.201506402

[8] Zhang Q, Peng W, Li Y, Zhang F, Fan X. Topochemical synthesis of low-dimensional nanomaterials.

Nanoscale. 2020;**12**:21971-21987. DOI: 10.1039/D0NR04763E

[9] Pazhamalai P, Krishnamoorthy K, Mariappan VK, Sathyaseelan A, Kim S-J. Solar driven renewable energy storage using rhenium disulfide nanostructure based rechargeable supercapacitors. Materials Chemistry Frontiers. 2020;**4**:3290-3301. DOI: 10.1039/D0QM00421A

[10] Pazhamalai P, Krishnamoorthy K, Mariappan VK, Sahoo S, Manoharan S, Kim S-J. A high efficacy self-charging MoSe₂ solid-state supercapacitor using electrospun nanofibrous piezoelectric separator with ionogel electrolyte. Advanced Materials Interfaces. 2018;**5**:1800055. DOI: 10.1002/admi.201800055

[11] Xiao X, Wang H, Urbankowski P, Gogotsi Y. Topochemical synthesis of 2D materials. Chemical Society Reviews. 2018;**47**:8744-8765. DOI: 10.1039/C8CS00649K

[12] Gopalakrishnan D, Damien D, Shaijumon MM. MoS₂ quantum dot-interspersed exfoliated MoS₂ nanosheets. ACS Nano. 2014;**8**:5297-5303. DOI: 10.1021/nn501479e

[13] Saha SK, Baskey M, Majumdar D. Graphene quantum sheets: A new material for spintronic applications. Advanced Materials. 2010;**22**:5531-5536. DOI: 10.1002/adma.201003300

[14] Chhowalla M, Shin HS, Eda G, Li L-J, Loh KP, Zhang H. The chemistry of two-dimensional layered transition metal dichalcogenide nanosheets. Nature Chemistry. 2013;**5**:263-275. DOI: 10.1038/nchem.1589

[15] Li S, Lin YC, Zhao W, Wu J, Wang Z, Hu Z, et al. Vapour-liquid-solid growth of monolayer MoS₂ nanoribbons. Nature Materials. 2018;**17**:535-542. DOI: 10.1038/s41563-018-0055-z

[16] Zhao X, Ning S, Fu W, Pennycook SJ, Loh KP. Differentiating polymorphs in molybdenum disulfide via electron microscopy. Advanced Materials. 2018;**30**:1-11. DOI: 10.1002/adma.201802397

[17] Ares P, Palacios JJ, Abellán G, Gómez-Herrero J, Zamora F. Recent progress on antimonene: A new bidimensional material. Advanced Materials. 2018;**30**:1703771. DOI: 10.1002/adma.201703771

[18] Mizuguchi R, Imai H, Oaki Y. Formation processes, size changes, and properties of nanosheets derived from exfoliation of soft layered inorganic–organic composites. Nanoscale Advances. 2020;**2**:1168-1176. DOI: 10.1039/D0NA00084A

[19] Krishnamoorthy K, Pazhamalai P, Kim S-JJ. Two-dimensional siloxene nanosheets: Novel high-performance supercapacitor electrode materials. Energy & Environmental Science. 2018;**11**:1595-1602. DOI: 10.1039/C8EE00160J

[20] Pazhamalai P, Krishnamoorthy K, Sahoo S, Mariappan VK, Kim SJ. Understanding the thermal treatment effect of two-dimensional siloxene sheets and the origin of superior electrochemical energy storage performances. ACS Applied Materials & Interfaces. 2019;**11**:624-633. DOI: 10.1021/acsami.8b15323

[21] Ambrosi A, Pumera M. Exfoliation of layered materials using electrochemistry. Chemical Society Reviews. 2018;**47**:7213-7224. DOI: 10.1039/C7CS00811B

[22] Yang S, Zhang P, Nia AS, Feng X. Emerging 2D materials produced via electrochemistry. Advanced Materials. 2020;**32**:1907857. DOI: 10.1002/adma.201907857

[23] Jeon J, Park Y, Choi S, Lee J, Lim SS, Lee BH, et al. Epitaxial synthesis of molybdenum carbide and formation of a $Mo_2 C/MoS_2$ hybrid structure via chemical conversion of molybdenum disulfide. ACS Nano. 2018;**12**:338-346. DOI: 10.1021/acsnano.7b06417

[24] Cao J, Li T, Gao H, Lin Y, Wang X, Wang H, et al. Realization of 2D crystalline metal nitrides via selective atomic substitution. Science Advances. 2020;**6**. DOI: 10.1126/sciadv.aax8784

[25] Li H, Jing L, Liu W, Lin J, Tay RY, Tsang SH, et al. Scalable production of few-layer boron sheets by liquid-phase exfoliation and their superior supercapacitive performance. ACS Nano. 2018;**12**:1262-1272. DOI: 10.1021/acsnano.7b07444

[26] Shen J, He Y, Wu J, Gao C, Keyshar K, Zhang X, et al. Liquid Phase Exfoliation of Two-Dimensional Materials by Directly Probing and Matching Surface Tension Components. Nano Letters. 2015;**15**:5449-5454. DOI: 10.1021/acs.nanolett.5b01842

[27] Navarro-Suárez AM, Van Aken KL, Mathis T, Makaryan T, Yan J, Carretero-González J, et al. Development of asymmetric supercapacitors with titanium carbide-reduced graphene oxide couples as electrodes. Electrochimica Acta. 2018;**259**:752-761. DOI: 10.1016/j.electacta.2017.10.125

[28] Wang F, Wu X, Yuan X, Liu Z, Zhang Y, Fu L, et al. Latest advances in supercapacitors: From new electrode materials to novel device designs. Chemical Society Reviews. 2017;**46**:6816-6854. DOI: 10.1039/c7cs00205j

[29] Nakano H, Ohtani O, Mitsuoka T, Akimoto Y, Nakamura H. Synthesis of amorphous silica nanosheets and their photoluminescence. Journal of the American Ceramic Society. 2005;**88**:3522-3524. DOI: 10.1111/j.1551-2916.2005.00618.x

[30] Ramachandran R, Leng X, Zhao C, Xu Z-X, Wang F. 2D siloxene sheets: A novel electrochemical sensor for selective dopamine detection. Applied Materials Today. 2020;**18**:100477. DOI: 10.1016/j.apmt.2019.100477

[31] Chiesa M, Livraghi S, Giamello E, Albanese E, Pacchioni G. Ferromagnetic Interactions in Highly Stable, Partially Reduced TiO2: The S=2 State in Anatase. Angewandte Chemie. 2017;**56**:2604-2607. DOI: 10.1002/anie.201610973

[32] Pazhamalai P, Krishnamoorthy K, Mariappan VK, Kim S-JJ. Blue TiO2 nanosheets as a high-performance electrode material for supercapacitors. Journal of Colloid and Interface Science. 2019;**536**:62-70. DOI: 10.1016/j.jcis.2018.10.031

[33] Krishnamoorthy K, Pazhamalai P, Sahoo S, Kim SJ. Titanium carbide sheet based high performance wire type solid state supercapacitors. Journal of Materials Chemistry A. 2017;**5**:5726-5736. DOI: 10.1039/C6TA11198J

[34] C.Z. Wen, H.B. Jiang, S.Z. Qiao, H.G. Yang, G.Q. (Max) Lu, Synthesis of high-reactive facets dominated anatase TiO2, Journal of Materials Chemistry 21 (2011) 7052. doi:10.1039/c1jm00068c.

[35] Righini L, Gao F, Lietti L, Szanyi J, Peden CHF. Performance and properties of K and TiO2 based LNT catalysts. Applied Catalysis B: Environmental. 2016;**181**:862-873. DOI: 10.1016/j.apcatb.2015.07.008

[36] Chiesa M, Paganini MC, Livraghi S, Giamello E. Charge trapping in TiO2 polymorphs as seen by Electron Paramagnetic Resonance spectroscopy. Physical Chemistry Chemical Physics. 2013;**15**:9435. DOI: 10.1039/c3cp50658d

[37] Zhu Q, Peng Y, Lin L, Fan C-M, Gao G-Q, Wang R-X, et al. Stable blue TiO2−x nanoparticles for efficient visible light photocatalysts. Journal of Materials Chemistry A. 2014;**2**:4429. DOI: 10.1039/c3ta14484d

[38] Liu G, Yang HG, Wang X, Cheng L, Lu H, Wang L, et al. Enhanced photoactivity of oxygen-deficient anatase TiO2 sheets with dominant {001} facets. Journal of Physical Chemistry C. 2009;**113**:21784-21788. DOI: 10.1021/jp907749r

[39] Qiu J, Li S, Gray E, Liu H, Gu Q-F, Sun C, et al. Hydrogenation Synthesis of Blue TiO2 for High-Performance Lithium-Ion Batteries. Journal of Physical Chemistry C. 2014;**118**:8824-8830. DOI: 10.1021/jp501819p

[40] Dong J, Han J, Liu Y, Nakajima A, Matsushita S, Wei S, et al. Defective black TiO2 synthesized via anodization for visible-light photocatalysis. ACS Applied Materials & Interfaces. 2014;**6**:1385-1388. DOI: 10.1021/am405549p

[41] Zhang C, Hua H, Liu J, Han X, Liu Q, Wei Z, et al. Enhanced Photocatalytic Activity of Nanoparticle-Aggregated Ag–AgX(X = Cl, Br)@TiO2 Microspheres Under Visible Light. Nano-Micro Letters. 2017;**9**:49. DOI: 10.1007/s40820-017-0150-8

[42] Knorr FJ, Mercado CC, McHale JL. Trap-State Distributions and Carrier Transport in Pure and Mixed-Phase TiO2: Influence of Contacting Solvent and Interphasial Electron Transfer. Journal of Physical Chemistry C. 2008;**112**:12786-12794. DOI: 10.1021/jp8039934

[43] Jiang Z, Wan W, Wei W, Chen K, Li H, Wong PK, et al. Gentle way to build reduced titanium dioxide nanodots integrated with graphite-like carbon spheres: From DFT calculation to experimental measurement. Applied Catalysis B: Environmental. 2017;**204**:283-295. DOI: 10.1016/j.apcatb.2016.11.044

[44] Jin C, Liu B, Lei Z, Sun J. Structure and photoluminescence of the TiO2 films grown by atomic layer deposition using tetrakis-dimethylamino titanium and ozone. Nanoscale Research Letters. 2015;**10**:95. DOI: 10.1186/s11671-015-0790-x

[45] Yan J, Wu G, Guan N, Li L, Li Z, Cao X. Understanding the effect of surface/bulk defects on the photocatalytic activity of TiO2: Anatase versus rutile. Physical Chemistry Chemical Physics. 2013;**15**:10978-10988. DOI: 10.1039/c3cp50927c

[46] Ohsaka T, Izumi F, Fujiki Y. Raman spectrum of anatase, TiO2. Journal of Raman Spectroscopy. 1978;**7**:321-324. DOI: 10.1002/jrs.1250070606

[47] Li L, Qiu F, Wang Y, Wang Y, Liu G, Yan C, et al. Crystalline TiB 2: An efficient catalyst for synthesis and hydrogen desorption/absorption performances of NaAlH 4 system. Journal of Materials Chemistry. 2012;**22**:3127-3132. DOI: 10.1039/c1jm14936a

[48] Erdem B, Hunsicker RA, Simmons GW, David Sudol E, Dimonie VL, El-Aasser MS. XPS and FTIR surface characterization of TiO2 particles used in polymer encapsulation. Langmuir. 2001;**17**:2664-2669. DOI: 10.1021/la0015213

[49] Bapna K, Phase DM, Choudhary RJ. Study of valence band structure of Fe doped anatase TiO 2 thin films. Journal of Applied Physics. 2011;**110**:043910. DOI: 10.1063/1.3624775

[50] Lin T, Yang C, Wang Z, Yin H, Lü X, Huang F, et al. Effective nonmetal incorporation in black titania with enhanced solar energy utilization. Energy & Environmental Science. 2014;7:967. DOI: 10.1039/c3ee42708k

[51] Kong L, Jiang Z, Wang C, Wan F, Li Y, Wu L, et al. Simple Ethanol Impregnation Treatment Can Enhance Photocatalytic Activity of TiO 2 Nanoparticles under Visible-Light Irradiation. ACS Applied Materials & Interfaces. 2015;7:7752-7758. DOI: 10.1021/acsami.5b00888

[52] Xia Y, Mathis TS, Zhao MQ, Anasori B, Dang A, Zhou Z, et al. Thickness-independent capacitance of vertically aligned liquid-crystalline MXenes. Nature. 2018;**557**:409-412. DOI: 10.1038/s41586-018-0109-z

[53] Lukatskaya MR, Kota S, Lin Z, Zhao M-QQ, Shpigel N, Levi MD, et al. Ultra-high-rate pseudocapacitive energy storage in two-dimensional transition metal carbides. Nature Energy. 2017;**2**:17105. DOI: 10.1038/nenergy.2017.105

[54] Wang J, Polleux J, Lim J, Dunn B. Pseudocapacitive contributions to electrochemical energy storage in TiO 2 (anatase) nanoparticles. Journal of Physical Chemistry C. 2007;**111**:14925-14931. DOI: 10.1021/jp074464w

[55] Jiang Q, Kurra N, Alhabeb M, Gogotsi Y, Alshareef HN. All Pseudocapacitive MXene-RuO2 Asymmetric Supercapacitors. Advanced Energy Materials. 2018;**8**:1703043. DOI: 10.1002/aenm.201703043

[56] Mariappan VK, Krishnamoorthy K, Pazhamalai P, Sahoo S, Kim SJ. Electrodeposited molybdenum selenide sheets on nickel foam as a binder-free electrode for supercapacitor application. Electrochimica Acta. 2018;**265**:514-522. DOI: 10.1016/j.electacta.2018.01.075

[57] Huang B, Wang W, Pu T, Li J, Zhu J, Zhao C, et al. Two-dimensional porous (Co, Ni)-based monometallic hydroxides and bimetallic layered double hydroxides thin sheets with honeycomb-like nanostructure as positive electrode for high-performance hybrid supercapacitors. Journal of Colloid and Interface Science.

2018;**532**:630-640. DOI: 10.1016/j.
jcis.2018.08.019

[58] Han J, Hirata A, Du J, Ito Y, Fujita T,
Kohara S, et al. Intercalation
pseudocapacitance of amorphous
titanium dioxide@nanoporous
graphene for high-rate and large-
capacity energy storage. Nano Energy.
2018;**49**:354-362. DOI: 10.1016/j.
nanoen.2018.04.063

[59] Wang J, Polleux J,
Brezesinski T, Tolbert S, Dunn B. The
Pseudocapacitance Behaviors of TiO2
(Anatase) Nanoparticles. ECS
Transactions. 2008:101-111.
DOI: 10.1149/1.2953511

[60] Fong KD, Wang T, Kim HK,
Kumar RV, Smoukov SK. Semi-
Interpenetrating Polymer Networks for
Enhanced Supercapacitor Electrodes.
ACS Energy Letters. 2017;**2**:2014-2020.
DOI: 10.1021/acsenergylett.7b00466

[61] Sahoo S, Pazhamalai P,
Krishnamoorthy K, Kim SJ.
Hydrothermally prepared α-MnSe
nanoparticles as a new pseudocapacitive
electrode material for supercapacitor.
Electrochimica Acta. 2018;**268**:403-410.
DOI: 10.1016/j.electacta.2018.02.116

[62] Rakhi RB, Ahmed B, Hedhili MN,
Anjum DH, Alshareef HN. Effect of
Postetch Annealing Gas Composition on
the Structural and Electrochemical
Properties of Ti2CTx MXene Electrodes
for Supercapacitor Applications.
Chemistry of Materials. 2015;**27**:5314-
5323. DOI: 10.1021/acs.chemmater.
5b01623

[63] Barai HR, Rahman MM, Joo SW.
Template-free synthesis of two-
dimensional titania/titanate nanosheets
as electrodes for high-performance
supercapacitor applications. Journal of
Power Sources. 2017;**372**:227-234.
DOI: 10.1016/j.jpowsour.2017.10.076

Chapter 10

Silk Fibroin Nanoparticles: Synthesis and Applications as Drug Nanocarriers

Guzmán Carissimi, Mercedes G. Montalbán,
Marta G. Fuster and Gloria Víllora

Abstract

The use of nanoparticles in biomedical fields is a very promising scientific area and has aroused the interest of researchers in the search for new biodegradable, biocompatible and non-toxic materials. This chapter is based on the features of the biopolymer silk fibroin and its applications in nanomedicine. Silk fibroin, obtained from the *Bombyx mori* silkworm, is a natural polymeric biomaterial whose main features are its amphiphilic chemistry, biocompatibility, biodegradability, excellent mechanical properties in various material formats, and processing flexibility. All of these properties make silk fibroin a useful candidate to act as nanocarrier. In this chapter, the structure of silk fibroin, its biocompatibility and degradability are reviewed. In addition, an intensive review on the silk fibroin nanoparticle synthesis methods is also presented. Finally, the application of the silk fibroin nanoparticles for drug delivery acting as nanocarriers is detailed.

Keywords: silk fibroin, structure, biocompatibility, nanoparticle, synthesis, nanocarrier

1. Introduction

Silk is an ancestral material used since 2450 BC [1] for making fabrics. After having been an economic engine of several empires and coining the name to the trade route that linked Asia, Europe and Africa, silk suffered a debacle in the early 20th century when the much cheaper synthetic polymers derived from hydrocarbons were introduced. However, today, motivated by its biocompatibility and excellent mechanical properties, researchers around the world are trying to produce biomaterials based on this biopolymer for a variety of biomedical applications: films with a surface roughness that increase cell adhesion, 3D structures for bone implants, hydrogels for wound protection and nanoparticles for drug delivery, among others [2–5].

Silk fibroin probably receives a lot of attention from the general public due to its mechanical properties, so a brief text will be devoted to comparing them with other natural fibers and engineering materials. **Table 1** shows values of stress at break, elasticity and percentage of nominal deformation at the break of silk fibroin together with the values of other biomaterials and synthetic materials. Excluding

Material	Tension at break (MPa)	Elasticity (GPa)	% nominal strain at break	Ref.
Silk B. Mori (with sericin)	500	5–12	19	[6]
Silk fibroin B. mori	610–690	15–17	4–16	[6]
Spider silk N. clavipes	875–972	11–13	17–18	[7]
Collagen[a]	0.9-7.4	0.0018-0.046	24–68	[8]
Collagen[b]	42-72	0.4-0.8	12–16	[8]
PLA[c]	28-50	1.2-3.0	2–6	[9]
Tendon (mostly collagen)	150	1.5	12	[10]
Bone	160	20	3	[10]
Kevlar (49 fibers)	3600	130	2.7	[10]
Carbon fiber	4000	300	1.3	[10]
Synthetic rubber	50	0.001	850	[10]

[a]Type I collagen fibers extruded from rat tail tested after stretching from 0–50%.
[b]Cross-linked rat tail collagen tested after stretching from 0–50%.
[c]Polylactic acid with molecular weights ranging from 50,000 to 300,000 units.
Adapted from reference (5) with permission of Elsevier.

Table 1.
Comparison of the mechanical properties of different natural and synthetic fibers.

mineralized biomaterials (bones), Kevlar and carbon fibers, *Bombyx mori* silk fibroin together with *Nephila clavipes* spider silk are the biomaterials with the highest stress at break. While the list of biomaterials is incomplete, it is fair to say that fibroins are among the strongest polymeric biomaterials known. However, the tensile strength of fibroin is substantially lower than that of Kevlar and carbon fiber, engineering materials that are commonly used to transmit and support tensile forces. At first glance, we could infer that fibroin is superior to other biomaterials, such as collagen, but not as "good" as Kevlar and carbon fibers. However, this interpretation is based on the assumption that "good" means stiff and strong. Looking closely at **Table 1**, it can be seen that fibroin is quite extensible, presenting a maximum deformation of approximately 18%, while engineering materials fail in deformations of the order of 1–3%. The great extensibility of fibroin makes it more resistant than engineering materials.

It is especially notable that silkworms can produce strong and stiff fibers at room temperature and from an aqueous solution, while synthetic materials with comparable properties must be processed at elevated high temperatures and/or with less benign solvents. Furthermore, synthetic polymer fibers typically require post-spin stretching to ensure the necessary degree of molecular orientation in their structure [11]. On the other hand, this is not necessary for silk in the natural spinning process. This is due to its impressive amino acid sequence which gives rise to an extraordinary polymorphic secondary structure that will be discussed below.

2. Silk structure

Silk is a protein biopolymer synthesized by a wide variety of lepidoptera and arachnids in specialized glands. However, it is important to note that this chapter focuses on silk fibroin from the silkworm *Bombyx mori* of the *Bombycidae* family, fed only with mulberry leaves (*Morus alba L*). This is important because the amino acid sequence

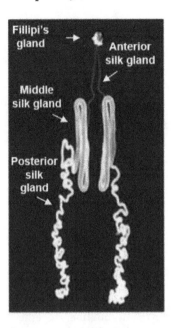

Figure 1.
Diagram of the silk fibroin gland. From reference [12] with permission of public library of science (PLOS).

varies from species to species and with it its mechanical and physicochemical properties, which can have great implications in different applications.

Silk proteins are synthesized in a gland that extends through the abdomen of the worm and is divided into three sections, posterior, middle and anterior, as illustrated in **Figure 1**. The posterior gland cells secrete silk fibroin, reaching a concentration of approximately 12% by weight. At this time, the protein is in a water-soluble state [13], with a partially ordered secondary structure composed of irregular structures and type II β-turns [14], commonly known as *silk I*. This protein is pushed into the gland media, where cells lining the lumen secrete sericin along with other flavonoids (assimilated by worms in the diet) [15]. Fibroin and sericin are concentrated to 25% by weight and are driven to the anterior gland where they experience pH gradients (maintained through the secretion of carbonic anhydrase) and ionic strength gradients. These factors contribute to the elongation of fibroin (at this point 30% by weight) into two thin filaments and promote the crystallization of repetitive domains. Lastly, during the spinning process, the non-Newtonian protein solution is subjected to crystallization induced by changes in pH and ionic strength through the gland [16] and by the shear stress generated by an opposing pressure front to the flow, generating a velocity gradient from the inlet (0.334 mm/s) to the outlet (13.8 mm/s) of the spinning organ [17]. Throughout the process, the silk fibroin initially secreted with a partially ordered structure (*silk I*) undergoes a transition to one composed mainly (58%) [18] by antiparallel β sheets, adopting an insoluble crystalline structure known as *silk II* [18, 19].

To understand the formidable properties of this biopolymer, its structure must be studied in detail. A silk cocoon is composed of a single silk fiber between 1000 and 1500 m in length with a diameter of between 10 and 25 μm [20]. This fiber is composed of a core of two fibroin filaments, each one of approximately 10 μm covered by a layer of sericin that hold the fibers together, as illustrated in **Figure 2** left, providing greater resistance to the assembly of the fiber. In turn, fibroin fibers are

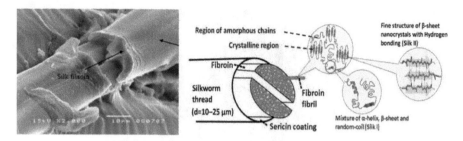

Figure 2.
Left, scanning electron microscope image (2,000X magnification) of silk fiber, containing two fibroin fibers coated by sericin. From reference [21] with permission of Elsevier. Right, schematic representation of the structure of silk fibroin. Insets show the general structure of the fibrils and the alignment of the antiparallel β-sheets. From reference [22] with permission of John Wiley and Sons.

composed of coiled nanofibrils of between 20 and 25 nm, which gives them greater tensile strength (**Figure 2**, right) [23].

Fibroin, representing approximately 75% of the weight of the cocoon, is a linear, water-insoluble protein with high tensile strength. On the other hand, sericin represents approximately 25% of the cocoon weight, it is a globular, water-soluble protein whose function is to keep the fibroin fibers together [24]. Silk fibroin is made up of three components, a heavy chain (391 kDa) and a light chain (26 kDa) linked by a disulfide bridge, and a glycoprotein, P25 (25–30 kDa) in a 6: 6: 1 molar ratio to yield a 6.3 MDa megastructure [25]. The primary structure of the silk fibroin heavy chain is schematically represented in **Figure 3A**. This chain is composed of 5,263 amino acids divided into N- and C-terminal domains, both hydrophilic, and

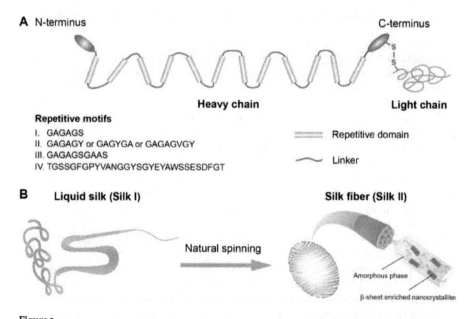

Figure 3.
Organization of the amino acid sequence within silk fibroin. A) In green and magenta, the N- and C-terminal domains are shown, respectively. The repetitive sequences of the GAGAGS type that give rise to the crystalline domains of the silk fibroin are represented by orange cylinders and in blue, the hydrophilic sequences that flank the crystalline regions. B) Diagram representing the transition from water-soluble silk (silk I) to crystalline silk fibers after the bio-spinning process. In silk fiber, amorphous regions (44%) and crystalline regions rich in antiparallel β sheets (56%) are represented. Reprinted with permission from reference [26]. Copyright 2020 American Chemical Society.

12 hydrophobic domains flanked by 11 short and hydrophilic domains. The hydrophobic domains contain highly conserved and repetitive sequences of the GAGAGS, GAGAGY, and GAGAGVGY types that form the β-sheet structures that, in addition, are packaged in crystalline areas [16]. 86% of the amino acids of the heavy chain of fibroin are Glycine (45%), Alanine (29%) and Serine (12%) [27], which are mostly found in the hydrophobic and highly repetitive regions. The great bias that the primary sequence of fibroin presents towards amino acids with small residues such as Glycine and Alanine, promotes the formation of antiparallel β-sheets, which are mostly packed in the crystalline areas.

The secondary structure of fibroin contains approximately a total of 58% of β-sheet [28], of which approximately 33% correspond to antiparallel β-sheets organized in crystalline structures. Fibroin fibers are generally described as a matrix of disordered structures with β-sheet crystals embedded in it [26], as represented in **Figure 3B**. The intra and intermolecular hydrogen bond network provide strength and tensile strength to the biopolymer, while amorphous regions provide flexibility and elasticity [29]. In the literature, there is great variability in the sizes reported for these crystals [30–33]. To illustrate the size of these, reference can be made to X-ray diffraction measurements and low voltage transmission electron microscopy performed by Drummy et al. [30]. They have determined that the crystals within the fibroin fibers have dimensions of 21 x 6 x 2 nm and their major axis is aligned parallel to the axis of the fibers.

The secondary structure of fibroin with *silk II* conformation is extremely stable thanks to a large number of hydrogen bonds which makes it insoluble in most solvents, including under moderate acidic and alkaline conditions. As the content of acidic and basic groups is low in fibroin, the electrostatic factor is not decisive in the formation of the secondary structure, however, it can be decisive in the dissolution of fibroin. The secondary structure of fibroin is not only relevant in the biomaterial synthesis process due to the need for its dissolution, but also due to its influence on the mechanical and physicochemical properties of the resulting biomaterials. For example, Wang et al. [34] prepared silk and polyvinylpyrrolidone micro- and silk fibroin nanoparticles for controlled drug release and concluded that release profiles can be adjusted by modulating the number of β-sheets in the secondary structure of fibroin.

3. Silk biocompatibility

Silk fibroin is an attractive material for numerous biomedical applications as, due to its mechanical and physicochemical properties, it encompasses applications such as drug delivery, tissue engineering, and implantable devices. However, in addition to the functionalities necessary for specific applications, a key factor necessary for the clinical success of any biomaterial is the appropriate in vivo interactions with the body or biocompatibility. Among them, (i) the immune and inflammatory response and (ii) the biodegradability can be studied.

3.1 Immune and inflammatory response

As already mentioned, silk fiber is essentially made up of two proteins, fibroin and sericin. While fibroin is highly biocompatible [3, 4] with a low immune response [35, 36], sericin can present unwanted adverse allergic reactions [37, 38]. For this reason, sericin is normally removed by different procedures, known as degumming [39]. Depending on the format of the material and the location of implantation, silk fibroin can induce a mild inflammatory response that diminishes within a few hours/days after implantation [40]. The response involves the recruitment and activation of macrophages and may include the activation of a mild foreign body response with the formation of multinucleated giant cells, again depending on the format of the material and the location of implantation [36]. The number of immune cells decreases with time, and granular tissue, if formed, is replaced by endogenous non-fibrous tissue, although these responses are reserved for films, hydrogels, and bone implants [36].

The study carried by Meinel et al. [40] indicated that collagen films implanted in rats produce a greater inflammatory reaction in the tissue than equivalent films prepared with fibroin after 6 weeks. In another study comparing fibroin membranes and poly (styrene) and poly (2-hydroxyethyl methacrylate) membranes, Santin et al. [41] demonstrated that the former has a milder immune response than the latter. The results indicated that lower levels of fibrinogen were bound to the fibroin membrane than to the two synthetic polymers, while the same amounts of C_3 human plasma complement fragment and adsorbed IgG were detected. The activation of mononuclear cells by fibroin, measured as production of interleukin 1β, was lower than that of synthetic materials. Another study indicated that the braided silk fibroin used for the reconstruction of the anterior cruciate ligament produces a mild inflammatory response after seven days of implantation in vivo, while an equivalent implant made with the biodegradable polymer polyglycolic acid (PGA) produced a more acute response [42]. In this case, although the breaking load for the PGA implant was twice that for the fibroin graft, the initial attachment and growth of cells in the prosthetic ligament was higher in the latter.

In the case of silk fibroin nanoparticles, the literature is not as extensive as for other formats of the same material. Tan et al. [43] showed that nanoparticles coated with fibroin hardly produced an immune response and the adaptive immune system was not activated. In another study, Totten et al. [44] used in vitro and nuclear magnetic resonance-based metabolomics assays to examine the inflammatory phenotype and metabolic profiles of macrophages after exposure to PEGylated and unmodified silk fibroin nanoparticles. The macrophages internalized both types of nanoparticles but showed different phenotypic and metabolic responses to each type of nanoparticle. Unmodified silk fibroin nanoparticles induced upregulation of several processes, including the production of pro-inflammatory mediators (such as cytokines), the release of nitric oxide, and the promotion of antioxidant activity. These responses were accompanied by changes in macrophage metabolomic profiles that were consistent with a pro-inflammatory state and indicated an increase in glycolysis and reprogramming of the tricarboxylic acid cycle and the creatine kinase/phosphocreatine pathway. In contrast, PEGylated silk fibroin nanoparticles induced milder changes in inflammatory and metabolic profiles, suggesting that immunomodulation of macrophages with silk fibroin nanoparticles is dependent on PEGylation. This would indicate that the PEGylation of silk fibroin nanoparticles reduces the inflammatory and metabolic responses initiated by macrophages. In the case of silk fibroin microparticles (10–200 μm) prepared by enzymatic digestion, Panilaitis et al. [38] found that suspension of the particles induced a significant release of TNF cytokines. In contrast, macrophages grown in the presence of silk fibroin fibers did not upregulate transcription levels for a wide range of pro-inflammatory cytokines to a significant degree. The combination of results from these two studies could indicate that the immune response is dependent on the size of the biomaterial, excluding materials at the nanoscale and macroscale, but not at the microscale. In a recent study carried out in our research group [45], the HeLa and EA.hy926 cell lines were incubated with up to 250 μg/mL of silk fibroin nanoparticles in vitro. Viability was studied by MTT tests, and the results did not show significant variations (p < 0.05) with respect to the controls.

Recent studies indicated that nanoparticles loaded with resveratrol have shown immunomodulatory properties and anti-inflammatory effects in murine models with inflammatory bowel disease [46] and periodontal infections [47]. In another similar study [48], treatments with RGD linear peptide-functionalized silk fibroin nanoparticles were performed and were found to improve colonic damage in rats, reduce neutrophil infiltration, and improve the compromised oxidative state of the colon. It was also found that only rats treated with RGD-silk fibroin nanoparticles showed a significant reduction in the expression of different pro-inflammatory cytokines (interleukin-1β, IL-6 and IL-12) and inducible nitric oxide synthase compared to the control group. Furthermore, the expression of both cytokine-induced neutrophil chemoattractant-1 and monocyte chemotactic protein-1 was significantly decreased with RGD-silk fibroin nanoparticle treatment.

3.2 In vivo degradation of silk fibroin

Fibroin fibers implanted in the human body retain more than 50% of their mechanical properties after 60 days, which is why the North American Pharmacopeia classifies this material as non-biodegradable [49]. However, the rate at which fibroin degrades depends on the size of the implanted material, its morphology, mechanical and biological conditions at the implantation site, the secondary structure of the protein, and the molecular weight distribution of the fibroin chains. In particular, for the application of silk fibroin nanoparticles to drug transport, three of these parameters must be taken into account mainly: (i) size, (i) molecular weight distribution and (iii) secondary structure. But before analyzing each of them, the possible degradation pathways of fibroin will be discussed to later mention how these parameters are able to influence degradation.

As a protein, fibroin exhibits degradation against proteases capable of degrading amide bonds including α-chymotrypsin, collagenase IA, protease XIV, and metalloproteases [50–52]. The residues of the degradation process are the corresponding amino acids of the proteins, so they are easily absorbed in vivo and do not generate toxicity. The partial hydrolysis of the protein by enzymes into small fragments is not a problem either, since these can be easily phagocytosed by macrophages [38]. Li et al. [51] observed that the mean molecular weight of fibroin film products after degradation with the three enzymes followed the order of protease XIV > collagenase IA > α-chymotrypsin. The degradation mechanism is based on a two-stage process, based on enzymes finding binding domains on the surface of materials and their subsequent hydrolysis [53]. In this manner, different enzymes have different results for the degradation of different structures within fibroin. For example, chymotrypsin has been used to degrade the amorphous regions of fibroin to obtain highly crystallized fibroin [51]. Collagenase preferentially degrades the content of β sheets in hydrogels [38]. On the other hand, after incubation of fibroin with protease XIV, it was found that the mass was significantly reduced [51]. Brown et al. [38] concluded that the ability of enzymes to break down a biomaterial not only depends on the cleavage sites being present in the primary structure of the protein but also the secondary structure and the format of the material play a fundamental role. This indicates that the degradability of fibroin can be modulated by controlling the relative abundance of its secondary structures. In this way, for example, by reducing the content of highly crystalline structures in stacked β-sheets, degradation can be accelerated, since both protease XIV and chymotrypsin can act simultaneously in these areas.

Horan et al. [49] concluded that the degradation of electrospun fibers exhibited a predictable degradation dependent on the diameter of the fibers. As expected, as the diameter decreases and, therefore, the surface/volume ratio increases, the degradation occurs at a higher rate. Decreasing the size from macroscopic fibers to nanoparticles will clearly increase this ratio, allowing greater access to enzymatic degradation and phagocytosis by macrophages. The degradation of PEG-functionalized and non-functionalized silk fibroin nanoparticles by proteolytic enzymes (protease XIV and α-chymotrypsin) and papain, as well as cysteine protease, were studied by Wongpinyochit et al. [54]. Both classes of particles presented similar degradation patterns in a period of 20 days, establishing the order of degradation of the particles by means of enzymes such as: Protease XIV > papain > > α-chymotrypsin. The authors reported that, after 1 day, silk fibroin nanoparticles and PEG-silk fibroin nanoparticles incubated with protease XIV lost 60 and 40% of their mass, respectively, a reduction in the amorphous content of the secondary structure and an increase in diameter. In contrast, 10 days of incubation were required for similar degradations with papain and 20 with α-chymotrypsin. Silk fibroin nanoparticles were also exposed to a complex mixture of rat liver lysosomal enzymes ex vivo, finding that they lost 45% of their mass in 5 days.

Lastly, it should be noted that modifications in the molecular weight distribution of fibroin chains can alter the rate of degradation [55]. A decrease in this can alter the order of crosslinking between polymers and potentially result in faster degradation [56]. For this reason, the purification and subsequent processing of fibroin not only affects the mechanical and physicochemical properties of the resulting biomaterials [57] but can also be used to modulate their biodegradability.

4. Synthesis of silk fibroin nanoparticles

Silk fibroin particles can be produced from the top-down as well as the bottom-up approach. The first of these involves grinding the fibroin fibers to reduce their size. This can be achieved through ball milling [58, 59], bead milling [60], air-jet [61] as well as irradiating the material with an electron beam [62]. However, these tend to produce particles in the micrometer range and with a wide size distribution, so in the rest of this section, we will focus on the bottom-up approach which offers more control over the particles produced. This approach is based on the self-assembly of the smallest units that constitute a nanoparticle. In the particular case of silk fibroin nanoparticles, the fibroin fibers are firstly dissolved to obtain their individual constituent units and subsequently regenerated into nanoparticle format. This is normally achieved through a desolvation process, which can be achieved in different ways that will be discussed in this section along with some examples; but first, the dissolution of silk fibroin will be discussed, which is not an easy task due to its high structural stability and deserves a detailed analysis that will be exposed below.

4.1 Solubilization of silk fibroin

The preparation of most biomaterials depends on achieving a complete dissolution of fibroin. This process is referred to by some authors as reverse engineering [63], since it attempts to obtain water-soluble fibroin with *silk I* structure, starting from fibers with *silk II* structure. However, only a limited number of solvents are able to dissolve silk fibroin. Examples of these are strong acidic solutions (phosphoric, formic, sulfuric, hydrochloric) or aqueous/organic solutions concentrated in salts (LiCNS, LiBr, CaCl$_2$, Ca (CNS)$_2$, ZnCl$_2$, NH$_4$CNS, CuSO$_4$, NH$_4$OH, Ca(NO$_3$)$_2$) those that are capable of completing the dissolution of fibroin fibers [64–70]. This is due to the presence of the great network of intra- and intermolecular hydrogen bonds and the high crystallinity derived from its secondary structure.

In general, the literature describes the dissolution of fibroin in concentrated salt solutions as a two-step process [64, 67]. In the first step, the compact, crystalline structure of fibroin fibers swells due to the diffusion of solvent molecules. In the second, the dispersion of the silk fibroin molecules begins due to the collapse of the intermolecular interaction and their consequent dissolution [67]. During dissolution, amorphous sections with a higher content of massive amino acid residues or polar groups are firstly dissolved. The cations (Ca^{2+}, Zn^{2+}, Cu^{2+}, NH_4^+, Li^+) form stable chelate complexes with hydroxyl groups of the serine and tyrosine side chains and also with the oxygen of the carbonyls, breaking the hydrogen bonds and the van der Waals forces between polypeptide chains resulting in the dissolution of the protein that adopts *silk I* structure [68, 70, 71].

The described dissolution process depends on reaching a high concentration of salts. This entails certain disadvantages since extensive dialysis (72 h) is required to eliminate these, which requires 6 liters of distilled water for every 12 mL of solution [64]. Furthermore, the solutions obtained are unstable and tend to gel within a period of days. Alternatively, when long-term storage of fibroin is desired, aqueous fibroin solutions can be lyophilized and subsequently dissolved in organic solvents such as hexafluoroisopropanol [64] at the time they are to be used. However, these solvents are toxic and extremely corrosive so they require special care when handling [72].

Recently, ionic liquids have emerged as an alternative for the dissolution of silk fibroin [73] providing numerous advantages over traditional methods [64, 65]. Firstly, the negligible vapor pressure and easy recyclability of ionic liquids make them a more "green" alternative to organic solvents [74–76]. Secondly, the possibility of obtaining high concentrations of fibroin in a stable solution (up to 25% w/w in some ionic liquids [77]). Fibroin solutions in ionic liquids are more stable because the hydrophobic regions (highly conserved GAGAGS or GAGAGAGS sequences) are stabilized by the alkyl chains of cations such as imidazolium [78]. On the other hand, the bulky charged imidazolium ring is oriented outwards providing electrostatic repulsive forces between the hydrophobic blocks and preventing the transition from *silk I* to *silk II* through the formation of the β sheets. According to Wang et al. [78], solutions of silk fibroin in 1-allyl-3-methylimidazolium chloride can be stable for periods longer than one and a half years. Thirdly, the ease with which silk fibroin can be dissolved. According to the method described by Lozano-Pérez et al. [77], by means of high-power ultrasound, the complete dissolution of fibroin can be achieved in a few minutes compared to several hours with traditional methods [64, 65].

4.2 Regeneration of fibroin into nanoparticles

Nanoparticle synthesis processes in a bottom-up approach involve the dissolution of fibroin in its constituent polymer units and its subsequent regeneration into nanoparticles. This regeneration is normally carried out by means of a desolvation process in an "antisolvent", a process commonly referred to as antisolvation. As shown in **Figure 4**, in the nanoparticle synthesis process by antisolvation there are three key components, the polymer, the solvent and the antisolvent. The necessary conditions are that: (i) the solvent and the antisolvent are miscible under the process conditions, while (ii) the solute must be insoluble in the solvent/antisolvent mixture. In this way, when mixing the polymer solution, the antisolvent will seize the molecules that solvate it, leading to their aggregation. Employing kinetic and thermodynamic controls, a limited number of polymer units can be made to aggregate, thus forming nanoparticles. In practice, the preparation of nanoparticles by antisolvation can be achieved by different techniques that vary in the methodology

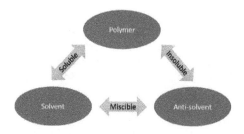

Figure 4.
Key components in the antisolvation process.

of mixing the solution of the polymer and the antisolvent or the nature of the latter. We will describe some of the most representative methods in the literature below.

4.2.1 Antisolvation in organic solvents

Probably one of the most widely used methods due to its simplicity and good results is the addition of the silk fibroin solution to water-miscible polar organic solvents, which act as an antisolvent when initiating the transition from *silk I* to *silk II* through the β-sheet formation [44, 77, 79–83]. It should be noted that the inverse variant, where the antisolvent is added to the fibroin solution, is also frequently found in the literature [84–86].

As an example, Wongpinyochit et al. [83] dissolved the fibroin fibers in a 9.3 M LiBr solution, keeping them stirred for 4 hours at 60°C. Subsequently, the solution is dialyzed for 72 hours and centrifuged to remove insoluble residues. Then, the fibroin concentration is adjusted to 5% w/v and added dropwise (10 μl/drop at a speed of 50 drops/min) to acetone under strong agitation, the volume of acetone being greater than 75% of the final volume of both liquids. A white suspension is immediately formed upon contact of both liquids, marking the formation of the nanoparticles. The particles are washed and collected by centrifugation. The overall process is illustrated in **Figure 5**. An average diameter of ca. 100 nm and a Z-potential of -50 mV (in distilled water) are provided for the particles obtained.

One way to optimize the nanoparticle synthesis process is to reduce the mixing time between the fibroin solution and the antisolvent [87]. This can be achieved by reducing the size of the fibroin solution droplets (increased surface/volume ratio) that come into contact with the antisolvent, favoring mass transfer [87]. In our research group, a method has been developed that uses a coaxial injector where the fibroin solution flows through the center and nitrogen under pressure runs through the concentric cylinder, this manages to produce an aerosol with very small droplets [45, 88].

4.2.2 Antisolvation in supercritical fluids

A supercritical fluid is any substance that is under conditions of pressure and temperature above its critical point. Under these conditions, the substance has hybrid properties between a liquid and a gas, that is, it can diffuse like a gas, and dissolve substances like a liquid [89]. Assuming that the compound to be used as a supercritical fluid meets the conditions described at the beginning of this section, it can act as an antisolvent in an antisolvation process. In this case, the process is known as supercritical antisolvation (SAS). In particular, carbon dioxide (CO_2) is one, if not the most used substance as a supercritical fluid due to its moderate critical conditions (T = 304 K and P = 7.38 MPa), harmless to the operator and the environment, as well as its economic obtaining and operation [89].

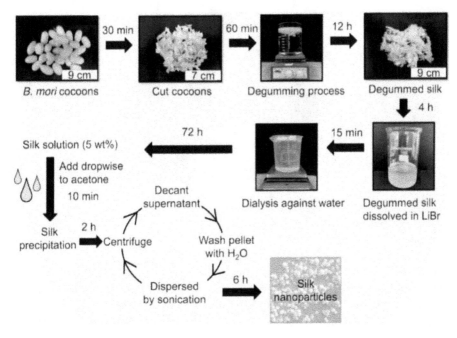

Figure 5.
Scheme illustrating the key steps to generate a solution and obtain silk fibroin nanoparticles described by Wongpinyochit et al. [83]. Reprinted with permission from reference (83). Copyright 2016 MyJoVE corporation.

The SAS process is well known and has been used for the preparation of silk fibroin nanoparticles [90–92]. SAS has some variants by other acronyms: Aerosol Solvent Extraction System (ASES), Solution Enhanced Dispersion by Supercritical Fluids (SEDS), Supercritical AntiSolvent with Enhanced Mass transfer (SAS-EM). The main difference between these processes is in the device for injecting the solution and CO_2. In the case of the SAS and ASES processes, the liquid solution is injected into the precipitation reactor through a micrometric nozzle, in the case of the SEDS process, the nozzle is coaxial; whereas, the SAS-EM process uses a baffle surface that vibrates at ultrasonic frequencies to improve the atomization of the solution [93].

This method can be exemplified by the process outlined by Xie et al. [90] for the preparation of curcumin-loaded silk fibroin nanoparticles by SEDS. Briefly, lyophilized silk fibroin in the *silk I* state is dissolved in hexafluoroisopropanol together with curcumin. The solution is subsequently injected into a precipitation reactor containing CO_2 at 20 MPa which will act as an antisolvent for fibroin. After the complete injection of the solution, a constant flow of CO_2 is maintained to remove the hexafluoroisopropanol from the precipitation reactor. Finally, the reactor is depressurized and opened for the collection of the nanoparticles. The process produces nanoparticles with a mean diameter of less than 100 nm.

4.2.3 Electrospray

Electrospray is a technique in which an electrical potential difference is applied between the nozzle of an injector and a manifold, which can contain a liquid that acts as an antisolvent [94]. In this technique, the surface of the liquid emerging from a capillary subjected to electrical stress is deformed into an elongated injector that initially produces a series of micrometer-sized drops. Because the drops are charged, the repulsive forces break each drop into a group of smaller drops in a process called coulombic explosion [94]. Using this technique to spray a silk fibroin solution onto

Figure 6.
Microfluidic chip made of glass. Channels are 50 μm deep and 150 μm wide (image by IX-factory STK, 2014, CC BY-SA 3.0; no changes were made to the original image).

aluminum foil, Gholami et al. [95] succeeded in synthesizing silk fibroin nanoparticles of up to 80 nm on average. The authors showed that lower concentrations, lower feed rates, and longer distances between the needle and the collector led to a decrease in mean particle size. Increasing the voltage to 20 kV decreased the size of the particles but voltages higher than this produced an increase in the particle size.

4.2.4 Microfluidics

Microfluidic equipment can also be used for the preparation of nanoparticles. Microfluidic kits are devices, generally the size of millimeters/centimeters, that contain microcapillaries specially designed for mixing fluids [96, 97]. A representative image of these is shown in **Figure 6**. These types of equipment allow precise manipulation of liquids (like the dissolution of silk and the antisolvent) by means of the control of the parameters of the process such as the total flow, the relations of speed between different lines of injection, etc. The greatest advantages provided by this equipment are the possibility of producing particles in continuous flow and with a narrow size distribution.

Wongpinyochit et al. [63] have used a commercial microfluidic kit to mix a 3% wt solution of silk fibroin with acetone or isopropanol as an antisolvent. Through the use of different mixing conditions, the authors were able to control the final size of the nanoparticles obtained, which varied between 110 and 310 nm, with polydispersity and Z-potential indices between 0.1/0.25 and − 20/−30 mV, respectively.

4.2.5 Salting-out

Phase separation has also been used for the preparation of nanoparticles using the salting-out method. For example, by adding potassium phosphate to a solution of silk fibroin, Lammel et al. [98] prepared fibroin particles with sizes varying between 500 and 2000 nm depending on the initial concentration of fibroin in solution. The authors revealed that, for the formation of nanoparticles, a salt concentration greater than 750 mM is required, otherwise, the solution gels.

5. Mechanism of nanoparticle formation in antisolvation processes

After having explained some of the synthesis methods, it is appropriate to mention the mechanism by which antisolvation can generate nano-sized particles. Upon mixing the fibroin solution and the antisolvent, supersaturation occurs, leading to phase separation (precipitation). The mechanism can be divided into two steps, with a nucleation stage driven by supersaturation occurring first and then a growth stage. Growth can occur by two coagulation mechanisms: i) the nuclei of particles converge to form a larger particle or condensation and ii) the polymer units add to growing nuclei. This is exemplified in **Figure 7**. Condensation decreases supersaturation by reducing the mass of solute in the mixture and therefore competes with nucleation. Coagulation can reduce the rate of condensation by reducing the total number of particles and therefore the surface area [99]. Supersaturation influences nucleation and growth rates to different degrees. The nucleation rate depends more strongly on supersaturation than the condensation rate. High nucleation rates offer the potential to produce a large number of submicron particles in the final suspension if growth can be controlled. This process can be compared with the formation of crystals in the order of millimeters for X-ray crystallography when it is sought to obtain a low number of nuclei and greater growth in slow nucleation processes.

The key to generate rapid nucleation is to achieve rapid supersaturation. This process will be directly influenced by mixing and phase separation, which can be represented by the Damkohler number (Da) defined as the relationship between the mixing time (τ_{mix}) and the total precipitation time (τ):

$$Da = \frac{\tau_{mix}}{\tau}. \tag{1}$$

Under poor mixing conditions, τ_{mix} is large (as is Da) and the nucleation rate is slow relative to the growth rate, resulting in large particles and wide size distributions. As the τ_{mix} is reduced with respect to the τ , greater supersaturation and faster nucleation are achieved, resulting in smaller particles with narrow size distribution [87].

The τ_{mix} can be reduced by reducing the size of the droplets (increasing the surface/volume ratio) of fibroin solution that meet the antisolvent, favoring mass transfer. Reduction in droplet size is typically achieved by increasing Reynolds number which produces turbulent flow and thus results in dissolution jet

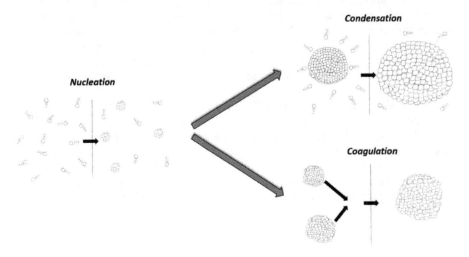

Figure 7.
Scheme representing the mechanism of precipitation by nucleation and growth of particles by coagulation and condensation.

fragmentation. In fact, this is the goal of previously proposed methods such as the coaxial injector [45, 88], SEDS [90], electrospray [95] and microfluidic equipment [63].

According to Eq. (1), another strategy to reduce Da is to increase the τ. This can be achieved by adding stabilizers that interact with the polymer units generating steric hindrances that retard growth by condensation and coagulation [87]. According to Matteucci et al. [87], adding the stabilizers to the antisolvent is more effective in preventing the growth of the particles than adding them to the polymer solution. This is because, when placed in the antisolvent, the stabilizing agents are more available to interact with the polymeric units, as they do not need to diffuse across the interface from the overall aqueous phase to the organic phase.

Temperature is another important thermodynamic control factor for the preparation of nanoparticles by antisolvation that affects the formation of particles in different ways. Firstly, an increase in temperature leads to an increase in the solubility of the polymer and therefore reduces the degree of supersaturation when mixing the solutions, favoring slow nucleation. Secondly, elevated temperatures increase diffusion and growth kinetics at the interface of the particle boundary layer. And thirdly, higher solubility also increases Oswald's maturation rate [100]. For these reasons, a reduced temperature in the precipitation stage is preferable for the formation of small nanoparticles.

6. Nanoparticle drug loading

The loading of drugs can be carried out mainly by two approaches, (i) during the nanoparticle formation process or (ii) a posteriori, by adsorption of the drug on the surface of the nanoparticle. The first approach can be achieved by adding the drug to the polymer solution (nanoparticle matrix) [88] or the antisolvent [101] before mixing both. This approach is often referred to as coprecipitation because the polymer and drug precipitate together. In contrast to this method, adsorption of the drug to the surface of the nanoparticle can be achieved by incubating the nanoparticles in a solution of the drug. These methods have advantages and disadvantages. On the one hand, the first method is usually simpler, since the loading and preparation of the nanoparticle are carried out in a single step. However, this process could affect the formation of the nanoparticle and therefore the second method could be preferential. On the other hand, drug release profiles must be considered. As demonstrated by Montalbán et al. [88], particles charged by coprecipitation have slower release profiles compared to particles charged by absorption. This is to be expected since, in the latter, the drug is on the surface of the nanoparticle and readily available to the medium.

Drug loading and encapsulation efficiency depend on drug-polymer interactions and the presence of functional groups (i.e, hydroxyl, carboxyl, etc.) in both. Montalbán et al. [102] used computational methods such as blind docking and molecular dynamics simulations to study the interactions of different drugs with silk fibroin nanoparticles. The authors found a strong correlation between drug-fibroin interactions and their loading content. Similarly, drugs with weaker interactions had a higher release rate and a higher percentage of loaded drug release.

7. Conclusions

Silk fibroin of the Bombyx mori silkworm is a natural, protein polymer that presents an interesting combination of mechanical properties, such as flexibility and resistance which are still difficult to achieve with synthetic polymers. Furthermore, fibroin is biodegradable and biocompatible, which makes it an excellent material for the production of nanoparticles for drug delivery.

Although the remarkable mechanical resistance of fibroin is one of the attractive properties of the biomaterial, the nanoparticle synthesis process is hampered by its high stability, due to the high number of hydrogen bonds in its secondary structure, mostly in the form of antiparallel β-sheets. A recent method, developed by our research group, is based on the use of ionic liquids to dissolve native fibroin and has allowed the production of nanoparticles in an easy and scalable process for industry.

The silk fibroin nanoparticle synthesis process comprises several stages. Firstly, the fibroin is purified by removing the sericin (a method known as degumming). Secondly, fibroin must be dissolved in its monomeric units and, subsequently, regenerated into nanoparticles. Each of these steps can have significant effects on the secondary structure of the protein and given the implications that it has on the resistance, degradability and biocompatibility of the final product, their study is essential.

Acknowledgements

This publication is part of the grant ref. CTQ2017-87708-R funded by MCIN/AEI/ 10.13039/501100011033 and by "ERDF A way of making Europe" and grant ref. PID2020-113081RB-I00 funded by MCIN/AEI/ 10.13039/501100011033. In addition, the publication is also part of the grant ref. 20977/PI/18 funded by the research support program of the Seneca Foundation of Science and Technology of Murcia, Spain. Marta G. Fuster acknowledges support from FPI grant (Ref. PRE2018-086441) funded by MCIN/AEI/ 10.13039/501100011033 and by "ESF Investing in your future". Mercedes G. Montalbán acknowledges support from the University of Murcia and Santander Bank through the research project (Ref. RG2020-002UM) associated with her postdoctoral contract.

Author details

Guzmán Carissimi, Mercedes G. Montalbán*, Marta G. Fuster and Gloria Víllora
Faculty of Chemistry, Chemical Engineering Department, Regional Campus
of International Excellence "Campus Mare Nostrum", University of Murcia,
Murcia, Spain

*Address all correspondence to: mercedes.garcia@um.es

IntechOpen

References

[1] Good IL, Kenoyer JM, Meadow RH. New evidence for early silk in the Indus civilization. Archaeometry. 2009;51(3): 457-466.

[2] Vepari C, Kaplan DL, Charu Vepari; David L. Kaplan. Silk as a Biomaterial. Prog Polym Sci. 2009;32(8-9):991-1007.

[3] Janani G, Kumar M, Chouhan D, Moses JC, Gangrade A, Bhattacharjee S, et al. Insight into Silk-Based Biomaterials: From Physicochemical Attributes to Recent Biomedical Applications. ACS Appl Bio Mater. 2019;2(12):5460-5491.

[4] Aramwit P. Biomaterials for Wound-Healing Applications Complimentary Contributor Copy. 2015;(January): 49-104.

[5] Altman GH, Diaz F, Jakuba C, Calabro T, Horan RL, Chen J, et al. Silk-based biomaterials. Biomaterials. 2003;24(3):401-416.

[6] Pérez-Rigueiro J, Viney C, Llorca J, Elices M. Mechanical properties of single-brin silkworm silk. J Appl Polym Sci. 2000;75(10):1270-1277.

[7] Cunniff PM, Fossey SA, Auerbach MA, Song JW, Kaplan DL, Adams WW, et al. Mechanical and thermal properties of dragline silk from the spider Nephila clavipes. Polym Adv Technol. 1994 Aug;5(8):401-410.

[8] Pins GD, Christiansen DL, Patel R, Silver FH. Self-assembly of collagen fibers. Influence of fibrillar alignment and decorin on mechanical properties. Biophys J. 1997;73(4):2164-2172.

[9] Engelberg I, Kohn J. Physico-mechanical properties of degradable polymers used in medical applications: A comparative study. Biomaterials. 1991 Apr;12(3):292-304.

[10] Gosline JM, Guerette PA, Ortlepp CS, Savage KN. The mechanical design of spider silks: From fibroin sequence to mechanical function. J Exp Biol. 1999;202(23):3295-3303.

[11] Tesoro G. Textbook of polymer science, 3rd ed., Fred W. Billmeyer, Jr., Wiley-Interscience, New York, 1984, 578 pp. No price given. J Polym Sci Polym Lett Ed. 1984 Dec;22(12):674-674.

[12] Wang W, Huang M-H, Dong X-L, Chai C-L, Pan C-X, Tang H, et al. Combined Effect of Cameo2 and CBP on the Cellular Uptake of Lutein in the Silkworm, Bombyx mori. PLoS One. 2014;9(1):e86594.

[13] Sehnal F, Sutherland T. Silks produced by insect labial glands. Prion. 2008;2(4):145-153.

[14] Asakura T, Ashida J, Yamane T, Kameda T, Nakazawa Y, Ohgo K, et al. A repeated β-turn structure in poly(Ala-Gly) as a model for silk I of Bombyx mori silk fibroin studied with two-dimensional spin-diffusion NMR under off magic angle spinning and rotational echo double resonance. J Mol Biol. 2001;306(2):291-305.

[15] Kumar JP, Mandal BB. Antioxidant potential of mulberry and non-mulberry silk sericin and its implications in biomedicine. Free Radic Biol Med. 2017;108(February):803-818.

[16] He YX, Zhang NN, Li WF, Jia N, Chen BY, Zhou K, et al. N-terminal domain of Bombyx mori fibroin mediates the assembly of silk in response to pH decrease. J Mol Biol. 2012;418(3-4):197-207.

[17] Jin H-J, Kaplan DL. Mechanism of silk processing in insects and spiders. Nature. 2003 Aug;424(6952):1057-1061.

[18] Asakura T, Yao J. 13C CP/MAS NMR study on structural heterogeneity in

Bombyx mori silk fiber and their generation by stretching. Protein Sci. 2002;11(11):2706-2713.

[19] Vollrath F, Knight DP. Liquid crystalline spinning of spider silk. Nature. 2001;410(6828):541-548.

[20] Hao LC, Sapuan SM, Hassan MR, Sheltami RM. Natural fiber reinforced vinyl polymer composites. Natural Fibre Reinforced Vinyl Ester and Vinyl Polymer Composites. Elsevier Ltd; 2018. 27-70 p.

[21] Aramwit P. Introduction to biomaterials for wound healing. In: Wound Healing Biomaterials. 2016, Woodhead Publishing; p. 3-38.

[22] Volkov V, Ferreira A V., Cavaco-Paulo A. On the Routines of Wild-Type Silk Fibroin Processing Toward Silk-Inspired Materials: A Review. Macromol Mater Eng. 2015;300(12): 1199-1216.

[23] Greving I, Cai M, Vollrath F, Schniepp HC. Shear-induced self-assembly of native silk proteins into fibrils studied by atomic force microscopy. Biomacromolecules. 2012;13(3):676-682.

[24] Padamwar MN, Pawar AP. Silk sericin and its application : A review. J Sci Ind Res (India). 2004;63(10): 323-329.

[25] Inoue S, Tanaka K, Arisaka F, Kimura S, Ohtomo K, Mizuno S. Silk fibroin of *bombyx mori* is secreted, assembling a high molecular mass elementary unit consisting of H-chain, L-chain, and P25, with a 6:6:1 molar ratio. J Biol Chem. 2000;275(51): 40517-40528.

[26] Guo C, Li C, Kaplan DL. Enzymatic Degradation of Bombyx mori Silk Materials: A Review. Biomacromolecules. 2020;21(5): 1678-1686.

[27] Lotz B, Colonna Cesari F. The chemical structure and the crystalline structures of *bombyx mori* silk fibroin. Biochimie. 1979;61(2):205-214.

[28] Carissimi G, Baronio CM, Montalbán MG, Víllora G, Barth A. On the Secondary Structure of Silk Fibroin Nanoparticles Obtained Using Ionic Liquids: An Infrared Spectroscopy Study. Polymers (Basel). 2020 Jun;12(6): 1294.

[29] Holland C, Numata K, Rnjak-Kovacina J, Seib FP. The Biomedical Use of Silk: Past, Present, Future. Adv Healthc Mater. 2019;8(1).

[30] Drummy LF, Farmer BL, Naik RR. Correlation of the β-sheet crystal size in silk fibers with the protein amino acid sequence. Soft Matter. 2007;3(7): 877-882.

[31] Martel A, Burghammer M, Davies RJ, Riekel C. Thermal Behavior of Bombyx mori Silk: Evolution of Crystalline Parameters, Molecular Structure, and Mechanical Properties. Biomacromolecules. 2007 Nov;8(11): 3548-3556.

[32] Pérez-Rigueiro J, Elices M, Plaza GR, Guinea G V. Similarities and Differences in the Supramolecular Organization of Silkworm and Spider Silk. Macromolecules. 2007 Jul;40(15): 5360-5365.

[33] Liang K, Gong Y, Fu J, Yan S, Tan Y, Du R, et al. Microstructural change of degummed Bombyx mori silk: An in situ stretching wide-angle X-ray-scattering study. Int J Biol Macromol. 2013 Jun;57:99-104.

[34] Wang X, Yucel T, Lu Q, Hu X, Kaplan DL. Silk nanospheres and microspheres from silk/pva blend films for drug delivery. Biomaterials. 2010; 31(6):1025-1035.

[35] Catto V, Farè S, Cattaneo I, Figliuzzi M, Alessandrino A, Freddi G,

et al. Small diameter electrospun silk fibroin vascular grafts: Mechanical properties, in vitro biodegradability, and in vivo biocompatibility. Mater Sci Eng C. 2015;54:101-111.

[36] Thurber AE, Omenetto FG, Kaplan DL. In vivo bioresponses to silk proteins. Biomaterials. 2015;71:145-157.

[37] Soong HK, Kenyon KR. Adverse Reactions to Virgin Silk Sutures in Cataract Surgery. Ophthalmology. 1984 May;91(5):479-483.

[38] Panilaitis B, Altman GH, Chen J, Jin HJ, Karageorgiou V, Kaplan DL. Macrophage responses to silk. Biomaterials. 2003;24(18):3079-3085.

[39] Carissimi G, Lozano-Pérez AA, Montalbán MG, Aznar-Cervantes SD, Cenis JL, Víllora G. Revealing the Influence of the Degumming Process in the Properties of Silk Fibroin Nanoparticles. Polymers (Basel). 2019; 11(12).

[40] Meinel L, Hofmann S, Karageorgiou V, Kirker-Head C, McCool J, Gronowicz G, et al. The inflammatory responses to silk films in vitro and in vivo. Biomaterials. 2005 Jan;26(2):147-155.

[41] Santin M, Motta A, Freddi G, Cannas M. In vitro evaluation of the inflammatory potential of the silk fibroin. J Biomed Mater Res. 1999 Sep;46(3):382-389.

[42] Seo YK, Choi GM, Kwon SY, Lee HS, Park YS, Song KY, et al. The Biocompatibility of Silk Scaffold for Tissue Engineered Ligaments. Key Eng Mater. 2007 Jul;342-343:73-76.

[43] Tan M, Liu W, Liu F, Zhang W, Gao H, Cheng J, et al. Silk fibroin-coated nanoagents for acidic lysosome targeting by a functional preservation strategy in cancer chemotherapy. Theranostics. 2019;9(4):961-973.

[44] Totten JD, Wongpinyochit T, Carrola J, Duarte IF, Seib FP. PEGylation-Dependent Metabolic Rewiring of Macrophages with Silk Fibroin Nanoparticles. ACS Appl Mater Interfaces. 2019;11(16):14515-14525.

[45] Fuster MG, Carissimi G, Montalbán MG, Víllora G. Improving Anticancer Therapy with Naringenin-Loaded Silk Fibroin Nanoparticles. Nanomater (Basel). 2020;10(4).

[46] Lozano-Pérez AA, Rodriguez-Nogales A, Ortiz-Cullera V, Algieri F, Garrido-Mesa J, Zorrilla P, et al. Silk fibroin nanoparticles constitute a vector for controlled release of resveratrol in an experimental model of inflammatory bowel disease in rats. Int J Nano medicine. 2014;9:4507-4520.

[47] Giménez-Siurana A, García FG, Bernabeu AP, Lozano-Pérez AA, Aznar-Cervantes SD, Cenis JL, et al. Chemoprevention of experimental periodontitis in diabetic rats with silk fibroin nanoparticles loaded with resveratrol. Antioxidants. 2020;9(1):1-13.

[48] Rodriguez-Nogales A, Algieri F, De Matteis L, Lozano-Perez AA, Garrido-Mesa J, Vezza T, et al. Intestinal anti-inflammatory effects of RGD-functionalized silk fibroin nanoparticles in trinitrobenzenesulfonic acid-induced experimental colitis in rats. Int J Nanomedicine. 2016;11:5945-5958.

[49] Horan RL, Antle K, Collette AL, Wang Y, Huang J, Moreau JE, et al. In vitro degradation of silk fibroin. Biomaterials. 2005;26(17):3385-3393.

[50] Brown J, Lu CL, Coburn J, Kaplan DL. Impact of silk biomaterial structure on proteolysis. Acta Biomater. 2015;11(1):212-221.

[51] Li M, Ogiso M, Minoura N. Enzymatic degradation behavior of

porous silk fibroin sheets. Biomaterials. 2003;24(2):357-365.

[52] Nair LS, Laurencin CT. Biodegradable polymers as biomaterials. Prog Polym Sci. 2007;32(8-9):762-798.

[53] Cao Y, Wang B. Biodegradation of silk biomaterials. Int J Mol Sci. 2009; 10(4):1514-1524.

[54] Wongpinyochit T, Johnston BF, Seib FP. Degradation Behavior of Silk Nanoparticles - Enzyme Responsiveness. ACS Biomater Sci Eng. 2018;4(3):942-951.

[55] Wang L, Luo Z, Zhang Q, Guan Y, Cai J, You R, et al. Effect of Degumming Methods on the Degradation Behavior of Silk Fibroin Biomaterials. Fibers Polym. 2019;20(1):45-50.

[56] Zuo B, Dai L, Wu Z. Analysis of structure and properties of biodegradable regenerated silk fibroin fibers. J Mater Sci. 2006;41(11): 3357-3361.

[57] Wang Z, Yang H, Li W, Li C. Effect of silk degumming on the structure and properties of silk fibroin. J Text Inst. 2018;5000(May):1-7.

[58] Bhardwaj N, Rajkhowa R, Wang X, Devi D. Milled non-mulberry silk fibroin microparticles as biomaterial for biomedical applications. Int J Biol Macromol. 2015;81:31-40.

[59] Rajkhowa R, Wang L, Wang X. Ultra-fine silk powder preparation through rotary and ball milling. Powder Technol. 2008;185(1):87-95.

[60] Kazemimostaghim M, Rajkhowa R, Tsuzuki T, Wang X. Production of submicron silk particles by milling. Powder Technol. 2013;241:230-235.

[61] Rajkhowa R, Wang L, Kanwar J, Wang X. Fabrication of ultrafine powder from eri silk through attritor

and jet milling. Powder Technol. 2009;191(1-2):155-163.

[62] Jacobs M, Heijnen N, Bastiaansen C, Lemstra P. Production of fine powder from silk by radiation. Macromol Mater Eng. 2000;283:126-131.

[63] Wongpinyochit T, Totten JD, Johnston BF, Seib FP. Microfluidic-assisted silk nanoparticle tuning. Nanoscale Adv. 2019;1(2):873-883.

[64] Rockwood DN, Preda RC, Yücel T, Wang X, Lovett ML, Kaplan DL. Materials fabrication from Bombyx mori silk fibroin. Nat Protoc. 2011; 6(10):1612-1631.

[65] Ajisawa A. Dissolution aqueous of silk fibroin with calciumchloride / ethanol solution. J Sericultural Sci Japan. 1997;67(2):91-94.

[66] Zheng Z, Guo S, Liu Y, Wu J, Li G, Liu M, et al. Lithium-free processing of silk fibroin. J Biomater Appl. 2016;31(3): 450-463.

[67] Shen T, Wang T, Cheng G, Huang L, Chen L, Wu D. Dissolution behavior of silk fibroin in a low concentration CaCl2-methanol solvent: From morphology to nanostructure. Int J Biol Macromol. 2018;113:458-463.

[68] Mathur AB, Tonelli A, Rathke T, Hudson S. The dissolution and characterization of bombyx mori silk fibroin in calcium nitrate-methanol solution and the regeneration of films. Biopolym - Nucleic Acid Sci Sect. 1997;42(1):61-74.

[69] Ha SW, Park YH, Hudson SM. Dissolution of bombyx mori silk fibroin in the calcium nitrate tetrahydrate-methanol system and aspects of wet spinning of fibroin solution. Biomacromolecules. 2003;4(3):488-496.

[70] Cheng G, Wang X, Tao S, Xia J, Xu S. Differences in regenerated silk

fibroin prepared with different solvent systems: From structures to conformational changes. J Appl Polym Sci. 2015 Jun;132(22):1-8.

[71] Ajisawa A. Dissolution of silk fibroin with calciumchloride/ethanol aqueous solution. J Seric Sci Jpn. 1997;67(2):91-94.

[72] Kanbak M, Karagoz AH, Erdem N, Oc B, Saricaoglu F, Ertas N, et al. Renal Safety and Extrahepatic Defluorination of Sevoflurane in Hepatic Transplantations. Transplant Proc. 2007;39(5):1544-1548.

[73] Phillips DM, Drummy LF, Conrady DG, Fox DM, Naik RR, Stone MO, et al. Dissolution and Regeneration of *Bombyx mori* Silk Fibroin Using Ionic Liquids. J Am Chem Soc. 2004;126(10):14350-14351.

[74] Hernández-Fernández FJ, de los Ríos AP, Tomás-Alonso F, Gómez D, Rubio M, Víllora G. Integrated reaction/ separation processes for the kinetic resolution of rac-1-phenylethanol using supported liquid membranes based on ionic liquids. Chem Eng Process Process Intensif. 2007;46(9):818-824.

[75] Hernández-Fernández FJ, de los Ríos AP, Tomás-Alonso F, Gómez D, Víllora G. On the development of an integrated membrane process with ionic liquids for the kinetic resolution of rac-2-pentanol. J Memb Sci. 2008;314(1-2):238-246.

[76] Rogers RD, Seddon KR. Ionic Liquids - Solvents of the Future? Vol. 302, Science. 2003. p. 792-3.

[77] Lozano-Pérez AA, Montalbán MG, Aznar-Cervantes SD, Cragnolini F, Cenis JL, Víllora G. Production of silk fibroin nanoparticles using ionic liquids and high-power ultrasounds. J Appl Polym Sci. 2014;132, 41702.

[78] Wang Q, Yang Y, Chen X, Shao Z. Investigation of Rheological Properties

and Conformation of Silk Fibroin in the Solution of AmimCl. Biomacro molecules. 2021;13(6):1875-1881.

[79] Zhang Y-Q, Shen W-D, Xiang R-L, Zhuge L-J, Gao W-J, Wang W-B. Formation of silk fibroin nanoparticles in water-miscible organic solvent and their characterization. J Nanoparticle Res. 2007;9(5):885-900.

[80] ZhuGe DL, Wang LF, Chen R, Li XZ, Huang ZW, Yao Q, et al. Cross-linked nanoparticles of silk fibroin with proanthocyanidins as a promising vehicle of indocyanine green for photo-thermal therapy of glioma. Artif Cells, Nanomedicine Biotechnol. 2019;47(1):4293-4304.

[81] Kundu J, Chung Y-I, Kim YH, Tae G, Kundu SC. Silk fibroin nanoparticles for cellular uptake and control release. Int J Pharm. 2010;388(1):242-250.

[82] Perteghella, Sottani, Coccè, Negri, Cavicchini, Alessandri, et al. Paclitaxel-Loaded Silk Fibroin Nanoparticles: Method Validation by UHPLC-MS/MS to Assess an Exogenous Approach to Load Cytotoxic Drugs. Pharmaceutics. 2019 Jun;11(6):285.

[83] Wongpinyochit T, Johnston BF, Philipp Seib F. Manufacture and drug delivery applications of silk nanoparticles. J Vis Exp. 2016;2016 (116):1-9.

[84] Li H, Tian J, Wu A, Wang J, Ge C, Sun Z. Self-assembled silk fibroin nanoparticles loaded with binary drugs in the treatment of breast carcinoma. Int J Nanomedicine. 2016;11:4373-4380.

[85] Yu S, Yang W, Chen S, Chen M, Liu Y, Shao Z, et al. Floxuridine-loaded silk fibroin nanospheres. RSC Adv. 2014;4(35):18171-18177.

[86] Wu P, Liu Q, Li R, Wang J, Zhen X, Yue G, et al. Facile preparation of paclitaxel loaded silk fibroin

nanoparticles for enhanced antitumor efficacy by locoregional drug delivery. ACS Appl Mater Interfaces. 2013;5(23): 12638-12645.

[87] Matteucci ME, Hotze MA, Johnston KP, Williams RO. Drug nanoparticles by antisolvent precipitation: Mixing energy versus surfactant stabilization. Langmuir. 2006;22(21):8951-8959.

[88] Montalbán M, Coburn J, Lozano-Pérez A, Cenis J, Víllora G, Kaplan D. Production of Curcumin-Loaded Silk Fibroin Nanoparticles for Cancer Therapy. Nanomaterials. 2018;8(2):126.

[89] Kiran E, Debenedetti PG, Peters CJ. Supercritical Fluids. 2000, Springer.

[90] Xie M, Fan D, Li Y, He X, Chen X, Chen Y, et al. Supercritical carbon dioxide-developed silk fibroin nanoplatform for smart colon cancer therapy. Int J Nanomedicine. 2017;12:7751-7761.

[91] Zhao Z, Chen A, Li Y, Hu J, Liu X, Li J, et al. Fabrication of silk fibroin nanoparticles for controlled drug delivery. J Nanoparticle Res. 2012;14(4).

[92] Chen BQ, Kankala RK, He GY, Yang DY, Li GP, Wang P, et al. Supercritical Fluid-Assisted Fabrication of Indocyanine Green-Encapsulated Silk Fibroin Nanoparticles for Dual-Triggered Cancer Therapy. ACS Biomater Sci Eng. 2018;4(10): 3487-3497.

[93] Reverchon E, De Marco I, Torino E. Nanoparticles production by supercritical antisolvent precipitation: A general interpretation. J Supercrit Fluids. 2007;43(1):126-138.

[94] Jaworek A, Krupa A. Jet and drops formation in electrohydrodynamic spraying of liquids. A systematic approach. Exp Fluids. 1999;27(1):43-52.

[95] Gholami A, Tavanai H, Moradi AR. Production of fibroin nanopowder through electrospraying. J Nanoparticle Res. 2011;13(5):2089-2098.

[96] Composition P, Droplets V, Shishulin A V, Fedoseev VB. Microfluidics-a review. 1993;D.

[97] Suh YK, Kang S. A review on mixing in microfluidics. Micromachines. 2010;1(3):82-111.

[98] Lammel AS, Hu X, Park SH, Kaplan DL, Scheibel TR. Controlling silk fibroin particle features for drug delivery. Biomaterials. 2010;31(16): 4583-4591.

[99] Weber M, Thies M. Understanding the RESS Process. In: Supercritical Fluid Technology in Materials Science and Engineering. CRC Press; 2002.

[100] Ostwald ripening. In: IUPAC Compendium of Chemical Terminology. Research Triangle Park, NC: IUPAC; 2007. p. 1824.

[101] Perteghella S, Crivelli B, Catenacci L, Sorrenti M, Bruni G, Necchi V, et al. Stem cell-extracellular vesicles as drug delivery systems: New frontiers for silk/curcumin nanoparticles. Int J Pharm. 2017; 520(1-2):86-97.

[102] Montalbán MG, Chakraborty S, Peña-García J, Verli H, Villora G, Pérez-Sánchez H, et al. Molecular insight into silk fibroin based delivery vehicle for amphiphilic drugs: Synthesis, characterization and molecular dynamics studies. J Mol Liq. 2020;299: 112156.

Chapter 11

Nanoparticles as Drug Delivery Systems

Guzmán Carissimi, Mercedes G. Montalbán,
Marta G. Fuster and Gloria Víllora

Abstract

This chapter presents a review on the design of nanoparticles which have been proposed as drug delivery systems in biomedicine. It will begin with a brief historical review of nanotechnology including the most common types of nanoparticles (metal nanoparticles, liposomes, nanocrystals and polymeric nanoparticles) and their advantages as drug delivery systems. These advantages include the mechanism of increased penetration and retention, the transport of insoluble drugs and the controlled release. Next, the nanoparticle design principles and the routes of administration of nanoparticles (parental, oral, pulmonary and transdermal) are discussed. Different routes of elimination of nanoparticles (renal and hepatic) are also analyzed.

Keywords: nanoparticle, drug delivery, insoluble drug, controlled release, route of administration

1. Introduction

Nanomedicine is a relatively new discipline that arises from the intersection between nanotechnology and medicine. It is based on the control of matter at the nanometer scale for applications in the field of human health. The use of materials in this range has been a great advance for the pharmacology by modifying fundamental properties of the drugs such as solubility, diffusivity, half-life in the bloodstream and drug release and distribution profiles [1–4]. Although the production and use of nano-sized matter dates from hundreds of years [5, 6], nanomedicine as a modern interdisciplinary science was first established at the end of the last century. Many authors consider the beginning of nanotechnology in the famous lecture of the physicist and Nobel laureate Richard P. Feynman in 1959 for the American Physical Society entitled: "There's Plenty of Room at the Bottom: An Invitation to Enter a New Field of Physics" [7]. In it, Feynman presented a futuristic vision of technology that leads towards the atomic scale and towards the final limits established by physical laws. Revolutionary ideas were put forward, such as reducing the integrated circuits of a computer to diameters between 10 and 100 atoms. To understand the scope of his predictions, suffice it to remember that, at the time of presenting these ideas, a computer occupied an entire room if not several. However, the word (or the prefix) nano was not mentioned even once in his presentation, Feynman stuck to describing the miniaturization of machines and its possible applications. The honor of having coined the term "nano" is awarded to Norio Taniguchi, for his presentation "On the basic concept of nanotechnology" in 1974 [8].

It is important to emphasize that the term nanotechnology applied to the study of nanoparticles simply consists in renaming the study of colloidal dispersions, in which field the contributions of renowned scientists such as Michael Faraday stand out, who in 1857 disseminated the first synthesis of gold nanoparticles and other metals [9]. In the paper, Faraday reveals his amazement at the changes in the optical properties of metallic colloidal dispersions. These properties were later explained in

Drug	Company	Application	Date of approval
Lipid-based			
Doxil	Janssen	Kaposi's sarcoma, ovarian cancer, multiple myeloma	1995
DaunoXome	Galen	Kaposi's sarcoma	1996
AmBisome	Gilead Sciences	Fungal/protozoal infections	1997
Visudyne	Bausch and Lomb	Wet age- related macular degeneration, myopia, ocular histoplasmosis	2000
Marqibo	Acrotech Biopharma	Acute lymphoblastic leukemia	2012
Onivyde	Ipsen	Metastatic pancreatic cancer	2015
Vyxeos	Jazz Pharmaceuticals	Acute myeloid leukemia	2017
Onpattro	Alnylam Pharmaceuticals	Transthyretin- mediated amyloidosis	2018
Polymer-based			
Oncaspar	Servier Pharmaceuticals	Acute lymphoblastic leukemia	1994
Copaxone	Teva	Multiple sclerosis	1996
PegIntron	Merck	Hepatitis C infection	2001
Eligard	Tolmar	Prostate cancer	2002
Neulasta	Amgen	Neutropenia, chemotherapy induced	2002
Abraxane	Celgene	Lung cancer, metastatic breast cancer, metastatic pancreatic cancer	2005
Cimiza	UCB	Crohn's disease, rheumatoid arthritis, psoriatic arthritis, ankylosing spondylitis	2008
Plegridy	Biogen	Multiple sclerosis	2014
ADYNOVATE	Takeda	Hemophilia	2015
Inorganic			
INFeD	Allergan	Iron-deficient anemia	1992
DexFerrum	American Regent	Iron-deficient anemia	1996
Ferrlecit	Sanofi	Iron deficiency in chronic kidney disease	1999
Venofer	American Regent	Iron deficiency in chronic kidney disease	2000
Feraheme	AMAG	Iron deficiency in chronic kidney disease	2009
Injectafer	American Regent	Iron-deficient anemia	2013

Table 1.
FDA-approved nanomedicines for drug delivery. Adapted from reference [16] with permission of Springer Nature.

1908 by Gustav Mie who would give a solution to Maxwell's equations for particles with a finite volume [10]. In 1925, Richard Zsigmondy would be awarded the Nobel Prize in Chemistry for his demonstration of the heterogeneous nature of colloidal dispersions [11]. His contributions in methodological terms have become fundamental for the study of modern colloidal chemistry and nanotechnology.

In 1981, Eric Drexler [12] proposed what is now known as a bottom-up approach, where atoms are self-assembled to create higher-order structures. It contrasts with the approach proposed by Feynman, who conceives the beginning of nanotechnology from a top-down approach, building smaller and smaller machines that, ultimately, are used to manipulate matter with atomic precision. In particular, the bottom-up approach is of great interest in nanoparticle synthesis, where self-assembly properties, the product of natural chemical and physical interactions between molecules, can be exploited to produce defined characteristics. It is this concept that opens a wide range of possibilities towards the synthesis of nanoparticles with a wide variety of functionalities. From this point of view, nanoparticle engineering is based on "programming" with predetermined instructions the self-assembly of atoms or molecules in such a way that the desired nanoparticles are the final product.

Different authors see the paths of nanotechnology and medicine intertwined in 1986 when Matsumura and Maeda [13] observed that an anticancer protein bound to polymeric nanoparticles exhibited greater accumulation in tumor tissues than in healthy tissues. This discovery led to the theory of enhanced permeability and retention (EPR) as a consequence of tumor physiology and the size of nanoparticles (<200 nm), which are capable of penetrating tumor cells due to their reduced size and, at the same time, being retained [13]. The discovery lays the foundation for the development of different theories on targeted delivery via passive transport to tumor tissues and a large cascade of advances in the design of drug nanocarriers. In 1995, the first liposome-based nanostructure for the delivery of doxorubicin, an important anticancer drug, was approved by the FDA (Food and Drug Administration, USA) under the trade name Doxil® [14]. Since then and until April 2016, more than 50 nanomedicines of different kinds have been approved by the FDA and this is expected to be only the beginning of the near future [15]. **Table 1**, which has been adapted from reference [16], summarizes the FDA-approved nanomedicines used for drug delivery up to date. At the time of writing this chapter, one of the most conservative market capitalization estimates that the value of all nanomedicines is comprised of $ 47.5 billion and is expected to rise to $ 164 billion by 2027 driven by the crisis SARS-CoV-2 [17].

2. Types of nanoparticles

Modern and advanced synthesis techniques have led to the preparation of a great variety of nanoparticles with different shapes and sizes, together with the use of a great variety of materials. The classification of nanoparticles can be based on different physical and/or chemical parameters. This is a brief summary of the most important characteristics and functions of different types of nanoparticles used in biomedicine, classified based on the materials used in their synthesis.

2.1 Metal nanoparticles

Metallic nanoparticles have attracted great interest for use in medicine as anticancer agents [18], imaging contrast agents [19], and drug carriers [20]. One of the most exploited properties of these nanoparticles is the increase in molar absorptivity that colloidal dispersions present due to the intensity of their surface plasmon resonance [21], classic examples of this being nanoparticles of metals such as gold, silver or copper. Plasmon resonance can radiate light (Mie scattering), a process that finds great utility in the optical and imaging fields, or it can be rapidly converted to heat (absorption). The latter mechanism can be used to convert metallic nanoparticles into light-activated heat sources for use in medicine in selective laser photo-thermolysis of cancer cells [22–24]. The properties of the resonance plasmon can be tuned by modifying the size, morphology and nature of the metals used for the synthesis of nanoparticles, thus being able to serve different purposes [18]. Their optical properties and high capacity to catalyze reactions and electron transfer also give them applications as biosensors [21] which, through ingenious modifications, are capable of significantly amplifying signals [25].

Metallic nanoparticles are also of interest as vehicles for the administration of drugs and other active principles due to their high surface-volume ratio, stability, functionality through chemical modifications of their surface and relative harm-lessness. For example, Libutti et al. [26] functionalized the surface of 27 nm gold nanoparticles with tumor necrosis factor-α and polyethylene glycol. The nanoparticles managed to passively accumulate in the cancerous tissues avoiding healthy tissues. This allowed the researchers to administer doses of tumor necrosis factor-α that were previously considered toxic. Iron oxide nanoparticles have been approved by the United States Food and Drug Administration (FDA) for the treatment of anemia [27]. Recently, molecular docking studies propose the reuse of these nanoparticles to combat the current global pandemic of SARS-CoV-2 [28]. The studies revealed that both Fe_2O_3 and Fe_3O_4 nanoparticles interact effectively with the different proteins and glycoproteins of the virus. These interactions associated with conformational changes in proteins are expected to result in the inactivation of the virus.

However, despite the great boom in metallic nanoparticles due to their long history and simplicity in terms of their synthesis, they present toxicity problems in prolonged use as they cannot be biodegraded [29–31]. In addition, different authors have already expressed their concerns regarding the neurotoxicity of these particles as they are capable of crossing the blood–brain barrier [32, 33].

2.2 Liposomes

Liposomes are spherical vesicles composed of one or more concentric mem-branes of lipid bilayers with an internal compartment that normally contains water. Liposomes have the ability to encapsulate both lipophilic molecules in their membrane and hydrophilic in their internal cavity. The size of these vesicles can vary from a few nanometers to several microns. However, liposomes applied for medical use range between 50 and 450 nm [34]. Liposomes were discovered in the 1960s [35] and claim to be the first nanoparticles to be used for the delivery of nanomedicines after Doxil® was approved by the FDA in 1995 [14]. At present, there have been technological advances that have managed to use various natural or synthetic lipids, as well as surfactants to modify the physicochemical properties of liposomes, giving rise to the second and third generation of them [36]. Changes in the physicochemical properties of liposomes influence their interaction with cells, their half-life in circulation, their ability to penetrate tissues, and their final fate in vivo [36]. For example, through the exchange of a phospholipid bilayer in liquid phase for a bilayer in solid phase in liposomes, by incorporating cholesterol (bilayer tightening effect) or sphingomyelin, the retention of the drug loaded in the lipo-somes increases, delaying the release.

Despite all the hopes for conventional liposomes, they have presented various problems and pharmacological implications over the years. A major drawback of conventional liposomes is their rapid capture by the reticuloendothelial system [37]. Liposomes accumulate mainly in the liver and spleen, due to their abundant blood supply and the abundance of phagocytic cells resident in these tissues [38]. The marked increase in the retention and accumulation of liposomal drugs in these organs may delay the clearance of lipophilic anticancer drugs from the circulation [39]. Furthermore, during chemotherapy, it can lead to partial depletion of macrophages and interfere with important host defense functions in these cell types [40].

2.3 Nanocrystals

Nanocrystals are perhaps the simplest forms of nanomedicine, i.e., nanoparticles made up of 100% of the drug. The large surface/area ratio offered by the nanometric scale increases the dissolution rate, allowing improved pharmacokinetic profiles. The small size of the nanoparticles increases the penetration of the nanocrystals to biological barriers such as the digestive tract, thus increasing the bioavailability of insoluble drugs. The production of crystalline nanoparticles has been applied to both organic drugs and inorganic materials [41–43]. Although the inorganic crystalline nanoparticles approved by the FDA (year 2016) are limited to hydroxyapatite and calcium phosphate for use as substitutes for bone grafts and iron oxide, iron oxide nanoparticles have been used for the treatment of glioblastoma and anemia, due to iron deficiency in kidney diseases [15]. Solubility problems associated with several pharmacological compounds have been improved by conversion to nanocrystals and are marketed for a variety of indications [43]. The pearl mill developed by Elan Nanosystems was used to produce the first three FDA-approved nanocrystals: Rapamune®, Tricor® and Emend®, and is expected to be almost universally applicable to a variety of drugs with low solubility, estimated to be 70–90% of potential drug compounds [41].

2.4 Polymeric nanoparticles

Polymeric nanoparticles are colloidal particles of solid nature that, depending on the preparation method, can form two types of structures: nanospheres or nanocapsules [44]. Nanospheres consist of a matrix system in which the drug can be adsorbed on the surface or co-precipitated with the polymer [45], while in nanocapsules the drug is contained in an internal cavity surrounded by a polymeric membrane [46]. Natural polymers like carbohydrates and proteins vary in their properties between hydrophilic, hydrophobic and even amphiphilic. On the other hand, synthetic polymers are mostly hydrophilic in nature and can be present in a prepolymerized form or be polymerized during the nanoparticle synthesis process. Synthetic polymers, in turn, can be subdivided into two classes, biodegradable and non-biodegradable. Polylactic-co-glycolic acid (PLGA) is a biodegradable polymer widely used for drug delivery [47, 48]. On the other hand, polyacrylates are non-biodegradable polymers that have also been studied for drug delivery [49, 50], although to a lesser extent compared to biodegradable polymers for clear biocompatibility reasons [51].

Polymeric nanoparticles have immense potential as drug carriers, since they can deliver them in different organs, they protect drugs against degradation in vitro and in vivo, they release the drug in a controlled manner and also offer the possibility of passively targeting drugs to tumors or other tissues actively [44]. The use of polymeric nanoparticles for drug delivery is a universal approach to increase the therapeutic performance of those that are poorly soluble in any route of administration.

3. Advantages of nanoparticles for drug delivery

Nanoparticles bring a new level of engineering and control to the field of medicine by being able to modify parameters such as solubility, diffusivity, half-life, toxicity, pharmacokinetics and biodistribution of drugs and diagnostic agents. The applications of nanoparticles are very diverse and are expected to increase with the advancement of technology. In recent years, numerous studies have demonstrated their ability to act as sensors [52], drug carriers [53–55], and diagnostic agents [1, 56]. Recent efforts have managed to integrate treatments and diagnoses in a single application, giving rise to the procedures known as "theranostic".

The justification for the use of nanoparticles as drug delivery systems lies in at least three mechanisms: (i) Enhanced Penetration and Retention (EPR) of nanoparticles in solid tumors; (ii) The possibility of transporting insoluble drugs in the blood through stable colloidal systems and (iii) the controlled release thereof. In this section, the possible advantages of each of these points will be developed.

3.1 Enhanced permeability and retention (EPR)

The term EPR was coined by Matsumura and Maeda in 1986 [13]. In their work, the researchers observed that the anticancer protein neocarzinostatin, conjugated to a polymeric matrix, exhibited greater accumulation in tumor tissues than free neocarzinostatin. By applying labeled macromolecules to tumor-bearing mice, they observed that their concentration was up to 5 times higher in tumor areas than in blood over a period of 19 to 72 hours [13]. The authors affirm that the passive accumulation of these macromolecules in tumors is due to the abnormal physiology associated with tumor masses: fenestrated hypervascularization with increased permeability to macromolecules (or nanoparticles) and poor recovery through blood vessels or lymphatic vessels [57]. Subsequently, it was shown that other plasma proteins greater than 40 kDa are capable of passively and selectively accumulating in tumor areas [58]. The EPR effect can be demonstrated in mice with the intravenous injection of the Evans Blue marker, which binds to plasma albumin forming a complex that demonstrates differential accumulation in tumor areas [59], as shown in **Figure 1**.

Figure 1.
Image of a metastatic lung cancer originating from 26 colon tumors implanted in the dorsal skin of a mouse. The mouse was sacrificed 3 months after implantation and 10 hours before sacrificing, a solution of Evans blue (5%) was injected intravenously to allow the EPR effect to become visible. Albumin-Evans blue complex (70 kDa) preferentially accumulated in metastatic tumor nodules, as in primary tumors. Arrows point to metastatic tumor nodules. From reference [60] with permission of Elsevier.

For passive accumulation through the EPR effect to be important, different requirements are needed. On the one hand, the nanoparticles must remain in circulation for a time greater than 6 hours [60, 61]. This can generally be achieved by functionalizing the nanoparticles with polyethylene glycol (PEG) [62]. On the other hand, the mechanism also depends on the particles being small enough to penetrate biological membranes but large enough to be retained. Yuan et al. [63] measured the microvascular permeability of several macromolecules in human colon adenocarcinoma LSI74T transplanted in mice with immunodeficiency and the results indicated that the cut-off size of the pore is around 400-600 nm, depending on physicochemical properties such as charge and hydrophobicity of the nanoparticles. Regarding the minimum size, Maeda et al. [58] estimated that the nanoparticle size must be greater than 40 kDa to show significant retention in the tumor area.

Vascular extravasation is also highly dependent on the morphology and the specific type of tumor. Scanning electron micrographs of normal vascular epithelium and two epithelia associated with different tumors are shown in **Figure 2**. As can be seen, tumor-associated epithelia have significant pores (fenestrations) and their size depends on the type of tumor. Smith et al. [64] studied the extravasation capacity of quantum-dots (20-25 nm) and single-walled carbon nanotubes (2-3 x 200 nm) in tumors implanted in mouse ears. The surface of both types of nanostructures was modified by PEG to avoid differences in charge or surface chemistry and that the results were only due to the morphology of the particles. The authors found that spherical quantum-dots are capable of extravasation of the endothelium of LS174T tumors, whereas cylindrical nanotubes are capable of extravasation in U87MG tumors. Surprisingly, the authors were not able to see the extravasation of the nanomaterials in normal endothelium. This suggests that the morphology of the nanoparticles may be a determining factor for penetrating certain tumors, while healthy endothelium could prevent nanoparticle transfer.

Although the EPR model has been tested in rodents with large induced tumor masses [59, 65], these models differ widely in morphology and physiology of possible human tumors and, for these reasons, there is still much controversy regarding it [55, 60, 66]. Firstly, tumors of up to 10% of body weight have been reported in mice. If we make an analogy with a 70 kg human, the tumor would be the size of a basketball [67], when they actually have a size between millimeters and centimeters at the time of diagnosis and treatment [68]. Such tumor masses filter out a significant proportion of the injected drug dose and act as a reservoir, enhancing efficacy while mitigating toxicity. In addition to this, the

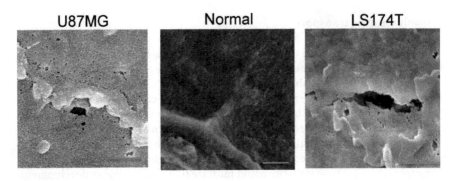

Figure 2.
Scanning emission microscopy (SEM) of tumors and normal blood vessels. SEM images show pores in the U87MG and LS174T tumor vasculature at the apparent border between endothelial cells. No pores are seen at the border of the vasculature of a tumor-free mouse ear. Scale bars: U87MG (500 nm), Normal (1 µm) and LS174T (1 µm). Reprinted with permission from reference [64]. Copyright 2012 American Chemical Society.

tumor microenvironment in humans presents important physiological differences compared to murine tumors: (i) lack of fenestrations in the tumor endothelium for the entry of nanoparticles, (ii) heterogeneity of blood flow through tissues, which causes the regions to become acidic or hypoxic [69], (iii) lower pericyte coverage, (iv) heterogeneous basement membrane and (v) higher and heterogeneous density of the extracellular matrix. This leads to high interstitial pressure and therefore the main mechanism of matter transport is by diffusion and not by convective transport, which is more efficient [69, 70].

For these reasons, it is not possible to directly transfer the results obtained in rodents to humans, mainly because cell penetration depends on the nanoparticles go from the point of application to the tumor mass and be able to interact with cells to be internalized. Currently, different methods are being investigated to increase the EPR effect. For example, Fang et al. [71] developed agents which can selectively generate vasodilator molecules (carbon monoxide) in tumor areas, achieving an increase in the concentration of the nanocarrier between 2 and 3 times higher in these, while an increase in tissues healthy was not detected. Similar results have been achieved with nanocarriers that can release nitric oxide [72, 73]. The increase in blood pressure results in an increase in the osmotic pressure, which promotes the filtration of the particles towards the tumor areas so that when angiotensin II is co-administered with the nanocarriers, an increase in the transfer and accumulation in the tumor areas can be observed [74].

In contrast to the passive accumulation of drug nanocarriers in tumor areas by the EPR mechanism, active targeting is presented, which is based on the functionalization of the nanoparticle surface with recognition molecules such as antibodies [75, 76] or ligands [77, 78] which can specifically bind to molecules overexpressed at the target site [79]. In the active targeting strategy, two cellular targets can be distinguished: (i) targeting cancer cells, which present overexpression of molecules such as transferrin, folate, epidermal growth factor receptor or glycoprotein receptors, and (ii) targeting tumor endothelium, which have overexpression of vascular endothelial growth factors (VEGF), $\alpha_v\beta_3$ integrins, vascular cell adhesion molecule-1 (VCAM-1), or matrix metalloproteinases [66, 80]. In some cases, both receptors are overexpressed in cancer cells and endothelium and can be exploited simultaneously [80]. In addition, the design of nanocarriers as active targeting systems may involve the coupling of recognition molecules as surface receptors which are able to initiate endocytosis, and hence to increase cell internalization in contrast to simple accumulation [81]. Not only would this increase the antitumor efficacy of many drugs, but it could also be used for the delivery of genetic material [82].

3.2 Insoluble drug transport

Most orally administered drugs that are soluble in water and capable of penetrating biological membranes during the passage of the gastrointestinal tract will eventually become bioavailable in the body. In contrast, water-insoluble drugs will generally not be bioavailable after oral ingestion as they cannot dissolve and pass through the gastrointestinal barrier. Along the same lines, due to their low solubility, they cannot be administered intravenously and parenteral administration does not always increase bioavailability [83]. It is estimated that 90% of drugs in development are insoluble in water, while only 40% of drugs on the market share this characteristic [84]. These statistics could indicate that many drugs in development do not reach their administration to patients due to their low solubility in water. This not only means less capital invested in research and development but also lost treatment opportunities. The development of a drug in 2011 was estimated at between 92 million and 1.8 billion dollars [85], lasting for a period of between 11.4

and 13.5 years on average [86]. Considering these, we can see that low water solubility represents a formidable challenge and opportunity for nanotechnology.

Three factors govern the speed and degree of absorption of orally administered drugs: (i) dissolution rate, (ii) solubility and (iii) intestinal permeability, which are grouped according to the biopharmaceutical classification system (BCS, Biopharmaceutical Classification System) in the categories [87]:

Class I: High Solubility - High Permeability.
Class II: Low Solubility - High Permeability.
Class III: High Solubility - Low Permeability.
Class IV: Low Solubility - Low Permeability.

The criterion established by the BCS classifies a drug as soluble when it is capable of dissolving an entire therapeutic dose in 250 mL of water, being this volume equivalent to the average amount of water found in the stomach [87].

As can be deduced, the possibilities of entering the market for a class I drug are substantially greater than that of the rest of the categories, however, a possible solution to these problems lies in the development of drug carriers which can transport them in a stable colloidal dispersion and with particles capable of crossing biological membranes [88]. As an example, Atovaquone (Wellvone®) is an antibiotic used for the treatment of *Pneumocystis carinii*, leishmaniasis and *P. falciparum* malaria, however, its low solubility limits its absorption. By formulating a dispersion of nanoparticles of this drug, it was possible to increase absorption from 15 to 40% with a 3-fold lower drug dose [89]. Xie et al. [90] prepared curcumin-loaded silk fibroin nanoparticles (SFN) to increase the dissolution rate of the drug and the mass of the drug in dispersion. Recent results from our research group revealed that SFN are an excellent vehicle for the transport of the natural drug naringenin, with anti-cancer properties [91], which has low solubility in water. The results indicated that this drug loaded in the nanoparticles is 1.7 times more effective in reducing the viability of HeLa cells than by itself. These results can be attributed to the low solubility and slow dissolution of free naringenin which, when loaded in the SFN, remains stable in dispersion, increasing its cellular penetration and improving the dissolution profile.

3.3 Controlled release

Nanoparticles can be used as drug reservoirs for their controlled release over time, which offers numerous advantages compared to conventional administration of multiple doses. Among them, it can be highlighted the improvement in efficacy and reduction of toxicity and patient cooperation [92]. The former can be considered as the increase in therapeutic activity compared to the intensity of the side effects, while the latter offers the advantage of reducing the number of applications required during treatment.

Controlled release is especially beneficial for those drugs whose half-life in the blood is relatively low due to a high rate of metabolism and elimination by the body. This effect can be observed in **Figure 3**, where the concentration of a drug in blood applied by a conventional method (red line) is represented against a controlled release system (blue line). As can be seen, the drug administered in a conventional manner is only a fraction of the time in the zone considered therapeutic, while fluctuating between subtherapeutic concentrations and above the maximum tolerable level. On the other hand, the controlled release system takes longer to reach the therapeutic concentration window but remains stable within it. The goal of the system is to match the rate of clearance to that of release in the

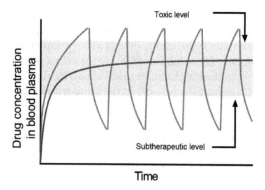

Figure 3.
Diagram of the blood concentration of a drug after multiple administrations as a conventional injection (red line) and as a controlled release system (blue line).

therapeutic concentration zone. In the clinic, this translates into numerous benefits, for example, in the case of administration of analgesics, the concentration could be prevented from falling to subtherapeutic levels and therefore the patient feeling pain. This is transferable to a large number of drugs including anti-inflammatories, antibiotics, anesthesia, hormones, chemotherapeutics, etc.

There are different mechanisms by which polymer nanoparticles can allow controlled drug release. On the one hand, the release can be delayed by using a water-soluble polymer as a matrix, whose dissolution rate is slow and consequently releases the drug at the rate of dissolution of the polymer. In the case of insoluble polymers, they can act as a diffusion barrier, slowing down the release of the drug from inside the nanoparticle to the medium. The release can also be controlled by an osmotic flow generated by a semipermeable membrane, which is itself the nanosystem, as is the case with liposomes. Finally, a delivery system that responds to internal or external stimuli could be achieved, which would be very useful, for example, in diabetic patients in which the nanosystem would release insulin on demand of the blood glucose concentration [93]. Volpatti et al. [94] have succeeded in synthesizing nanoparticles whose insulin release is sensitive to glucose levels by adding glucose oxidase and catalase to them. These researchers demonstrated that a single subcutaneous injection provides 16 h of glycemic control in diabetic mice. Cheng et al. [95] developed SFN capable of loading the antitumor drug paclitaxel (3%) and delivering it sustainably for 14 days.

4. Nanoparticle design

From the point of application to the site of action, nanoparticles face a host of challenges. In the first place, they are diluted in approximately 5 L of blood that circulates at 5 L/min through the circulatory system about 106 km long, where the velocity in each blood vessel can be between 1.5-33 cm/s [96] hindering the interaction between nanoparticles and the target tissue. Interstitial fluids have a much lower speed, just a few µm/s, where interactions would be favored. However, reaching them means crossing biological barriers, which is not an easy task. Finally, to all of the above, it is added that when nanoparticles enter the body they are treated hostilely by the immune system. For these reasons, different design principles are applied to nanoparticles to try to get around different obstacles depending on their final application.

As mentioned above, as soon as the nanoparticles enter the body, they are exposed to the mononuclear phagocyte system which consists of a system of phagocytic cells, predominantly macrophages resident in the spleen, lymph nodes and liver, which sequester the nanoparticles immediately after administration [97]. This process begins with the opsonization of the nanoparticles based on the adsorption of plasma proteins, including albumin, complementary system proteins, pattern recognition receptors and immunoglobulins. This process is relatively fast and can occur in a period as short as 30 seconds [98]. This "natural functionalization" is known as the formation of the protein crown and clearly can alter the function or fate of nanoparticles by disturbing different parameters such as size, charge and surface chemistry, as well as hydrophobicity. This protein crown can even mask the receptors or ligands attached to the nanoparticles [99].

Different design strategies have been developed to avoid opsonization and subsequent clearance by the immune system. This evasion of the immune system tries to increase the circulation time of the nanoparticles in the body and, consequently, the chances that they find the target tissue while they circulate through the bloodstream. One of the easiest and most direct strategies is PEGylation, based on the functionalization with polyethylene glycol (PEG) molecules on the surface of the nanoparticles where the polymer units form very strong associations with the water molecules, generating a hydration layer and a steric barrier to opsonization [100, 101]. An alternative strategy may be to functionalize the nanoparticles with endogenous signals normally present in healthy cells. Rodríguez et al. [102] functionalized viral particles with the CD47 membrane protein, which acts as a "non-phagocytizing" signal [103], thus prolonging the circulation time. Another similar strategy is to cover the particles with biomimetic molecules such as cell membranes, to hide the particles from the immune system [104, 105]. Another way to increase circulation time is the one proposed by Nikitin et al. [106], which is based on a slight and transient suppression of the mononuclear phagocyte system through the administration of anti-erythrocyte antibodies. They were able to increase the circulating half-life of different nanosystems up to 32 times through the suppression of ca. 5% of hematocrits.

Silk fibroin exhibits unique low immune response properties, allowing it to evade the immune system. This can be exemplified by the study by Catto et al. [107], who implanted tubular matrices based on silk fibroin in mice, detecting few macrophages labeled with anti-ED1 antibodies, which was indicative of a low inflammatory response. The absence of T lymphocytes (anti-CD4 antibodies) demonstrated that there was no cell-mediated immune response. Recently, under a state-of-the-art design, Tan and colleagues [108] have designed a doxorubicin delivery nanosystem using silk fibroin as a Trojan horse. The researchers synthesized drug-loaded amorphous calcium carbonate nanoparticles and coated them with silk fibroin. It prevents the premature release of doxorubicin and helps evade the immune system. Thanks to the EPR mechanism, nanoparticles are accumulated in cancerous tissues and, finally, internalized by lysosomes. The acidic pH of the latter promotes the generation of CO_2 from calcium carbonate, resulting in the bursting of the lysosome due to the expansion of the gas and the release of doxorubicin inside the target cell. Results in mice revealed that silk fibroin-coated nanoparticles are more effective in reducing tumor mass and preventing side effects in mice compared to free doxorubicin or uncoated calcium carbonate nanoparticles. In addition, the immunotoxicity tests indicated that the nanoparticles did not initiate an immune response by not increasing the amount of T cells ($CD4^+$ and $CD8^+$) or IgM, IgG and IgA compared to the control group. More information on intracellular drug release can be found in the review by Fenghua et al. [109].

5. Routes of administration of nanoparticles

SFN have proven to be extremely versatile for the transport of therapeutic compounds such as small drugs, proteins and DNA molecules [110]. The functionality of these compounds is closely related to the route of administration. For example, nanoparticles can be injected into the bloodstream and make use of the EPR effect for passive accumulation in metastatic tumors or they can be injected directly into the tumor mass [111]. On the other hand, they can be applied topically for the treatment of skin cancer [112] and in a similar way for lung treatments [113, 114]. SFN have been used in a wide variety of routes of administration [54]. For the sake of simplicity, only the main routes of administration will be mentioned: parental, oral, transdermal and pulmonary, and some of the studies that address the use of SFN for these different routes of administration will be cited as examples.

5.1 Parenteral

Parenteral administration forms are intended for administration by injection, which can be subdivided into intravenously (into a vein), intramuscular (into the muscle), subcutaneous (under the skin), or intradermal (into the skin). Parenteral administration acts faster than topical or enteral administration, and the onset of action often occurs in a range of seconds to minutes. Essentially, the bioavailability of the injected drug is 100% and its distribution is systemic, which means that it is potentially capable of reaching the entire body. This last concept paired with the EPR effect as mentioned above is of special interest for the treatment of tumor masses. For example, ZhuGe et al. [115] prepared SFN with their surface functionalized by proanthocyanidins and loaded with indocyanine green. Indocyanine can absorbe near-infrared light (650-900 nm) and producing a thermal effect both in vitro and in vivo. This photothermic compound is approved by the FDA and can be used to kill cells by photothermolysis. To test their functionality, the researchers injected the loaded nanoparticles intravenously into mice bearing C6 glioma. The pharmacokinetic study showed that the nanoparticles managed to reach the gliomas after intravenous administration in vivo, while the pharmacological study demonstrated inhibition of tumor growth after irradiation with near-infrared light. On the other hand, nanoparticles also offer temporary release control. Recently, in another study, Zhan et al. [116] administered Celastrol-loaded SFN to rats intravenously. The results showed that an increase in the total exposure time to the drug is reduced by increasing its residence time and reducing its metabolism.

5.2 Oral

Oral administration is the most common route and probably the one preferred by patients when receiving medications. However, conventional formulations, such as tablets and capsules, can release drugs in a rapid and poorly controlled manner, which can result in degradation and alteration of the drug due to the environment of the gastrointestinal tract (variations in pH and the presence of digestive enzymes and microbiota). Furthermore, the common mechanism of drug absorption through the gastrointestinal tract is passive diffusion. Consequently, most of the initial dose is not absorbed but is metabolized and excreted. SFN possess favorable characteristics to overcome the aforementioned problems and become candidates of interest for the oral administration of therapeutic compounds. Firstly, due to their mucoadhesive capacity, SFN can adhere firmly to the gastrointestinal mucosa or intestinal epithelial cells (Peyer's lymphatic M cells), followed by cell internalization via endocytosis [117]. Thus, encapsulated drugs can enter the bloodstream effectively and intact. Zhan et al. [116] increased more than doubled the absolute bioavailability of Celastrol from 3.14% to 7.56% by loading the drug in SFN and administering it orally to rodents.

5.3 Pulmonary

The lung is a potential target for drug delivery for both local and systemic treatments. Locally, lung and respiratory diseases, e.g., lung cancer or tuberculosis, can be treated with a reduced dose and fewer side effects compared to conventional dosage forms. At the systemic level, due to the large surface area of the lung, the drug can be absorbed quickly and efficiently without being degraded by the first-pass metabolism as in oral administration [118]. In 2015, Kim et al. [113] prepared cisplatin-loaded SFN for the treatment of lung cancer. The particles showed compatibility with the human lung epithelial cell line A549. The results indicated that the cisplatin loaded in the particles increases the cytotoxicity concerning the drug applied alone. The researchers concluded that the particles showed a high aerosolization performance through in vitro lung deposition measurement, which is at the level of commercially available dry powder inhalers.

5.4 Transdermal

The transdermal administration of drugs improves their bioavailability and is useful for systemic and local treatment as in pulmonary application. Takeuchi et al. [119] evaluated the in vivo permeability of 40 nm SFN through the skin using mice and demonstrated that the particles are capable of reaching the dermis in 6 hours in addition to the stratum corneum, hair follicles and epidermis that surrounds them.

6. Nanoparticle elimination

The use of nanoparticles in humans raises great doubts about their safety and their elimination capacity. If the removal is very fast, the nanoparticles will not reside long enough to fulfill their function. On the contrary, if the retention is very high, the concentration of nanoparticles can increase to the point of becoming harmful. Consequently, a relevant question in the use of nanoparticles in humans is how these biological systems can eliminate nanoparticles once their functions have been fulfilled. The properties of nanoparticles that affect their removal are mainly based on size, shape, composition, charge, and surface chemistry. These aspects will be briefly discussed within the two main elimination routes, (i) renal and (ii) hepatic elimination, to obtain a global vision of the process.

6.1 Renal elimination

The kidneys have the potential for rapid removal of particles from the vascular system without the need for decomposition. Renal elimination involves the mechanisms of glomerular filtration and tubular secretion to end in urinary excretion [120]. The nanoparticles reach the nephrons through the afferent arteriole, where they meet three endothelial barriers: the fenestrated endothelium; the highly negatively charged glomerular basement membrane; and the podocyte extensions of glomerular epithelial cells. The fenestrated epithelium has pores with a functional physiological diameter of between 9 and 10 nm, and a few (ca. 1%) pores of 15 to 23 nm in diameter [121], which means that nanoparticles with diameters less than 10 nm can spread freely regardless of the charge of the particle. The second barrier presented by the glomerular basement membrane filters particles between 6 and 8 nm depending on the electrostatic interactions between the nanoparticle and the membrane [122]. In this way, low-charged or positively charged nanoparticles can diffuse more freely. After glomerular filtration, the nanoparticles enter the lumen of Bowman's capsule, where they can be reabsorbed. Because the proximal tube epithelium is negatively charged, positively charged nanoparticles can be more easily reabsorbed.

Choi et al. [123] administered quantum-dots (inorganic nanoparticles) intravenously to rodents to study their renal elimination. The results indicated that particles with a hydrodynamic diameter less than 5.5 nm present rapid elimination and the increase in this diameter is inversely proportional to the retention time of the quantum-dots in the body.

6.2 Hepatic clearance and the reticulum endothelial system

Those nanoparticles that are too large to be excreted by the renal system must be eliminated by the hepatobiliary system. In 1924, Karl Albert Ludwig Aschoff coined the term reticuloendothelial system (RES) to describe a functional cellular system widely distributed in the body, composed of sessile and circulating macrophages of mesenchymal origin. These cells have a marked phagocytic capacity towards particulate matter. Macrophages stored in the RES can be found in the central nervous system (microglia), in the spleen, lymph nodes, tonsils, in the bone marrow (reticular cells) and, particularly, in the liver (90% of all macrophages) [124]. The exogenous structures are subjected to very intensive phagocytosis by the RES as well as the foreign proteins of higher molecular weight. Total blood flow must pass through the liver, making it a central organ to monitor the blood for endogenous, foreign substances and particles that must be removed for physiological reasons. In order to perform their functions, RES cells have special abilities such as: phagocytosis, pinocytosis, the release of signaling substances (cytokines, eicosanoids) and elimination of endotoxins, among others [124]. In addition, these are equipped with numerous pores of various diameters, depending on their different functions, which gives them the ability to filter larger molecules and particles, keeping them away from the liver parenchyma. The Kupffer cells and the endothelial sinus are in a privileged position to engulf any colloid foreign to the body. For this purpose, Kupffer cells are equipped with a branched and ciliated surface that act as capture mechanics. Besides, they possess specific receptors for carbohydrate components, as well as for the Fc region of IgG and for complement C_3, allowing them to differentiate the opsonized matter. They also possess lysosomal enzymes, although in much lower amounts than sinus endothelial cells.

In a very complete study, Poon et al. [125] proposed an algorithm to infer how nanoparticles can be eliminated in vivo (**Figure 4**). Most of the nanoparticles with diameters smaller than the glomerular filtration size limit (~5.5 nm) are eliminated by the kidneys and leave the body through the urine [123] although fecal elimination of small nanoparticles is also observed [125]. Biodegradable nanocarriers or nanoparticles larger than 5.5 nm can be decomposed [126, 127] or metabolized [128] and can be returned to the systemic circulation. Most non-biodegradable nanoparticles larger than 5.5 nm are retained long-term in Kupffer cells [129]. If the nanoparticles can evade Kupffer cells or if Kupffer cells are incapacitated, the nanoparticles can undergo hepatobiliary clearance. Similar to the glomerular filtration size limit, the authors proposed that there is a filtration size limit in hepatic sinusoidal endothelium. Nanoparticles larger than the fenestra of sinusoidal endothelium in the liver have restricted access to hepatocytes, whereas nanoparticles smaller than the

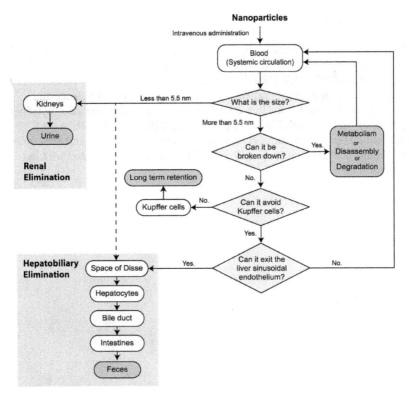

Figure 4.
Flow diagram for removal of nanoparticles in vivo. Reprinted with permission from reference [125]. Copyright 2019 American Chemical Society.

fenestrae have better access through the fenestra to enter the perisinusoidal space. In general, nanoparticles must escape these barriers established by non-parenchymal cells in the liver before they have the potential to enter the perisinusoidal space and interact with hepatocytes for elimination. Once the nanoparticles successfully interact with them, they can transit to enter the bile ducts. Finally, the nanoparticles enter the intestine and are eliminated from the body through the feces.

7. Conclusions

The academy and industry have made extraordinary advances in a wide variety of areas due to the development of nanotechnology and the control of structures at the nanoscopic levels. Particularly in the field of medicine, nanotechnology has the potential to generate a significant impact on human health, being able to improve the diagnosis, prevention and treatment of diseases. In this field, nanotechnology seeks to encapsulate drugs and/or tracer compounds in nanoparticles to increase their efficiency by allowing direct delivery to target tissues, while they reduce their toxicity avoiding accumulation and the consequent side effects in healthy tissues. The encapsulation of drugs also allows their controlled release, thus avoiding maximum levels of highly harmful or subtherapeutic concentrations. Moreover, nanoparticles are of great value in the transport of drugs with low solubility in water, which turns out to be the major problem when introducing new drugs to the market because it limits their bioavailability in the body. A wide variety of materials can be used for the preparation of nanoparticles depending on the intended function of the system.

Acknowledgements

This work has been partially supported by the European Commission (FEDER/ ERDF) and the Spanish Ministry of Science and Innovation (Refs. CTQ2017-87708-R and PID2020-113081RB-I00) through the Spanish State Research Agency, and the research support program of the Seneca Foundation of Science and Technology of Murcia, Spain (Ref. 20977/PI/18). Marta G. Fuster acknowledges support from Spanish MINECO (FPI grant, PRE2018-086441). Mercedes G. Montalbán acknowledges support from the University of Murcia and Santander Bank through the research project (Ref. RG2020-002UM) associated with her postdoctoral contract.

Author details

Guzmán Carissimi, Mercedes G. Montalbán*, Marta G. Fuster and Gloria Víllora
Faculty of Chemistry, Chemical Engineering Department, Regional Campus
of International Excellence "Campus Mare Nostrum", University of Murcia,
Murcia, Spain

*Address all correspondence to: mercedes.garcia@um.es

IntechOpen

References

[1] L Zhang, FX Gu, JM Chan, AZ Wang RL and OF. Nanoparticles in Medicine: Therapeutic Applications and Developments. Educ Policy Anal Arch. 2007;8(5):761-769.

[2] van der Meel R, Sulheim E, Shi Y, Kiessling F, Mulder WJM, Lammers T. Smart cancer nanomedicine. Nat Nanotechnol. 2019 Nov;14(11): 1007-1017.

[3] Jain RK, Stylianopoulos T. Delivering nanomedicine to solid tumors. Nat Publ Gr. 2010;7:653-664.

[4] Dogra P, Butner JD, Ruiz Ramírez J, Chuang Y, Noureddine A, Jeffrey Brinker C, et al. A mathematical model to predict nanomedicine pharmacokinetics and tumor delivery. Comput Struct Biotechnol J. 2020 Jan;18:518-531.

[5] Schaming D, Remita H. Nanotechnology: from the ancient time to nowadays. Found Chem. 2015;17(3): 187-205.

[6] Reibold M, Paufler P, Levin AA, Kochmann W, Pätzke N, Meyer DC. Carbon nanotubes in an ancient Damascus saber. Nature. 2006 Nov;444(7117):286-286.

[7] Feynman RP. There's plenty of room at the bottom. Gilbert ed. H, editor. New York: Reinhold Publishing Corp; 1961.

[8] Taniguchi N. On the basic concept of nanotechnology. Proceeding ICPE. 1974;

[9] Faraday M. Experimental relations of gold (and other metals) to light. Philos Trans R Soc London. 1857 Dec;147(0): 145-181.

[10] Mie G. Beiträge zur Optik trüber Medien, speziell kolloidaler Metallösungen. Ann Phys. 1908;330(3): 377-445.

[11] Richard Zsigmondy. Properties of colloids. Nobel Lect. 1926;1:1.

[12] Drexler KE. Molecular engineering: An approach to the development of general capabilities for molecular manipulation. Proc Natl Acad Sci U S A. 1981;78(9 I):5275-8.

[13] Matsumura Y, Maeda H. A New Concept for Macromolecular Therapeutics in Cancer Chemotherapy: Mechanism of Tumoritropic Accumulation of Proteins and the Antitumor Agent Smancs. Cancer Res. 1986;46(8):6387-6392.

[14] Youn YS, Bae YH. Perspectives on the past, present, and future of cancer nanomedicine. Adv Drug Deliv Rev. 2018;

[15] Bobo D, Robinson KJ, Islam J, Thurecht KJ, Corrie SR. Nanoparticle-Based Medicines: A Review of FDA-Approved Materials and Clinical Trials to Date. Pharm Res. 2016;33(10): 2373-2387.

[16] Mitchell MJ, Billingsley MM, Haley RM, Wechsler ME, Peppas NA, Langer R. Engineering precision nanoparticles for drug delivery. Nat Rev. Drug Discov. 2021;20(2):101-124.

[17] Global Industry Analysts I. Nanotechnology in Drug Delivery - Global Market Trajectory & Analytics. 2020.

[18] Jain PK, ElSayed IH, El-Sayed MA. Au nanoparticles target cancer. Nano Today. 2007;2(1):18-29.

[19] Anderson SD, Gwenin V V., Gwenin CD. Magnetic Functionalized Nanoparticles for Biomedical, Drug Delivery and Imaging Applications. Nanoscale Res Lett. 2019 Dec;14(1):1-16.

[20] Sharma A, Goyal AK, Rath G. Recent advances in metal nanoparticles

in cancer therapy. J Drug Target. 2018 Sep;26(8):617-632.

[21] Malekzad H, Sahandi Zangabad P, Mirshekari H, Karimi M, Hamblin MR. Noble metal nanoparticles in biosensors: recent studies and applications. Nanotechnol Rev. 2017 Jun;6(3): 301-329.

[22] O'Neal DP, Hirsch LR, Halas NJ, Payne JD, West JL. Photo-thermal tumor ablation in mice using near infrared-absorbing nanoparticles. Cancer Lett. 2004;209(2):171-176.

[23] Vines JB, Yoon J-H, Ryu N-E, Lim D-J, Park H. Gold Nanoparticles for Photothermal Cancer Therapy. Front Chem. 2019 Apr;7(APR):167.

[24] Jain PK, Huang X, El-Sayed IH, El-Sayed MA. Noble metals on the nanoscale: Optical and photothermal properties and some applications in imaging, sensing, biology, and medicine. Acc Chem Res. 2008;41(12): 1578-1586.

[25] Cao X, Ye Y, Liu S. Gold nanoparticle-based signal amplification for biosensing. Anal Biochem. 2011;417(1):1-16.

[26] Libutti SK, Paciotti GF, Byrnes AA, Alexander HR, Gannon WE, Walker M, et al. Phase I and pharmacokinetic studies of CYT-6091, a novel PEGylated colloidal gold-rhTNF nanomedicine. Clin Cancer Res. 2010;16(24):6139-6149.

[27] Coyne DW. Ferumoxytol for treatment of iron deficiency anemia in patients with chronic kidney disease. Expert Opin Pharmacother. 2009 Oct;10(15):2563-2568.

[28] Abo-zeid Y, Ismail NS, McLean GR, Hamdy NM. A molecular docking study repurposes FDA approved iron oxide nanoparticles to treat and control COVID-19 infection. Eur J Pharm Sci. 2020;153(April):105465.

[29] Schrand AM, Rahman MF, Hussain SM, Schlager JJ, Smith DA, Syed AF. Metal-based nanoparticles and their toxicity assessment. Wiley Interdiscip Rev. Nanomedicine Nanobiotechnology. 2010 Sep;2(5): 544-568.

[30] Park S, Lee YK, Jung M, Kim KH, Chung N, Ahn EK, et al. Cellular toxicity of various inhalable metal nanoparticles on human alveolar epithelial cells. In: Inhalation Toxicology. Taylor & Francis; 2007. p. 59-65.

[31] Sengupta J, Ghosh S, Datta P, Gomes A, Gomes A. Physiologically Important Metal Nanoparticles and Their Toxicity. J Nanosci Nanotechnol. 2014 Jan;14(1):990-1006.

[32] Yildirimer L, Thanh NTK, Loizidou M, Seifalian AM. Toxicological considerations of clinically applicable nanoparticles. Nano Today. 2011 Dec;6(6):585-607.

[33] Teleanu D, Chircov C, Grumezescu A, Volceanov A, Teleanu R. Impact of Nanoparticles on Brain Health: An Up to Date Overview. J Clin Med. 2018 Nov;7(12):490.

[34] Bozzuto G, Molinari A. Liposomes as nanomedical devices. Int J Nanomedicine. 2015;10:975-999.

[35] Bangham AD, Horne RW. Negative staining of phospholipids and their structural modification by surface-active agents as observed in the electron microscope. J Mol Biol. 1964 Jan;8(5):660-IN10.

[36] Cattel L, Ceruti M, Dosio F. From conventional to stealth liposomes: A new frontier in cancer chemotherapy. J Chemother. 2004;16(SUPPL. 4): 94-97.

[37] Torchilin V. Tumor delivery of macromolecular drugs based on the EPR

effect. Adv Drug Deliv Rev. 2011;63(3):131-135.

[38] Zamboni WC. Concept and Clinical Evaluation of Carrier-Mediated Anticancer Agents. Oncologist. 2008 Mar;13(3):248-260.

[39] Juliano RL, Stamp D. Pharmacokinetics of liposome-encapsulated anti-tumor drugs. Biochem Pharmacol. 1978 Jan;27(1):21-27.

[40] Daemen T, Hofstede G, Ten Kate MT, Bakker-Woudenberg IAJM, Scherphof GL. Liposomal doxorubicin-induced toxicity: Depletion and impairment of phagocytic activity of liver macrophages. Int J Cancer. 1995 May;61(5):716-721.

[41] Junghanns JUAH, Müller RH. Nanocrystal technology, drug delivery and clinical applications. Int J Nanomedicine. 2008;3(3):295-309.

[42] Sato M, Sambito MA, Aslani A, Kalkhoran NM, Slamovich EB, Webster TJ. Increased osteoblast functions on undoped and yttrium-doped nanocrystalline hydroxyapatite coatings on titanium. Biomaterials. 2006 Apr;27(11):2358-2369.

[43] Shegokar R, Müller RH. Nanocrystals: Industrially feasible multifunctional formulation technology for poorly soluble actives. Int J Pharm. 2010;399(1-2):129-139.

[44] Jawahar N, Meyyanathan S. Polymeric nanoparticles for drug delivery and targeting: A comprehensive review. Int J Heal Allied Sci. 2012;1(4):217.

[45] Montalbán M, Coburn J, Lozano-Pérez A, Cenis J, Víllora G, Kaplan D. Production of Curcumin-Loaded Silk Fibroin Nanoparticles for Cancer Therapy. Nanomaterials [Internet]. 2018 Feb 24 [cited 2018 Apr 27];8(2):126. Available from: http://www.ncbi.nlm.nih.gov/pubmed/29495296

[46] Nagaich U. Polymeric nanocapsules: An emerging drug delivery system. J Adv Pharm Technol Res. 2018;9(3):65.

[47] Zabihi F, Xin N, Jia J, Chen T, Zhao Y. High yield and high loading preparation of curcumin-PLGA nanoparticles using a modified supercritical antisolvent technique. Ind Eng Chem Res. 2014;53(15):6569-6574.

[48] Teixeira M, Pedro M, Nascimento MSJ, Pinto MMM, Barbosa CM. Development and characterization of PLGA nanoparticles containing 1,3-dihydroxy-2-methylxanthone with improved antitumor activity on a human breast cancer cell line. Pharm Dev Technol. 2019 Oct;24(9):1104-1114.

[49] Zhang Z, Tsai P-C, Ramezanli T, Michniak-Kohn BB. Polymeric nanoparticles-based topical delivery systems for the treatment of dermatological diseases. Wiley Interdiscip Rev. Nanomedicine Nanobiotechnology. 2013 May;5(3): 205-218.

[50] Zhang Z, Wang J, Chen C. Near-infrared light-mediated nanoplatforms for cancer thermo-chemotherapy and optical imaging. Adv Mater. 2013 Jul;25(28):3869-3880.

[51] Ren H, Huang X. Polyacrylate nanoparticles: toxicity or new nanomedicine? Eur Respir J. 2010 Jul;36(1):218-221.

[52] El-Ansary A, Faddah LM. Nanoparticles as biochemical sensors. Nanotechnol Sci Appl. 2010;3(1):65-76.

[53] Etheridge ML, Campbell SA, Erdman AG, Haynes CL, Wolf SM, McCullough J. The big picture on nanomedicine: The state of

investigational and approved nanomedicine products. Nanomedicine Nanotechnology, Biol Med. 2013;9(1):1-14.

[54] Pham DT, Tiyaboonchai W. Fibroin nanoparticles: a promising drug delivery system. Drug Deliv. 2020;27(1):431-448.

[55] Lammers T, Ferrari M. The success of nanomedicine. Nano Today. 2020 Apr;31:100853.

[56] Chen H, Zhen Z, Todd T, Chu PK, Xie J. Nanoparticles for improving cancer diagnosis. Vol. 74, Materials Science and Engineering R: Reports. Elsevier Ltd.; 2013. p. 35-69.

[57] Maeda H, Matsumura Y. Tumoritropic and lymphotropic principles of macromolecular drugs. Crit Rev. Ther Drug Carrier Syst. 1989;6(3):193-210.

[58] Maeda H, Bharate GY, Daruwalla J. Polymeric drugs for efficient tumor-targeted drug delivery based on EPR-effect. Eur J Pharm Biopharm. 2009;71(3):409-419.

[59] Greish K, Fang J, Inutsuka T, Nagamitsu A, Maeda H. Macromolecular Therapeutics. Clin Pharmacokinet. 2003;42(13):1089-1105.

[60] Maeda H. Towards a full understanding of the EPR effect in primary and metastatic tumors as well as issues related to its heterogeneity. Adv Drug Deliv Rev. 2015;91:3-6.

[61] Maeda H, Nakamura H, Fang J. The EPR effect for macromolecular drug delivery to solid tumors: Improvement of tumor uptake, lowering of systemic toxicity, and distinct tumor imaging in vivo. Adv Drug Deliv Rev. 2013;65(1):71-79.

[62] Jokerst J V., Lobovkina T, Zare RN, Gambhir SS. Nanoparticle PEGylation for imaging and therapy. Vol. 6,

Nanomedicine. NIH Public Access; 2011. p. 715-728.

[63] Yuan F, Dellian M, Fukumura D, Leunig M, Berk DA, Jain RK, et al. Vascular Permeability in a Human Tumor Xenograft: Molecular Size Dependence and Cutoff Size. Cancer Res. 1995;55(17):3752-3756.

[64] Smith BR, Kempen P, Bouley D, Xu A, Liu Z, Dai H, et al. Shape Matters: Intravital Microscopy Reveals Surprising Geometrical Dependence for Nanoparticles in Tumor Models of Extravasation. 2013;12(7):3369-3377.

[65] Iwai K, Maeda H, Konno T. Use of Oily Contrast Medium for Selective Drug Targeting to Tumor: Enhanced Therapeutic Effect and X-Ray Image. Cancer Res. 1984;44(5):2115-2121.

[66] Danhier F. To exploit the tumor microenvironment: Since the EPR effect fails in the clinic, what is the future of nanomedicine? J Control Release. 2016;244:108-121.

[67] Nichols JW, Bae YH. Odyssey of a cancer nanoparticle: From injection site to site of action. Nano Today. 2012;7(6):606-618.

[68] Carter CL, Allen C, Henson DE. Relation of tumor size, lymph node status, and survival in 24,740 breast cancer cases. Cancer. 1989;63(181).

[69] Gillies RJ, Schornack PA, Secomb TW, Raghunand N. Causes and Effects of Heterogeneous Perfusion in Tumors. Neoplasia. 1999;1(3):197-207.

[70] Jain RK. Transport of Molecules in the Tumor Interstitium: A Review. Cancer Res. 1987;47(12):3039-3051.

[71] Fang J, Islam R, Islam W, Yin H, Subr V, Etrych T, et al. Augmentation of EPR Effect and Efficacy of Anticancer Nanomedicine by Carbon Monoxide

Generating Agents. Pharmaceutics. 2019;11(7):343.

[72] Tahara Y, Yoshikawa T, Sato H, Mori Y, Zahangir MH, Kishimura A, et al. Encapsulation of a nitric oxide donor into a liposome to boost the enhanced permeation and retention (EPR) effect. Medchemcomm. 2017;8(2):415-421.

[73] Islam W, Fang J, Imamura T, Etrych T, Subr V, Ulbrich K, et al. Augmentation of the enhanced permeability and retention effect with nitric oxide–generating agents improves the therapeutic effects of nanomedicines. Mol Cancer Ther. 2018;17(12):2643-2653.

[74] Nagamitsu A, Greish K, Maeda H. Elevating blood pressure as a strategy to increase tumor-targeted delivery of macromolecular drug SMANCS: Cases of advanced solid tumors. Jpn J Clin Oncol. 2009;39(11):756-766.

[75] Balasubramanian V, Domanskyi A, Renko JM, Sarparanta M, Wang CF, Correia A, et al. Engineered antibody-functionalized porous silicon nanoparticles for therapeutic targeting of pro-survival pathway in endogenous neuroblasts after stroke. Biomaterials. 2020 Jan;227:119556.

[76] Bian X, Wu P, Sha H, Qian H, Wang Q, Cheng L, et al. Anti-EGFR-iRGD recombinant protein conjugated silk fibroin nanoparticles for enhanced tumor targeting and antitumor efficiency. Onco Targets Ther. 2016;9:3153-3162.

[77] Heo DN, Yang DH, Moon HJ, Lee JB, Bae MS, Lee SC, et al. Gold nanoparticles surface-functionalized with paclitaxel drug and biotin receptor as theranostic agents for cancer therapy. Biomaterials. 2012;33(3):856-866.

[78] Bhattacharya D, Das M, Mishra D, Banerjee I, Sahu SK, Maiti TK, et al. Folate receptor targeted, carboxymethyl chitosan functionalized iron oxide nanoparticles: A novel ultradispersed nanoconjugates for bimodal imaging. Nanoscale. 2011 Apr;3(4):1653-1662.

[79] Lammers T, Kiessling F, Hennink WE, Storm G. Drug targeting to tumors: Principles, pitfalls and (pre-) clinical progress. J Control Release. 2012;161(2):175-187.

[80] Danhier F, Feron O, Préat V. To exploit the tumor microenvironment: Passive and active tumor targeting of nanocarriers for anti-cancer drug delivery. J Control Release. 2010;148(2):135-146.

[81] Kirpotin DB, Drummond DC, Shao Y, Shalaby MR, Hong K, Nielsen UB, et al. Antibody targeting of long-circulating lipidic nanoparticles does not increase tumor localization but does increase internalization in animal models. Cancer Res. 2006;66(13): 6732-6740.

[82] Jiménez Blanco JL, Benito JM, Ortiz Mellet C, García Fernández JM. Molecular nanoparticle-based gene delivery systems. J Drug Deliv Sci Technol. 2017;42:18-37.

[83] Wong J, Brugger A, Khare A, Chaubal M, Papadopoulos P, Rabinow B, et al. Suspensions for intravenous (IV) injection: A review of development, preclinical and clinical aspects. Adv Drug Deliv Rev. 2008;60(8):939-954.

[84] Loftsson T, Brewster ME. Pharmaceutical applications of cyclodextrins: Basic science and product development. J Pharm Pharmacol. 2010;62(11):1607-1621.

[85] Morgan S, Grootendorst P, Lexchin J, Cunningham C, Greyson D. The cost of drug development: A systematic review. Health Policy (New York). 2011;100(1):4-17.

[86] Paul SM, Mytelka DS, Dunwiddie CT, Persinger CC,

Munos BH, Lindborg SR, et al. How to improve RD productivity: The pharmaceutical industry's grand challenge. Nat Rev. Drug Discov. 2010;9(3):203-214.

[87] Services H. Waiver of In-vivo Bioavailability and Bioequivalence Studies for Immediate Release Solid Oral Dosage Forms Based on a Biopharmaceutics Classification System. US Dep Heal Hum Serv Food Drug Adm Cent Drug Eval Res. 2017;3-5.

[88] Kalepu S, Nekkanti V. Insoluble drug delivery strategies: Review of recent advances and business prospects. Acta Pharm Sin B. 2015;5(5):442-453.

[89] Müller RH, Jacobs C, Kayser O. Nanosuspensions as particulate drug formulations in therapy: Rationale for development and what we can expect for the future. Adv Drug Deliv Rev. 2001;47(1):3-19.

[90] Xie M-B, Li Y, Zhao Z, Chen A-Z, Li J-S, Hu J-Y, et al. Solubility enhancement of curcumin via supercritical CO2 based silk fibroin carrier. J Supercrit Fluids [Internet]. 2015 Aug [cited 2016 Apr 26];103:1-9. Available from: http://www.sciencedirect.com/science/article/pii/S0896844615001679

[91] Fuster MG, Carissimi G, Montalbán MG, Víllora G. Improving Anticancer Therapy with Naringenin-Loaded Silk Fibroin Nanoparticles. Nanomater (Basel, Switzerland). 2020 Apr;10(4).

[92] Uhrich KE, Cannizzaro SM, Langer RS, Shakesheff KM. Polymeric systems for controlled drug release. Chem Rev. 1999;99(11):3181-3198.

[93] Kost J, Horbett TA, Ratner BD, Singh M. Glucose-sensitive membranes containing glucose oxidase: Activity, swelling, and permeability studies. J Biomed Mater Res. 1985 Nov;19(9):1117-1133.

[94] Volpatti LR, Matranga MA, Cortinas AB, Delcassian D, Daniel KB, Langer R, et al. Glucose-Responsive Nanoparticles for Rapid and Extended Self-Regulated Insulin Delivery. ACS Nano. 2020;14(1):488-497.

[95] Chen M, Shao Z, Chen X. Paclitaxel-loaded silk fibroin nanospheres. J Biomed Mater Res - Part A. 2012;100 A(1):203-10.

[96] Jones RT. Blood Flow. Annu Rev. Fluid Mech. 1969 Jan;1(1):223-244.

[97] Moghimi SM, Patel HM. Serum-mediated recognition of liposomes by phagocytic cells of the reticuloendothelial system - The concept of tissue specificity. Adv Drug Deliv Rev. 1998;32(1-2):45-60.

[98] Tenzer S, Docter D, Kuharev J, Musyanovych A, Fetz V, Hecht R, et al. Rapid formation of plasma protein corona critically affects nanoparticle pathophysiology. Nat Nanotechnol. 2013;8(10):772-781.

[99] Salvati A, Pitek AS, Monopoli MP, Prapainop K, Bombelli FB, Hristov DR, et al. Transferrin-functionalized nanoparticles lose their targeting capabilities when a biomolecule corona adsorbs on the surface. Nat Nanotechnol. 2013;8(2):137-143.

[100] Milton Harris J, Chess RB. Effect of pegylation on pharmaceuticals. Nat Rev. Drug Discov. 2003;2(3):214-221.

[101] Totten JD, Wongpinyochit T, Carrola J, Duarte IF, Seib FP. PEGylation-Dependent Metabolic Rewiring of Macrophages with Silk Fibroin Nanoparticles. ACS Appl Mater Interfaces. 2019;11(16):14515-14525.

[102] Rodriguez PL, Harada T, Christian DA, Pantano DA, Tsai RK, Discher DE. Minimal "Self" Peptides That Inhibit Phagocytic Clearance and Enhance Delivery of Nanoparticles.

Science (80-). 2013 Feb;339(6122): 971-5.

[103] Wernig G, Chen SY, Cui L, Van Neste C, Tsai JM, Kambham N, et al. Unifying mechanism for different fibrotic diseases. Proc Natl Acad Sci U S A. 2017;114(18):4757-4762.

[104] Hu C-MJCMJ, Zhang LL, Aryal S, Cheung C, Fang RH, Zhang LL. Erythrocyte membrane-camouflaged polymeric nanoparticles as a biomimetic delivery platform. Proc Natl Acad Sci U S A. 2011;108(27):10980-10985.

[105] Parodi A, Quattrocchi N, Van De Ven AL, Chiappini C, Evangelopoulos M, Martinez JO, et al. Synthetic nanoparticles functionalized with biomimetic leukocyte membranes possess cell-like functions. Nat Nanotechnol. 2013;8(1):61-68.

[106] Nikitin MP, Zelepukin I V., Shipunova VO, Sokolov IL, Deyev SM, Nikitin PI. Enhancement of the blood-circulation time and performance of nanomedicines via the forced clearance of erythrocytes. Nat Biomed Eng. 2020;4(7):717-731.

[107] Catto V, Farè S, Cattaneo I, Figliuzzi M, Alessandrino A, Freddi G, et al. Small diameter electrospun silk fibroin vascular grafts: Mechanical properties, in vitro biodegradability, and in vivo biocompatibility. Mater Sci Eng C. 2015;54:101-111.

[108] Tan M, Liu W, Liu F, Zhang W, Gao H, Cheng J, et al. Silk fibroin-coated nanoagents for acidic lysosome targeting by a functional preservation strategy in cancer chemotherapy. Theranostics. 2019;9(4):961-973.

[109] Meng F, Cheng R, Deng C, Zhong Z. Intracellular drug release nanosystems. Mater Today. 2012;15(10):436-442.

[110] Zhao Z, Li Y, Xie M Bin. Silk fibroin-based nanoparticles for drug delivery. Int J Mol Sci. 2015;16(3): 4880-4903.

[111] Jordan A, Scholz R, Wust P, Fähling H, Krause J, Wlodarczyk W, et al. Effects of Magnetic Fluid Hyperthermia (MFH) on C3H mammary carcinoma in vivo. Int J Hyperth. 1997;13(6):587-605.

[112] Dianzani C, Zara GP, Maina G, Pettazzoni P, Pizzimenti S, Rossi F, et al. Drug delivery nanoparticles in skin cancers. Biomed Res Int. 2014;2014(July).

[113] Kim SY, Naskar D, Kundu SC, Bishop DP, Doble PA, Boddy A V., et al. Formulation of Biologically-Inspired Silk-Based Drug Carriers for Pulmonary Delivery Targeted for Lung Cancer. Sci Rep. 2015;5(April):1-13.

[114] Paranjpe M, Müller-Goymann CC. Nanoparticle-mediated pulmonary drug delivery: A review. Int J Mol Sci. 2014;15(4):5852-5873.

[115] ZhuGe DL, Wang LF, Chen R, Li XZ, Huang ZW, Yao Q, et al. Cross-linked nanoparticles of silk fibroin with proanthocyanidins as a promising vehicle of indocyanine green for photo-thermal therapy of glioma. Artif Cells, Nanomedicine Biotechnol. 2019;47(1):4293-4304.

[116] Zhan S, Paik A, Onyeabor F, Ding B, Prabhu S, Wang J. Oral Bioavailability Evaluation of Celastrol-Encapsulated Silk Fibroin Nanoparticles Using an Optimized LC–MS/MS Method. Molecules. 2020 Jul;25(15):3422.

[117] Brooks AE. The potential of silk and silk-like proteins as natural mucoadhesive biopolymers for controlled drug delivery. Front Chem. 2015;3(NOV):1-8.

[118] Gaul R, Ramsey JM, Heise A, Cryan SA, Greene CM. Nanotechnology

approaches to pulmonary drug delivery: Targeted delivery of small molecule and gene-based therapeutics to the lung. Design of Nanostructures for Versatile Therapeutic Applications. Elsevier Inc.; 2018. 221-253 p.

[119] Takeuchi I, Shimamura Y, Kakami Y, Kameda T, Hattori K, Miura S, et al. Transdermal delivery of 40-nm silk fibroin nanoparticles. Colloids Surfaces B Biointerfaces. 2019 Mar;175(November 2018):564-568.

[120] Deen WM, Lazzara MJ, Myers BD. Structural determinants of glomerular permeability. Am J Physiol - Ren Physiol. 2001;281(4 50-4).

[121] M. Ohlson, J. Sörensson BH. A gel-membrane model of glomerular charge and size selectivity in series. Am J Physiol. 2001;280:396-405.

[122] Longmire M, Choyke PL, Kobayashi H. Clearance Properties of Nano-sized Particles and Molecules as Imaging Agents: Considerations and Caveats. Nanomedicine. 2012;3(5):703-717.

[123] Soo Choi H, Liu W, Misra P, Tanaka E, Zimmer JP, Itty Ipe B, et al. Renal clearance of quantum dots. Nat Biotechnol. 2007;25(10):1165-1170.

[124] Kuntz E, Kuntz H-D. Hepatology Principles and Practices. Vol. 53, Journal of Chemical Information and Modeling. 2013. 1689-1699 p.

[125] Poon W, Zhang YN, Ouyang B, Kingston BR, Wu JLY, Wilhelm S, et al. Elimination Pathways of Nanoparticles. ACS Nano. 2019;13(5):5785-5798.

[126] Hwang KJ, Luk KFS, Beaumier PL. Hepatic uptake and degradation of unilamellar sphingomyelin/cholesterol liposomes: A kinetic study. Proc Natl Acad Sci U S A. 1980;77(7 II):4030-4.

[127] Mohammad AK, Reineke JJ. Quantitative detection of PLGA nanoparticle degradation in tissues following intravenous administration. Mol Pharm. 2013;10(6):2183-2189.

[128] Lu Y, Chen SC. Micro and nano-fabrication of biodegradable polymers for drug delivery. Adv Drug Deliv Rev. 2004;56(11):1621-1633.

[129] Sadauskas E, Wallin H, Stoltenberg M, Vogel U, Doering P, Larsen A, et al. Kupffer cells are central in the removal of nanoparticles from the organism. Part Fiber Toxicol. 2007;4(May 2014).

Chapter 12

Biological Synthesis of Metallic Nanoparticles from Different Plant Species

Kalyan Singh Kushwah and Deepak Kumar Verma

Abstract

Green chemistry for the synthesis of different nanoparticles (NPs) from metal has become a new and promising field of research in nanotechnology in recent years. The inspire applications of metal oxide NPs have attracted the interest of researchers around the world. Various physical, chemical and biological methods in material science are being adapted to synthesize different types of NPs. Green synthesis has gained widespread attention as a sustainable, reliable, and eco-friendly protocol for biologically synthesizing a wide range of metallic NPs. Green synthesis has been proposed to reduce the use of hazardous compounds and as a state of a harsh reaction in the production of metallic NPs. Plants extract used for biosynthesis of NPs such as silver (Ag), cerium dioxide (C_2O_2), copper oxide (CuO), Gold (Au), titanium dioxide (TiO_2), and zinc oxide (ZnO). This review article gives an overview of the plant-mediated biosynthesis of NPs that are eco-friendly and have less hazardous chemical effects.

Keywords: biosynthesis, metallic nanoparticles, plant extract

1. Introduction

In the last decade, novel synthesis methods for nanomaterials such as quantum dots (QDs), carbon nanotubes (CNTs), graphene, and their composites have been an interesting area in nanotechnology [1]. Despite the progress of the use of small metallic materials listed as nano, the debate still continues in many aspects associated with this new technological revolution. The conceptual beginnings of green chemistry and nanotechnology are among the great scientific developments that have influenced the design of experiments with the goal of environmental protection. Now the reduction in size of green chemistry and nanotechnology is one of the great scientific developments which are to preserve the environment with experiments [2]. Nanoparticles are the most fundamental component in the making of nanostructures. They are much smaller than the everyday objects around us governed by Newton's laws of motion but larger than an atom or a simple molecule that is the subject of quantum mechanics [3]. Nanoparticles exhibit specific properties that depend on their shape, size, and morphology and enable them to interact with plants, microbes, animals [4]. Nanoparticles are a subclass of ultrafine particles with a length of more than 1 nm and less than 100 nm in two or three dimensions and which cannot exhibit size-related depth properties

IntechOpen

that can vary the NP often their morphological or physical. Their application in controlling microbial growth in green synthesis of NPs and electronics, catalysts, drugs, and biological systems has made them eco-friendly [5]. 'Green Synthesis' is attracting a lot of attention in current research and development on materials science methods and technology. Science will basically make green synthesis of nanomaterials through regulation, control, cleaning and therapeutic process. Some basic principles of its environmental friendliness can thus be explained by several factors such as waste prevention, pollution reduction, and the use of safe non-toxic solvent as well as renewable [6]. Nanoparticles have many applications in several fields, such as microelectronics, hydrogen storage ferrofluids, catalytic systems, and chemical nanosensors as well as nanomedicine, agriculture, food science, and energy [7]. The metallic nanoparticles have unique properties that are different from fine-grained materials which use for many agricultural, industrial, and domestic applications, resulting in increased demand and production of nanoparticles. The list of nano-based commodities silica, iron, titania, alumina, and zinc oxide [8]. These types of NPs are the most white pigment and are being used in many products such as paints, plastics, paper, etc. [9] as well as chromosomal mutations in *Vicia faba* plants [10]. The number of multidrug-resistant bacteria and viral strains has been steadily increasing due to mutation, pollution, and changing environmental conditions are trying to develop drugs for the treatment of this microorganism infection to protect against this disease. Metal Nanoparticles have been found to be effective in inhibiting the growth of much infectious bacterial silver and occupy a prominent place in the category of Ag NPs metals used as antimicrobial agents [11]. Green synthesis methodologies based on biological precursors depend on various reaction parameters such as solvent, temperature, pressure, and pH conditions for the synthesis of various nanoparticles, broadly based on the availability of effective phytochemicals in various plant extracts. The leaves contain ketones, amides, terpenoids, carboxylic acids, aldehydes, flavones, phenols, and ascorbic acids that are capable of reducing the metal salts into metal nanoparticles [12]. *Chrysanthemum carinatum* is herbaceous perennial plant have deeply tapering leaves and large white flowers on the wall and extracts play a very important role in reducing and stabilizing agents that reduce cost production and environmental impact [13, 14].

2. Plant-based biosynthesis of metallic NPs

Many researchers have discovered many plant species and components that contain antioxidant compounds such as nitrogenous base amino acids, polyphenols, and sugars [15]. These compounds perform the function of a capping agent for the synthesis of nanoparticles [16]. Generally metal nanoparticles are manufactured in two ways such as bottom-up (fabrication of material from atom to bottom atom) and top to bottom (cutting of bulk material to obtain nano-sized particles). Metal and metal oxide NPs made from plant extract are generally stable even after 1 month and do not undergo any change [17]. Green synthesis of various metallic nanoparticles has reevaluated plants for their natural ability to reduce toxic and hazardous chemicals (**Figure 1**). Nanoparticles of plant-related parts such as leaves, stems, flowers, bark, roots, seeds, and their metabolites have been used for biosynthesis [19]. Generally, there are three aspects that need to be considered for green biosynthesis solvent medium, non-toxic ion reducing agents, and environmentally safe NPs stabilizers [20].

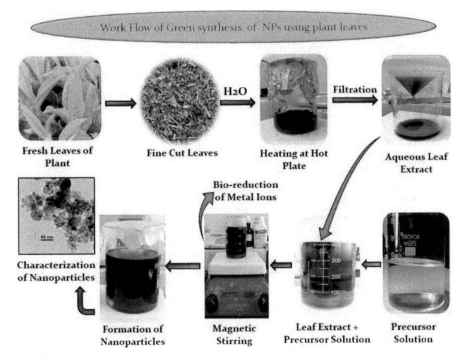

Figure 1.
Work flow of plant-based biosynthesis of metallic nanoparticles [18].

3. Biosynthesis of silver (Ag) nanoparticles

Plants have the ability to store heavy metals in their various parts. As a result, biosynthesis techniques employing plant extracts have produced a simple means to stabilize efficient growth, simple, cost-effective synthesis process, along with traditional preparation methods for the production of NPs [21]. Silver NPs have been synthesized and studied extensively due to their unique chemical, physical and

Biosynthesized NPs	Plant extract used	Size of NPs	Morphology of NPs	Reference
Ag NPs	*Erigeron bonariensis*	13	Spherical	[24]
Ag NPs	*Nigella sativa*	15	Spherical	[25]
Ag Nps	*Morinda tinctoria*	60–95	Spherical	[26]
Ag NPs	*Pedalium murex*	50	Spherical	[27]
Ag NPs	*Adhatoda vasica*	10–50	Spherical	[28]
CeO$_2$ NPs	*Gloriosa superba*	5	Spherical	[29]
CeO$_2$ NPs	*Centella asiatica*	19	Spherical	[30]
CeO$_2$ NPs	*Hibiscus sabdariffa*	4	Spherical	[31]
CeO$_2$ NPs	*Cymbopogon flexuosus*	10–40	Spherical	[32]
CeO$_2$ NPs	*Leucas aspera*	4–13	Spherical	[33]
CuO NPs	*Aloe barbadensis*	15–30	Spherical	[34]
CuO NPs	*Ixora coccinea*	80–110	Spherical	[35]

Biosynthesized NPs	Plant extract used	Size of NPs	Morphology of NPs	Reference
CuO NPs	Syzygium alternifolium	17	Spherical	[36]
CuO NPs	Leucaena leucocephala	10–25	Spherical	[37]
CuO NPs	Moringa oleifera	6–61	Spherical	[38]
Au NPs	Cinnamomum zeylanicum	25	Spherical	[39]
Au NPs	Scutellaria barbata	20–35	Spherical	[40]
Au NPs	Sesbania drummondii	6–20	Spherical	[41]
Au NPs	Avena sativa	5–20	Rod	[42]
Au NPs	Medicago sativa	2–40	Tetrahedral	[43]
TiO$_2$ NPs	Psidium guajava	32	Spherical	[44]
TiO$_2$ NPs	Solanum trilobatum	70	Spherical	[45]
TiO$_2$ NPs	Nyctanthes arbortristis	100–150	Spherical	[46]
TiO$_2$ NPs	Catharanthus roseus	25–110	Irregular	[47]
TiO$_2$ NPs	Annona squamosa	21–25	Spherical	[48]
ZnO NPs	Sedum alfredii	53	Hexagonal	[49]
ZnO NPs	Ruta graveolens	28	Hexagonal	[50]
ZnO NPs	Azadirachta indica	9–25	Spherical	[51]
ZnO NPs	Eichhornia crassipes	28–36	Crystalline	[52]
ZnO NPs	Plectranthus amboinicus	50–180	Hexagonal	[53]

Table 1.
Different type of metallic NPs biosynthesized from plants extracts.

biological properties [22]. Plants have biomolecules such as proteins, carbohydrates and coenzymes with the ability to reduce metal salts. Other biosynthesis processes of Ag NPs from herbal extracts perform extract-assisted biological synthesis and have yielded many advantages over chemical and physical methods of synthesis of NPs. It is also a fact that these routes are simple, cost-effective, eco-friendly for high production [23]. Synthesis of Ag NPs from *Arbutus unedo* leaf extracts and its organic compounds are responsible for the stabilization and reduction of ions. Various green-synthesized Ag NPs for natural capping agents can be used for drug delivery in pharmaceutical drugs. In most studies, the synthesis of Ag NPs by plant extracts is a simple method (**Table 1**) [54].

4. Biosynthesis of cerium dioxide (CeO$_2$) nanoparticles

Plants produce biologically active compounds to protect themselves. In plants, it is well organized and unique with various valuable metabolites that use phytochemical biologically active substances. These biocompounds are responsible for the reduction of metal ions during the synthesis of NPs [55]. Cerium dioxide has high catalytic properties due to their broad band gap energy and binding energy, indicating many applications [56]. Depending on the chemistry, the reaction kinetics are affected by the reaction temperature, and this determines the time that most of the temperature difference occurs as a result of changes in the synthesis of NPs [57]. Various compounds such as cerium nitrate, cerium acetate, and cerium chloride have been used as synthesis precursors, and extracts of several plant parts such as leaves, flowers, have been used as reducing agents. The synthesized CeO_2 NPs are spherical shape and crystalline in nature (**Table 1**) [58].

5. Biosynthesis of copper oxide (CuO) nanoparticles

The parameters of plant extracts such as phytochemicals, metal salt concentration, temperature and pH control the rate of formation of nanoparticles as well as their stability and yield [59]. Extracts of plant leaves are considered a good source for metal and metal oxide NPs synthesis. Additionally, plant leaf extracts play a greater role as reducing and stabilizing agents of the biosynthesis process, the phytochemical composition of plant leaf extract is also an important factor in the synthesis of CuO NPs [60]. Green synthesis of Cu NPs can reduce the use of many hazardous chemical substances; many methods use L-ascorbic acid as a reducing agent in synthesis, and the synthesized Cu NPs were highly stable (**Table 1**) [61].

6. Biosynthesis of gold (Au) nanoparticles

Water is the most accessible and cheapest solvent on earth. Since the advent of nanotechnology, water has been used as a solvent for the synthesis of various NPs [62]. Most research on the biosynthesis of NPs was done from angiosperms using plant extracts, mainly in gold NPs synthesis. Mechanisms of metal depletion and stabilization have also been explored by phytochemicals of plants. They can be helpful in forming NPs [63]. Plant extracts showed the green synthesized NPs to be effective in ions reduction. High variability suggests that this may be one of the reasons that produced large amounts of Au NPs using plant extract (**Table 1**) [64].

7. Biosynthesis of titanium dioxide (TiO₂) nanoparticles

The high prevalence of TiO_2 NPs is due to their many versatile applications resulting from the stability of chemical structure, optical, electrical, and physical properties. These characteristics offer a wide range of TiO_2 is present in three different mineral forms such as anatase, rutile, and brocite. It is generally preferred due to its photocatalytic activity [9]. The synthesis of TiO_2 NPs by plant leaf extracts is mixed with metal precursor solutions in various reactions and temperature conditions [65]. The high production and widespread use of TiO_2 NPs have led to little effort for biogenic production. Several plant species have been investigated for the production of TiO_2 NPs. Several processes have been developed in the last two decades to synthesize biogenic NPs. These methods effectively control the properties of NPs in which most of the potential toxicity materials are used (**Table 1**) [66].

8. Biosynthesis of zinc oxide (ZnO) nanoparticles

The biosynthesis of ZnO NPs an alternative to physical and chemical methods for the formation of NPs, it is representing an area of significant discovery for a wide range of applications (**Figure 2**) [73]. In recent times, the green method for the synthesis of NPs has become an area of great interest in this direction because the use of conventional chemical methods is very expensive, and as a reducing agent of chemical compounds or organic solvents use is required which produces toxicity [74]. Various studies where the plant-based synthesis of ZnO NPs from various zinc

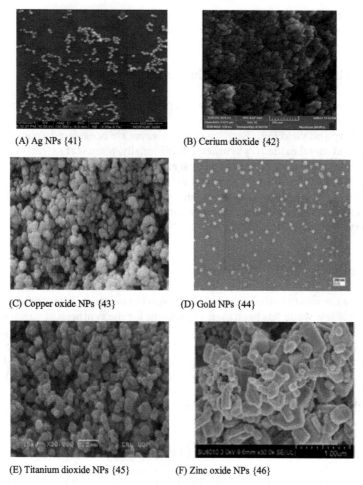

(A) Ag NPs {41} (B) Cerium dioxide {42}

(C) Copper oxide NPs {43} (D) Gold NPs {44}

(E) Titanium dioxide NPs {45} (F) Zinc oxide NPs {46}

Figure 2.
Scanning electron microscopic images of biosynthesized NPs: (A) Silver NPs [67], (B) Cerium dioxide NPs [68], (C) Copper oxide NPs [69], (D) Gold NPs [70], (E) Titanium dioxide NPs [71], (F) Zinc oxide NPs [72].

compounds was made using extracts from different parts of the plants are represented in **Table 1** [75].

XPS spectra of AgNPs were measured using a hemispherical analyzer (Physical Electronics 1257 system). For the XPS, a twin anode (Mg and Al) with X-ray source was operated at 400 W of constant power and using Al Kα radiation (1486.6 eV). The samples were placed in a sample stage with an emission angle of 45°. The measurements were carried out by suspending AgNPs on a gold film while gold was served as metallic reference. Au 4f binding energy was 84 eV for samples without any charging effect (**Figure 3**).

Raman spectra were measured using a Bruker Raman spectrometer; model Senterra with laser excitation at 633 nm, and laser power at 10 mW. Spectral data were collected using a 50× microscope objective (NA = 0.51) with 30 s integration time. The surface-enhanced Raman scattering (SERS) samples were prepared by mixing 360 µL of colloidal solution with 40 µL of aqueous solutions of the probe molecule, resulting in a final R6G concentration of 1.0×10^{-5} mol L^{-1} (**Figure 4**).

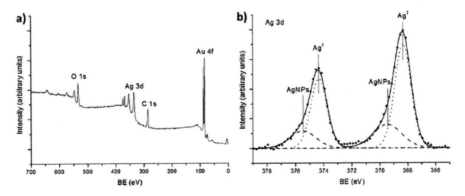

Figure 3.
X-ray photoelectron Spectroscopy of AgNPs [76].

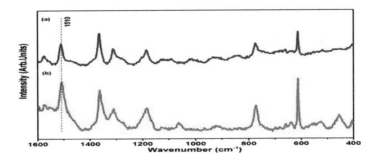

Figure 4.
Surface-enhanced Raman scattering spectrum of AgNPs [77].

9. Conclusion

Today, nanotechnology is considered to be very essential to help promote ultra-modern farming systems with very little environmental damage. Since the beginning, methods of synthesis of plant-based NPs using bioprocess have been considered eco-friendly and cost-effective for the last decades. Many types of natural extracts such as plant leaves have been employed as efficient resources for the synthesis of NPs. A large number of NPs are being explored in many areas of agriculture and biotechnology.

Conflict of interest

Authors declare that there is no conflict of interest for publication of this article.

Author details

Kalyan Singh Kushwah* and Deepak Kumar Verma
School of Studies in Botany, Jiwaji University, Gwalior, MP, India

*Address all correspondence to: kalyansinghkushwah85@gmail.com

IntechOpen

References

[1] Kim JS, Kuk E, Yu KN, Kim JH, Park SJ, Lee HJ, et al. Antimicrobial effects of silver nanoparticles. Nanomedicine (London). 2007;3:95-101

[2] Ahluwalia VK. Green Chemistry: Environmentally Benign Reaction. India: Ane Books Pvt Ltd.; 2009

[3] Horikoshi S, Serpone N, editors. Microwaves in Nanoparticle Synthesis: Fundamentals and Applications. Weinheim: John Wiley & Sons; 2013

[4] Husen A, Siddiqi KS. Phytosynthesis of nanoparticles: Concept, controversy and application. Nanoscale Research Letters. 2014;9(1):229

[5] Lokina S, Stephen A, Kaviyarasan V, Arulvasu C, Narayanan V. Cytotoxicity and antimicrobial activities of green synthesized silver nanoparticles. European Journal of Medicinal Chemistry. 2014;76:256-263

[6] Singh J, Dutta T, Kim KH, Rawat M, Samddar P, Kumar P. Green'synthesis of metals and their oxide nanoparticles: Applications for environmental remediation. Journal of Nanobiotechnology. 2018;16(1):84

[7] Pal SL, Jana U, Manna PK, Mohanta GP, Manavalan R. Nanoparticle: An overview of preparation and characterization. Journal of Applied Pharmaceutical Science. 2011;1(6):228-234

[8] Kouvaris P, Delimitis A, Zaspalis V, Papadopoulos D, Tsipas SA, Michailidis N. Green synthesis and characterization of silver nanoparticles produced using Arbutus Unedo leaf extract. Materials Letters. 2012;76:18-20

[9] Verma DK, Patel S, Kushwah KS. Synthesis of Titanium dioxide (TiO2) nanoparticles and Impact on morphological changes, seeds yield and

phytotoxicity of Phaseolus vulgaris L. Tropical Plant Research. 2020;7(1): 158-170

[10] Kushwah KS, Patel S. Effect of titanium dioxide nanoparticles (TiO2 NPs) on Faba bean (Vicia faba L.) and induced asynaptic mutation: A meiotic study. Journal of Plant Growth Regulation. 2019;7:1-12

[11] Jones SA, Bowler PG, Walker M, Parsons D. Controlling wound bioburden with a novel silver-containing Hydrofiber® dressing. Wound Repair and Regeneration. 2004;12(3):288-294

[12] Doble M, Rollins K, Kumar A. Green Chemistry and Engineering. Burlington: Academic Press; 2010

[13] Kushwah KS, Verma RC, Patel S, Jain NK. Colchicine induced polyploidy in Chrysanthemum carinatum L. Journal of Phylogenetics & Evolutionary Biology. 2018;6(193):2

[14] He Y, Du Z, Lv H, Jia Q, Tang Z, Zheng X, et al. Green synthesis of silver nanoparticles by Chrysanthemum morifolium Ramat. extract and their application in clinical ultrasound gel. International Journal of Nanomedicine. 2013;8:1809

[15] Vanathi P, Rajiv P, Narendhran S, Rajeshwari S, Rahman PK, Venckatesh R. Biosynthesis and characterization of phyto mediated zinc oxide nanoparticles: A green chemistry approach. Materials Letters. 2014;134:13-15

[16] Zayed MF, Eisa WH, Shabaka AA. Malva parviflora extract assisted green synthesis of silver nanoparticles. Spectrochimica Acta Part A: Molecular and Biomolecular Spectroscopy. 2012;98:423-428

[17] Huang L, Weng X, Chen Z, Megharaj M, Naidu R. Synthesis of

iron-based nanoparticles using oolong tea extract for the degradation of malachite green. Spectrochimica Acta Part A: Molecular and Biomolecular Spectroscopy. 2014;**117**:801-804

[18] Goutam SP, Saxena G, Roy D, Yadav AK, Bharagava RN. Green synthesis of nanoparticles and their applications in water and wastewater treatment. In: Saxena G, Bharagava RN, editors. Bioremediation of Industrial Waste for Environmental Safety. Singapore: Springer; 2020. pp. 349-379

[19] Reddy NJ, Vali DN, Rani M, Rani SS. Evaluation of antioxidant, antibacterial and cytotoxic effects of green synthesized silver nanoparticles by *Piper longum* fruit. Materials Science and Engineering: C. 2014;**34**:115-122

[20] Cruz D, Falé PL, Mourato A, Vaz PD, Serralheiro ML, Lino ARL. Preparation and physicochemical characterization of Ag nanoparticles biosynthesized by *Lippia citriodora* (Lemon Verbena). Colloids and Surfaces B: Biointerfaces. 2010;**81**(1):67-73

[21] Marchiol L. Synthesis of metal nanoparticles in living plants. Italian Journal of Agronomy. 2012;**7**(3):e37-e37

[22] Velmurugan P, Sivakumar S, Young-Chae S, Seong-Ho J, Pyoung-In Y, Jeong-Min S, et al. Synthesis and characterization comparison of peanut shell extract silver nanoparticles with commercial silver nanoparticles and their antifungal activity. Journal of Industrial and Engineering Chemistry. 2015;**31**:51-54

[23] Kumar B, Smita K, Cumbal L, Debut A. Green synthesis of silver nanoparticles using Andean blackberry fruit extract. Saudi Journal of Biological Sciences. 2017;**24**(1):45-50

[24] Kumar V, Singh DK, Mohan S, Hasan SH. Photo-induced biosynthesis of silver nanoparticles using aqueous extract of *Erigeron bonariensis* and its catalytic activity against Acridine Orange. Journal of Photochemistry and Photobiology B: Biology. 2016;**155**:39-50

[25] Amooaghaie R, Saeri MR, Azizi M. Catalytic properties and biomedical applications of cerium oxide nanoparticles. Synthesis, characterization and biocompatibility of silver nanoparticles synthesized from *Nigella sativa* leaf extract in comparison with chemical silver nanoparticles. Ecotoxicology and Environmental Safety. 2015;**120**:400-408

[26] Ramesh Kumar K, Nattuthurai GP, Mariappan T. Biosynthesis of silver nanoparticles from *Morinda tinctoria* leaf extract and their larvicidal activity against *Aedes aegypti* Linnaeus 1762. Journal of Nanomedicine & Nanotechnology. 2014;**5**(6):1-5

[27] Anandalakshmi K, Venugobal J, Ramasamy V. Characterization of silver nanoparticles by green synthesis method using *Pedalium murex* leaf extract and their antibacterial activity. Applied Nanoscience. 2016;**6**(3): 399-408

[28] Latha M, Priyanka M, Rajasekar P, Manikandan R, Prabhu NM. Biocompatibility and antibacterial activity of the *Adathoda vasica* Linn extract mediated silver nanoparticles. Microbial Pathogenesis. 2016;**93**:88-94

[29] Ting X, Wei S, Wang X, An-qi Y F. Positive effect of composite titanium on crop. Hunan Agricultural Science. 2018;**1**:51-54

[30] Sankar V, SalinRaj P, Athira R, Soumya RS, Raghu KG. Cerium nanoparticles synthesized using aqueous extract of *Centella asiatica*: Characterization, determination of free radical scavenging activity and evaluation of efficacy against cardiomyoblast hypertrophy. RSC Advances. 2015;**5**(27):21074-21083

[31] Thovhogi N, Diallo A, Gurib-Fakim A, Maaza M. Nanoparticles green synthesis by *Hibiscus sabdariffa* flower extract: Main physical properties. Journal of Alloys and Compounds. 2015;**647**:392-396

[32] Maensiri S, Labuayai S, Laokul P, Klinkaewnarong J, Swatsitang E. Structure and optical properties of CeO$_2$ nanoparticles prepared by using lemongrass plant extract solution. Japanese Journal of Applied Physics. 2014;**53**(6S):06JG14

[33] Malleshappa J, Nagabhushana H, Prashantha SC, Sharma SC, Dhananjaya N, Shivakumara C, et al. Eco-friendly green synthesis, structural and photoluminescent studies of CeO2: Eu3+ nanophosphors using *E. tirucalli* plant latex. Journal of Alloys and Compounds. 2014;**612**:425-434

[34] Gunalan S, Sivaraj R, Venckatesh R. *Aloe barbadensis Miller* mediated green synthesis of mono-disperse copper oxide nanoparticles: Optical properties. Spectrochimica Acta Part A: Molecular and Biomolecular Spectroscopy. 2012;**97**:1140-1144

[35] Vishveshvar K, Krishnan MA, Haribabu K, Vishnuprasad S. Green synthesis of copper oxide nanoparticles using Ixiro coccinea plant leaves and its characterization. BioNanoScience. 2018;**8**(2):554-558

[36] Yugandhar P, Vasavi T, Rao YJ, Devi PUM, Narasimha G, Savithramma N. Cost effective, green synthesis of copper oxide nanoparticles using fruit extract of *Syzygium alternifolium* (Wt.) Walp., characterization and evaluation of antiviral activity. Journal of Cluster Science. 2018;**29**(4):743-755

[37] Aher YB, Jain GH, Patil GE, Savale AR, Ghotekar SK, Pore DM, et al. Biosynthesis of copper oxide nanoparticles using leaves extract of Leucaena leucocephala L. and their promising upshot against diverse pathogens. International Journal of Molecular and Clinical Microbiology. 2017;**7**(1):776-786

[38] Galan CR, Silva MF, Mantovani D, Bergamasco R, Vieira MF. Green synthesis of copper oxide nanoparticles impregnated on activated carbon using *Moringa oleifera* leaves extract for the removal of nitrates from water. The Canadian Journal of Chemical Engineering. 2018;**96**(11):2378-2386

[39] Smitha SL, Philip D, Gopchandran KG. Green synthesis of gold nanoparticles using *Cinnamomum zeylanicum* leaf broth. Spectrochimica Acta Part A: Molecular and Biomolecular Spectroscopy. 2009;**74**(3):735-739

[40] Wang Y, He X, Wang K, Zhang X, Tan W. *Barbated Skullcup* herb extract-mediated biosynthesis of gold nanoparticles and its primary application in electrochemistry. Colloids and Surfaces B: Biointerfaces. 2009;**73**(1):75-79

[41] Sharma NC, Sahi SV, Nath S, Parsons JG, Gardea-Torresde JL, Pal T. Synthesis of plant-mediated gold nanoparticles and catalytic role of biomatrix-embedded nanomaterials. Environmental Science & Technology. 2007;**41**(14):5137-5142

[42] Gardea-Torresdey JL, Parsons JG, Gomez E, Peralta-Videa J, Troiani HE, Santiago P, et al. Formation and growth of Au nanoparticles inside live alfalfa plants. Nano Letters. 2002;**2**(4):397-401

[43] Armendariz V, Herrera I, Peralta VJR, Jose YM, Troiani H, Santiago P, et al. Size controlled gold nanoparticles formation by Biocatalytic Synthesis Pathways, Transformation, and Toxicity of NPs 1719 Avena sativa biomass: Use of plants in nanobiotechnology. Journal of Nanoparticle Research. 2004;**6**:377-382

[44] Santhoshkumar T, Rahuman AA, Jayaseelan C, Rajakumar G, Marimuthu S, Kirthi AV, et al. Green synthesis of titanium dioxide nanoparticles using *Psidium guajava* extract and its antibacterial and antioxidant properties. Asian Pacific Journal of Tropical Medicine. 2014;7(12):968-976

[45] Rajakumar G, Rahuman AA, Jayaseelan C, Santhoshkumar T, Marimuthu S, Kamaraj C, et al. Solanum trilobatum extract-mediated synthesis of titanium dioxide nanoparticles to control *Pediculus humanus capitis*, *Hyalomma anatolicum anatolicum* and *Anopheles subpictus*. Parasitology Research. 2014;113(2):469-479

[46] Sundrarajan M, Gowri S. Green synthesis of titanium dioxide nanoparticles by Nyctanthes arbor-tristis leaves extract. Chalcogenide Letters. 2011;8(8):447-451

[47] Velayutham K, Rahuman AA, Rajakumar G, Santhoshkumar T, Marimuthu S, Jayaseelan C, et al. Evaluation of *Catharanthus roseus* leaf extract-mediated biosynthesis of titanium dioxide nanoparticles against *Hippobosca maculata* and *Bovicola ovis*. Parasitology Research. 2012;111(6):2329-2337

[48] Roopan SM, Bharathi A, Prabhakarn A, Rahuman AA, Velayutham K, Rajakumar G, et al. Efficient phyto-synthesis and structural characterization of rutile TiO2 nanoparticles using Annona squamosa peel extract. Spectrochimica Acta Part A: Molecular and Biomolecular Spectroscopy. 2012;98:86-90

[49] Qu J, Luo C, Hou J. Synthesis of ZnO nanoparticles from Zn-hyperaccumulator (*Sedum alfredii* Hance) plants. Micro & Nano Letters. 2011;6(3):174-176

[50] Lingaraju K, Naika HR, Manjunath K, Basavaraj RB, Nagabhushana H, Nagaraju G, et al. Biogenic synthesis of zinc oxide nanoparticles using *Ruta graveolens* (L.) and their antibacterial and antioxidant activities. Applied Nanoscience. 2016;6(5):703-710

[51] Bhuyan T, Mishra K, Khanuja M, Prasad R, Varma A. Biosynthesis of zinc oxide nanoparticles from *Azadirachta indica* for antibacterial and photocatalytic applications. Materials Science in Semiconductor Processing. 2015;32:55-61

[52] Kumari M, Khan SS, Pakrashi S, Mukherjee A, Chandrasekaran N. Cytogenetic and genotoxic effects of zinc oxide nanoparticles on root cells of *Allium cepa*. Journal of Hazardous Materials. 2011;190(1-3):613-621

[53] Fu L, Fu Z. Plectranthus amboinicus leaf extract–assisted biosynthesis of ZnO nanoparticles and their photocatalytic activity. Ceramics International. 2015;41(2):2492-2496

[54] Mazumdar H, Haloi N. A study on biosynthesis of iron nanoparticles by Pleurotus sp. Journal of Microbiology and Biotechnology Research. 2011;1(3):39-49

[55] Arumugam A, Karthikeyan C, Hameed ASH, Gopinath K, Gowri S, Karthika V. Synthesis of cerium oxide nanoparticles using *Gloriosa superba* L. leaf extract and their structural, optical and antibacterial properties. Materials Science and Engineering: C. 2015;49:408-415

[56] Walkey C, Das S, Seal S, Erlichman J, Heckman VKK, Ghibelli L, et al. Environmental Science Nano. 2015;2:33-53

[57] Song JY, Kwon EY, Kim BS. Biological synthesis of platinum nanoparticles using *Diopyros kaki* leaf extract. Bioprocess and Biosystems Engineering. 2010;33(1):159

[58] Malleshappa J, Nagabhushana H, Sharma SC, Vidya YS, Anantharaju KS, Prashantha SC, et al. Leucas aspera mediated multifunctional CeO2 nanoparticles: Structural, photoluminescent, photocatalytic and antibacterial properties. Spectrochimica Acta Part A: Molecular and Biomolecular Spectroscopy. 2015;**149**:452-462

[59] Dwivedi AD, Gopal K. Biosynthesis of silver and gold nanoparticles using *Chenopodium album* leaf extract. Colloids and Surfaces A: Physicochemical and Engineering Aspects. 2010;**369**(1-3):27-33

[60] Malik P, Shankar R, Malik V, Sharma N, Mukherjee TK. Green chemistry based benign routes for nanoparticle synthesis. Journal of Nanoparticles. 2014;**2014**

[61] Xiong J, Wang Y, Xue Q, Wu X. Synthesis of highly stable dispersions of nanosized copper particles using L-ascorbic acid. Green Chemistry. 2011;**13**(4):900-904

[62] Yoosaf K, Ipe BI, Suresh CH, Thomas KG. In situ synthesis of metal nanoparticles and selective naked-eye detection of lead ions from aqueous media. The Journal of Physical Chemistry C. 2007;**111**(34):12839-12847

[63] Verma DK, Patel S, Kushwah KS. Green biosynthesis of silver nanoparticles and impact on growth, chlorophyll, yield and phytotoxicity of *Phaseolus vulgaris* L. Vegetos. 2020;**33**:648-657

[64] Narayanan KB, Sakthivel N. Green synthesis of biogenic metal nanoparticles by terrestrial and aquatic phototrophic and heterotrophic eukaryotes and biocompatible agents. Advances in Colloid and Interface Science. 2011;**169**(2):59-79

[65] Mittal AK, Chisti Y, Banerjee UC. Synthesis of metallic nanoparticles using plant extracts. Biotechnology Advances. 2013;**31**(2):346-356

[66] Ai J, Biazar E, Jafarpour M, Montazeri M, Majdi A, Aminifard S, et al. Nanotoxicology and nanoparticle safety in biomedical designs. International Journal of Nanomedicine. 2011;**6**:1117

[67] Pushkar DB, Sevak PI. Green synthesis of silver nanoparticles using Couroupita guianensis fruit pulp and its antibacterial properties. World Journal of Pharmaceutical Research. 2016;**5**(9): 1174-1187

[68] Miri A, Darroudi M, Sarani M. Biosynthesis of cerium oxide nanoparticles and its cytotoxicity survey against colon cancer cell line. Applied Organometallic Chemistry. 2020;**34**(1): e5308

[69] Buazar F, Sweidi S, Badri M, Kroushawi F. Biofabrication of highly pure copper oxide nanoparticles using wheat seed extract and their catalytic activity: A mechanistic approach. Green Processing and Synthesis. 2019;**8**(1):691-702

[70] Ali SG, Ansari MA, Alzohairy MA, Alomary MN, AlYahya S, Jalal M, et al. Biogenic gold nanoparticles as potent antibacterial and antibiofilm nano-antibiotics against *Pseudomonas aeruginosa*. Antibiotics. 2020;**9**(3):100

[71] Nabi G, Raza W, Tahir MB. Green synthesis of TiO 2 nanoparticle using cinnamon powder extract and the study of optical properties. Journal of Inorganic and Organometallic Polymers and Materials. 2020;**30**(4):1425-1429

[72] Ogunyemi SO, Abdallah Y, Zhang M, Fouad H, Hong X, Ibrahim E, et al. Green synthesis of zinc oxide nanoparticles using different plant extracts and their antibacterial activity against *Xanthomonas oryzae* pv. oryzae. Artificial Cells, Nanomedicine, and Biotechnology. 2019;**47**(1):341-352

[73] Makarov VV, Makarova SS, Love AJ, Sinitsyna OV, Dudnik AO, Yaminsky IV, et al. Biosynthesis of stable iron oxide nanoparticles in aqueous extracts of *Hordeum vulgare* and *Rumex acetosa* plants. Langmuir. 2014;**30**(20): 5982-5988

[74] Hussain I, Singh NB, Singh A, Singh H, Singh SC. Green synthesis of nanoparticles and its potential application. Biotechnology Letters. 2016;**38**(4):545-560

[75] Qu J, Yuan X, Wang X, Shao P. Zinc accumulation and synthesis of ZnO nanoparticles using *Physalis alkekengi* L. Environmental Pollution. 2011;**159**(7): 1783-1788

[76] Carmona ER, Benito N, Plaza T, Recio-Sánchez G. Green synthesis of silver nanoparticles by using leaf extracts from the endemic *Buddleja globosa* hope. Green Chemistry Letters and Reviews. 2017;**10**(4):250-256

[77] Soares MR, Corrêa RO, Stroppa PHF, Marques FC, Andrade GF, Corrêa CC, et al. Biosynthesis of silver nanoparticles using *Caesalpinia ferrea* (Tul.) Martius extract: Physicochemical characterization, antifungal activity and cytotoxicity. PeerJ. 2018;**6**:e4361

Chapter 13

Nanofibers: Production, Characterization, and Tissue Engineering Applications

Ece Bayrak

Abstract

Among all nanostructured materials, nanofibers (NFs) are the one class that is widely used in tissue engineering (TE) and regenerative medicine (RM) areas. NFs can be produced by a variety of different methods, so they can be used almost for any tissue engineering process with appropriate modifications. Also, the variety of materials that can form nanofibers, production methods, and application fields increase the value of NFs greatly. They are almost suitable for any tissue engineering applications due to their tunable properties. Hopefully, this chapter will provide brief information about the production methods (electrospinning, wet spinning, drawing, etc.), characterization methods (Scanning Electron Microscopy, Transmission Electron Microscopy, Atomic Force Microscopy, etc.), and tissue engineering applications (core-shell fibers, antibacterial fibers, nanoparticle-incorporated fibers, drug-loaded fibers, etc.) of NFs.

Keywords: nanomaterials, nanofibers, production, characterization, biomedical applications, regenerative medicine, tissue engineering

1. Introduction

Specific definition of nanofibers can vary from one discipline to another, but according to one of the most common descriptions, fibers with a diameter below 100 nm are referred as nanofibers. Nanofibers have two alike dimensions (diameter) in the nanoscale and a third dimension, which is significantly larger (length). Going back to its origins, nanofibers are produced for the first time by Formhals (1934) by electrospinning of cellulose acetate solution. Although electrospinning process was used before Formhals, no one was able to form long filaments due to the use of inelastic Newtonian fluids [1]. The use of viscoelasticity in the solutions led the formation of nanofibers because the applied electric field caused a considerable reduction of the fiber diameter due to the bending instability, which is later mentioned by Reneker [2]. Forming fibers in nanoscale was a major drawback at that time, and not much attention was paid to the topic until the breakthrough of nanotechnology in the late 1990s. After almost 60–70 years later, Formhals' work was appreciated, understood, and widened [3].

Nanofibers have many advantages because of their scales, which gave them high aspect ratio (length/diameter value) above 200 and high surface area. And because almost all their properties are tunable, one can select and use nanofibers in numerous applications. The vital point of nanofiber technology is the availability of a wide range of materials such as natural and synthetic polymers, composites, metals, metal oxides, carbon-based materials, etc., which can be used for fiber production process [4].

Types of nanofibers can vary due to their nature, structure, and composition. According to its nature, one can produce natural or engineered nanofibers while one can produce nonporous, porous, hollow, core-shell nanofibers due to its structure. It is also possible to blend fiber materials to acquire a composition that can be organic, inorganic, carbon-based, or a composite [5]. When all the advantages considered (high aspect ratio, tunable properties, ability to form 3D networks, etc.), nanofibers are perfect nominee for different biomedical applications, such as tissue engineering (TE), regenerative medicine, drug delivery, nanoparticle delivery, etc. [6].

This chapter mainly focuses on the different production methods of nanofibers, their characterization techniques, recent developments in tissue engineering applications.

2. Nanofiber production methods

Nanofiber production techniques can be divided into two main class: top-down and bottom-up approaches. Chemical and mechanical methods are considered in top-down approaches. In top-down techniques, nanofibers are formed from bulk materials. On the other hand, in bottom-up approaches such as electrospinning, drawing, phase separation, etc., nanofiber formation occurs from composing molecules. This chapter mainly focuses on bottom-up approaches since they are the widely used class of nanofiber production methods.

2.1 Electrospinning

Electrospinning (ES) directly emerged from electrospraying (Electrohydrodynamic spray (EHD)), which was discovered by Morton and Cooley in 1902. Both methods depend on dispersing fluids by using electrostatic forces. There is one important distinction between these methods. By using ES, continuous fibers can be produced, whereas only small droplets are formed in EHD. After the electrospinning method was found to be more suitable for producing nanofibers rather than EHD, this method received more attention, and more studies were carried out in this field. As a result of these studies over the years, electrospinning method has undergone many modifications. By using different types of ES, one can produce hollow fibers, core-shell fibers, nanoparticles, or drug-incorporated fibers, etc.

2.1.1 Traditional electrospinning

For traditional ES, three main components are needed: (i) a high voltage source, (ii) syringe pump (nozzle), and (iii) a grounded collector (**Figure 1a**). The nozzle is preferably a metallic needle with a blunt tip to proper observation of the Taylor cone. During the ES process, first certain amount of polymeric solution (preferably dissolved in a volatile solvent with a w:v ratio.) is placed into a proper syringe and then to the syringe pump. Then high voltage is applied to the tip of the nozzle, and the elongating conical shape of the droplet is observed. To form nanofibers, the electrostatic force has to overcome the surface tension of the droplet, then Taylor Cone occurs at the tip of the nozzle, and a charged jet ejects from the Taylor Cone, resulting in the formation of nanofibers following by the fast evaporation of the solvent [10].

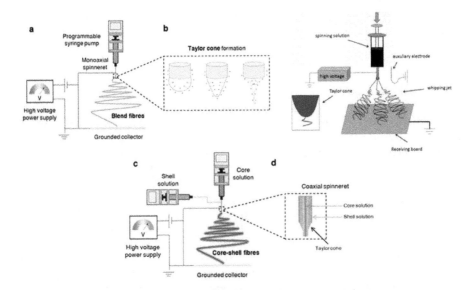

Figure 1.
(a) Traditional electrospinning setup, (b) multijet electrospinning setup, and (c) coaxial electrospinning setup. (a and c were reproduced with the permission from Gonçalves et al. [7], Ura et al. [8], and b was reproduced with the permission from Wu et al. [9]).

Morphology of the formed fibers can be controlled by many factors such as flow rate of the syringe pump, concentration of the polymer solution, collector type, solution viscosity, applied voltage, distance between the collector and the nozzle, diameter of the nozzle, etc. And each of these factors affects the fiber morphology significantly. For example, by increasing the voltage, fiber diameter can be decreased, low polymer concentrations can cause electrospraying rather than electrospinning, or increasing the flow rate can reduce the fiber diameter [11]. So, all these parameters have to be optimized before the ES process began in order to obtain fibers with maximum performance.

2.1.2 Multijet electrospinning

This method is also called "Multi Needle ES" in the literature. The reason for its development is to improve the productivity and produce composite fibers that cannot be dissolved in regular solvents (**Figure 1b**).

Needle diameter, needle number, and configuration play an important role in this approach such as the other ES methods. Unfortunately, this method holds one major drawback, which is a strong repulsion among the jets because of the multi needle system. This repulsion, which is generated by the coulomb force, may cause reduced fiber deposition and poor fiber quality. To avoid this problem, needles must be oriented at appropriate distance [12].

2.1.3 Coaxial electrospinning

Coaxial ES method is used to form core-shell nanofibers by using multiple syringe pumps or one syringe pump with multiple feeding systems. Mainly, a polymer and a composite solution, one is to form shell and the other is to form core parts, can be used individually, or two different polymer solutions can be employed as forerunner solutions (**Figure 1c**). Directed by the electrostatic repulsions between the surface charges, the polymer solution, which will form the shell part of the composite nanofibers, will be lengthened and will create viscous stress. After that this stress will be delivered to the core layer, and the polymer solution, which will form the core part, will be promptly stretched. As a result, composite jets will be formed, which will have coaxial structures [13].

2.1.4 Melt electrospinning

Addition to the conventional ES setup, melt ES technique requires a heating device such as heat guns, lasers, or electrical heating devices (**Figure 2a**). Polymer solution must stay in its molten state by a constant heat source. Main difference between melt ES and conventional ES method is the fiber formation process. In melt ES, instead of a solution, a molten polymer is used, and desired product is obtained on cooling; however, in conventional ES, fibers are formed with the help of solvent evaporation [14]. Other than this difference, the parameters that affect the fiber diameter, fiber quality, and the ES process are the same with the conventional ES method.

Main advantages of this method can be described as the absence of a solvent system and the high throughput rate of the polymer. This method can be used with the polymers that do not have a suitable solvent at room temperature. But in most of the cases, one of the major problems of melt ES is broad fiber diameter range due to the high viscosity of the melt polymer. Because of the high viscosity of the polymer, greater charge is required to initiate the jets. To reduce the fiber diameter or to obtain fibers with uniform diameters, some research groups used polymer blends or additives [17]. Requirement of high temperatures to melt the polymer can also be a disadvantage at this point. The melting temperature of the polymer can affect the structure and function of these additives (e.g., proteins, drugs, etc.) [18]. This situation makes selection and optimization steps critical.

2.1.5 Centrifugal electrospinning

Centrifugal ES is known by force spinning or rotary spinning as well. In this method, the electric field is replaced with a centrifugal force, which distinguishes centrifugal ES from conventional ES. Fiber formation is almost the same with conventional ES with a slight difference, instead of electric field, rotating speed surpasses the critical point to form a Taylor cone, and then the liquid jet gets ejected from the needle (**Figure 2b**) [19]. Therefore, rotating speed is one of the key parameters that determines the quality of the fibers along with the nozzle configuration, collector type, temperature, etc.

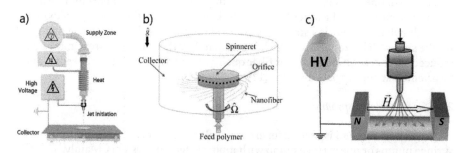

Figure 2.
(a) Melt electrospinning setup, (b) centrifugal electrospinning setup, and (c) magnetic-field-assisted electrospinning setup. (a was reproduced with the permission from Brown et al. [14]; b was reproduced with the permission from Taghavi et al. [15]; and c was reproduced with the permission from Blachowicz and Ehrmann [16]).

There are many advantages of this system due to the use of centrifugal force in place of electric field. Numerous conductive and nonconductive polymers can be electrospun with this method. Because no high voltage is needed, this method lightens safety-related concerns greatly. By adjusting the rotating speed, production efficiency can be improved, and large-scale production is allowed as well. Main limitations of conventional ES process (high voltage, misdirection of the jet, high cost, etc.) can be eliminated with this method. Aside of the advantages, the main disadvantage of this method is the spinneret design and material properties, which can highly affect the fiber quality and the yield of the process [20].

2.1.6 Magnetic-field-assisted electrospinning

In this method, magnetic properties are gained by incorporating magnetic nanoparticles to the polymer solution or using polymers that can respond to magnetic field (**Figure 2c**). This magnetic field can be obtained by two parallel permanent magnets, Helmholtz coils, or a magnetic field responder solution [21]. Besides mixing different polymers, adding non-polymeric materials (e.g., metals, ceramics, etc.) is another approach by which magnetic-field-assisted ES can be used. Fibers that are maintained with this method are reported to be more uniform. By using magnetic field fiber splitting from the jet can be prevented because of the magnetic field orientation. High velocity of the process supports smaller fiber diameter [22].

2.1.7 Needleless electrospinning

Some researchers proposed a new technique called "Needleless ES" to avoid the limitation caused by the capillaries and needles. Basis of this approach relies on a single principle, which is: Waves of an electrically conductive liquid self-organize on a mesoscopic scale and form jets when the intensity of the applied electrical field rises above a critical value (**Figure 3a**). Setup of the system can be divided in two groups: one of which is ES with a constrained feeding system and ES with an unconstrained feeding system. For the first system, a supply for the polymer solution, which is afterward injected into a closed nozzle, is preferred. On the contrary, for the second system, no nozzles are needed because the Taylor cones are formed on a free liquid surface. For both groups, high voltage source is a must to attract the polymer jets into nanofibers [25].

With the use of multiple jets without the needles, chances to increase the production rate of nanofibers are higher compared with the traditional ES systems. Some studies report an increase in polymer yield compared with single-needle

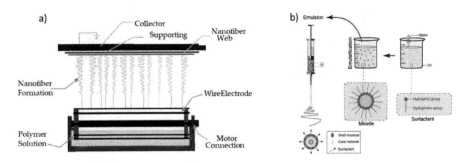

Figure 3.
(a) Needleless electrospinning setup, (b) emulsion electrospinning setup. (a was reproduced with the permission from Li et al. [23], and b was reproduced with the permission from Nikmaram et al. [24]).

solution and an improvement in fiber deposition, in opposition to multi-needle ES, which resulted from a reduction in mutual fiber repulsion [26].

2.1.8 Emulsion electrospinning

Emulsion ES is developed to produce fibers from two immiscible solutions. To blend these immiscible solutions and obtain an emulsion, vigorous stirring is required. Then this emulsion is loaded a glass syringe connected to a needle and a high voltage source (**Figure 3b**). Because this emulsion contains two immiscible solutions, fibers are difficult to produce due to properties and immiscible phases of these solutions. To overcome these difficulties, nanoparticles and surfactants such as detergents, sodium dodecyl sulfate, etc., are generally used. Even with this solution, there is a constant necessity for the emulsion to stay stable through the ES process, which is a major drawback [27, 28].

2.2 Wet spinning

Wet spinning (WS) is an alternative nanofiber fabrication method for polymers that are derived from natural sources. It is much cheaper and simpler in comparison to any ES method. Because there is no high voltage source, it is much easier to load therapeutic agents into fibers, which expands the range of polymers from natural or synthetic sources handled by means of WS [29]. It is also a developing approach, but it is possible to gather wet-spun nanofibers to produce biodegradable and biocompatible scaffolds with a 3D network for regenerative medicine approaches. This method is mainly based on extrusion of a polymeric solution into a coagulation bath where the solution in the coagulation bath contains a poor solvent or solvent/ non-solvent mixture (**Figure 4**).

Main goal here is to obtain coagulating fibers in the coagulation bath, which at the end solidifies as a constant fiber, as the extrusion process continues. Typical WS setup composed of a needle contains the polymeric solution, which is placed in a syringe pump and a coagulation medium. The needle must be immersed in the medium to initiate fiber coagulation. Different strategies have been developed for the collection or assembly of the fibers such as rotating drum, 3D assembly of the fibers by thermomechanical treatments, manually or computer-controlled motion of the coagulation bath or the needle, etc. After the WS setup is complete, quality and final morphology of the fibers still depend on several parameters, which include temperature, solvent system, properties of the selected polymer, needle diameter, flow rate of the syringe pump [31].

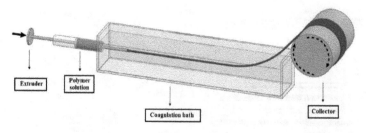

Figure 4.
Wet spinning setup. (Figure was reproduced and adapted with the permission from Wang et al. [30]).

2.3 Drawing

Drawing technique is usually used to produce continuous individual nanofibers. It is based on a sharp probe tip or a micropipette, which is soaked into the edge of the droplet deposited on a container. Then the sharp tip is withdrawn from the solution with a constant rate (usually 100 μm/s) to fabricate liquid fibers. The drawn nanofibers will be deposited on the surface by contacting the surface with the edge of the micropipette (**Figure 5a**). To form a 3D structure or a network, this process was repeated several times for every droplet [33], and continuous fibers in many adjustments can be fabricated with drawing method to use in biomedical applications.

In addition, specific control of the drawing process parameters such as drawing speed, viscosity, properties of the used polymer allows repeatability of the process, control of the fiber quality and fiber dimensions. Besides the advantages such as fabricating continuous fibers, simplicity, and the cost-effectiveness of the process, there are also some limitations. Because drawing causes nanofibers to be produced one single fiber at a time, productivity of this process is low. The only material type that can be used in this process is viscoelastic materials. Viscoelastic materials can resist the increased stress produced by drawing, and they can preserve their integration while going through a strong deformation [34].

2.4 Template synthesis

Template synthesis method allows to produce solid or hollow, discontinuous nanofibers with different properties such as polymeric, metallic, ceramic, semiconductor nanofibers. It is possible to convert multiple materials into fibrils or tubules in nanoscale diameter to use them in many applications, which include regenerative medicine, electronics, optoelectronics, gas sensors, etc. [35].

This method relies on the usage of a nanoporous membrane as a template/mold, containing cylindrical pores. The template/mold often refers to a metal oxide membrane such as aluminum oxide membranes or silica-based membranes, etc. Nanofibers are formed by passing through the polymer solution from the pores of the nanoporous membrane/template (**Figure 5b**). During the extrusion, polymer

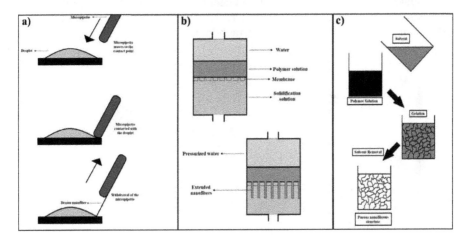

Figure 5.
(a) Drawing method, (b) template synthesis method, and (c) phase separation method. (Figures were adapted from Ramakrishna et al. [32]).

solution comes in contact with the solidifying solution and nanofibers are formed. The major disadvantage of this method is the continuity of the fibers. Only a few micrometers long fibers can be obtained with this method, and the diameter of these fibers depends on the pore size of the template [36]. By using template with different pore sizes, a variety of diameters can be achieved with template synthesis.

2.5 Phase separation

Phase separation method was developed by Zhang and Ma to mimic the 3D structure of collagen under the name of thermally induced liquid-liquid phase separation (**Figure 5c**). This method is mainly composed of five stages, which include preparing a homogeneous polymer solution, phase separation process, gelation, extraction of the solvent, freezing, and freeze-drying under the vacuum. Polymer solution is often prepared by dissolving the polymer at room temperature. Then the solution reaches the gelation temperature, which is the most critical step in this method because the duration of gelation depends on the concentration of the polymer and the gelation temperature. If the polymer acquires high gelation temperature, platelet-like structures are formed due to the nucleation of crystals so low gelation temperatures are required for this process. After gelation step was completed, solvent was extracted from the gel with water, and the freeze-drying stage was applied to the final product [37, 38].

For this method, minimum equipment is needed. Nanofiber matrix can be directly fabricated, and by adjusting the polymer concentration, properties of the matrix can be accustomed. Process parameters such as polymer concentration, polymer type, solvent type, etc., were found to influence the nanofiber quality, morphology, and the final nanofibrous matrix. The matrix fabricated by the phase separation method exhibits high porosity of almost 98% within the overall material. The major drawback of this method is that only a few polymers (e.g., polylactide, polyglycolide, etc.) can be used to obtain nanofibers by phase separation due to the fact that not all polymers are compatible with this process since it requires a certain gelation capability [39].

2.6 Self-assembly

This method relies on the idea of spontaneous organization of amphiphile compounds, which can be considered as active molecules (**Figure 6**). Because self-assembly is a bottom-up fabrication method, it is based on gathering small units

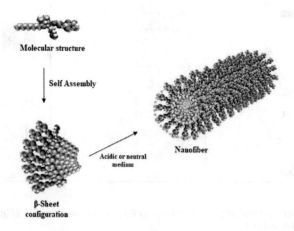

Figure 6.
Self-assembly method. (Figure was adapted from Xu et al. [40]).

together by the help of intermolecular forces such as hydrogen bonding, hydrophobic interactions, electrostatic reactions, biomolecule-specific interactions, etc. These units will organize and arrange themselves to form macromolecular nanofibers.

The overall shape of the nanofibers is determined by the shape of small units. With this method, it is possible to produce nanofibers smaller than 100 nm with a length of several micrometers, but the process is time-consuming. Also, low productivity, difficult control of the fiber dimensions, and limited active compound choices, which can self-assemble themselves, are the main disadvantages of self-assembly method [41].

2.7 Interfacial polymerization

This method depends on two different monomers, which can dissolve in different phases. Basically, this is a polycondensation reaction between two reactive monomers, which are dissolved in immiscible solvents. After these two different phases are prepared and mixed, polymerization will occur at the interface of the emulsion droplet. Homogeneous nucleated growth is the key factor in interfacial polymerization [42]. By separating the monomer precursors in different phases, localized reaction and nanofiber formation can be achieved (**Figure 7**).

By selecting different kinds of monomers, a variety of polymers can be synthesized. The properties and quality of the nanofibers are highly dependent on the reactivity and concentration of the monomers, reactive groups attached to the monomers, and the stability of the interface [44].

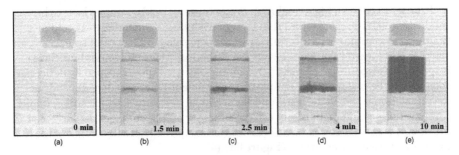

Figure 7.
Snapshots showing interfacial polymerization of aniline in a water/chloroform system. From a to e, the reaction times are 0, 1.5, 2.5, 4, and 10 min, respectively. (Figure was reproduced and adapted with the permission from Huang et al. [43]).

3. Nanofiber characterization methods

Nanofibers can be produced by many different methods according to the area in which they are to be used. After choosing the appropriate production method for the application area and producing the nanofibers, some characterization studies are required to examine the quality, composition, morphology, and structure of the nanofibers. Characterization methods are still improving, and request for the establishment of effective techniques is continuously increasing. Therefore, commonly used methods for the characterization of nanofibers are described below.

3.1 Morphological and structural characterization

3.1.1 Scanning electron microscopy (SEM)

Generally, microscopic imaging techniques are routinely used to observe fiber diameters, alignment, porous structure, fiber morphology, and orientation. With

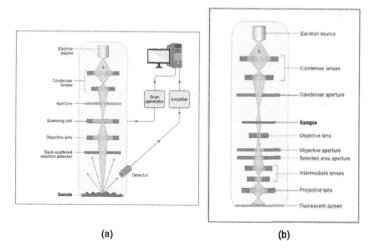

(a)   (b)

Figure 8.
(a) Schematic image of scanning electron microscope, and (b) schematic image of transmission electron microscope. (Figures are reproduced from https://biorender.com/).

scanning electron microscopy (SEM) imaging, high-resolution images of a scaffold surface can be obtained and surface properties (roughness, porosity, smoothness, etc.) can be determined.

In order to obtain a high-resolution image from scaffolds, samples have to be conductive, so sputtering with a thin layer of a conductive metal such as gold or titanium is a common modification for nonconductive samples. After sputtering, an electron gun is used to produce beams as a cathode source and focuses by electromagnetic lenses to an exact spot on the sample. Selected spot is shaped by deflection coils so that the surface of the sample can be scanned. This procedure depends on the interaction between the beam and the secondary electrons, which are produced from the sample. Interaction between the secondary electrons from the surface of the sample and the electron beam is monitored, amplified, and illustrated in the form of an image of the surface (**Figure 8a**) [45].

For the first evaluation of the nanofiber scaffolds, SEM is the most common characterization method due to its availability and the ease of use. It is possible to determine the porosity, the width, and length of pores on the surface, which can help understand the structure of the nanofibers [46]. To analyze the qualitative characteristics of the nanofibers, a convenient number of samples are needed to obtain statistical information about the materials.

Evaluation of nanofibers, cells, living organisms, or biological materials is also possible without any coating treatment required by using environmental SEM (ESEM). In this characterization method, the electron beam is wielded under water vapor environment. The ionization of water prevents the accumulation of the surface charges, which allows nonconductive materials to be evaluated without any modifications.

3.1.2 Transmission electron microscopy (TEM)

Transmission electron microscope (TEM) technology is considered one of the most important characterization techniques because of its ability to evaluate the interior structure of the samples. The pore structure of the scaffolds can be clearly seen by the images taken with TEM. The pore size and distribution of scaffolds are crucial parameters in tissue engineering field due to the fact that these parameters directly affect the ability of the cells to penetrate through the pores of the material. Similar to SEM imaging, TEM also yields two-dimensional (2D) images of nanofibers and pores as well [45–47].

The common method includes transmitting electron beams through the ultrathin part of the samples, which causes a phase shift in portion of the electrons. When the incoming electron beam descends from the microscope column, it interacts with the sample fluorescent screen. And then the electron beam hits the sample, which leads a large amount of radiation to be emitted from the sample. This interaction causes the elastic and inelastic scattering of the emitted electrons. Images that take origin from the elastically scattered electrons allow the observation of the structure of the scaffolds or the defects at a high resolution (**Figure 8b**). Ultrathin samples are required for TEM evaluation (~20–200 μm) because electron beams are absorbed completely by the thick samples and no image can be formed. It is a very common characterization technique, but it is also a detrimental technique as well because of the possibility to damage the samples, especially the biological samples, by the electron beam going through them [45, 46].

3.1.3 Atomic force microscopy (AFM)

Atomic force microscopy (AFM) technique is mostly used for the evaluation of surface topography. The analytical capabilities of AFM are limited to the uppermost atomic layer of a sample because its operation is based on the interactions with the electron clouds of atoms at the surface. This technique also gives information about morphology, surface roughness, fiber orientation, and particle/grain distribution from the surface of the samples [45].

In this technique, a small tip is attached to a cantilever, and when the tip encounters with the sample surface, Van der Waals and electrostatic interactions between atoms at the tip and those on the surface create a force profile and cause attraction of the tip to the surface. A photodiode detector detects the changes and converts them into data, which are later to be converted into images (**Figure 9**) [45, 49].

The operation of AFM can be carried out by three modes depending on the application: contact, noncontact, and tapping modes. The contact mode measures the repulsion between the tip and the surface of the sample where the force of the tip against sample surface remains constant. At this mode, sensitive samples can be damaged because scanning requires constant contact of the tip to the surface. The noncontact mode on the other hand measures the attractive forces between the tip

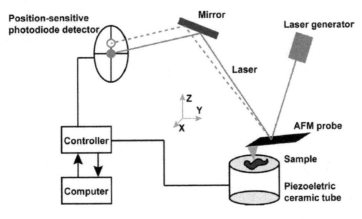

Figure 9.
Atomic force microscopy setup. (Figure is reproduced with the permission from Deng et al. [48]).

and the sample surface. Van der Waals forces between the tip and the sample surface are detected. Characterization of soft materials is often made with noncontact mode. At last, the tapping mode depends on the vertical oscillation of the tip. The tip contacts the surface of the sample and then lifts off at a certain frequency. Oscillation amplitude reduces as the tip contacts the surface due to the loss of energy. This mode overcomes problems with friction, adhesion, and electrostatic forces [49].

3.1.4 X-ray diffraction (XRD)

X-ray diffraction (XRD) spectroscopy is a safe non-damaging characterization technique, which can be performed on wide range of materials such as minerals, metals, semiconductors, ceramics, polymers, etc. This technique is mostly applied to evaluate structural properties of the samples such as phase formation, crystallite size, lattice strain, contents of each phase, and crystal structure. The wavelength of X-Rays (0.5–50 A°) is similar to the distance between atoms in a solid, they are ideal for exploring atomic arrangement in crystal structure [46].

XRD, rather than measuring how the absorbance of X-rays affects the sample, examines how X-rays are diffracted from the atoms in a sample. Diffraction occurs when incident rays are scattered by atoms in a way that reinforces the waves (**Figure 10**). Working principle of XRD is basically a filament is heated to produce electrons in a cathode tube. By applying voltage, electrons are accelerated toward the sample and the sample is bombarded with electrons. Characteristic X-ray spectra are produced when electrons have enough energy to remove the inner shell electrons of the target sample. These X-rays are adjusted and located onto the sample, and the intensity of the reflected X-rays is recorded. Then these recorded signals are processed and converted to a count rate and directed to a printer or a computer monitor as an output [51].

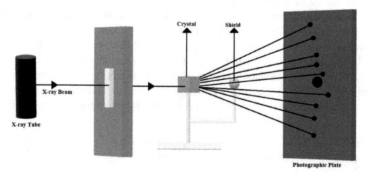

Figure 10.
Working principle of X-ray diffraction. (Figure is adapted from Kaur et al. [50]).

3.1.5 Thermogravimetric analysis (TGA)

Thermal methods can be examined under two categories: (a) differential thermal analysis and (b) thermogravimetric analysis (TGA). Differential analysis depends on the changes in heat content, which is measured as a function of increasing temperature. On the other hand, thermogravimetric analysis depends on the changes in weight, which is measured as a function of increasing temperature (**Figure 11**) [53].

TGA technique relies on the use of uniform heating to decompose all organic contents at high temperature, which eventually gives information about the

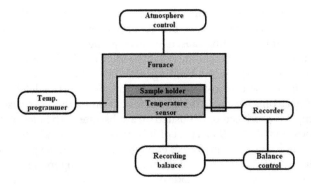

Figure 11.
Thermogravmetric analysis diagram. (Figure is adapted from Loganathan et al. [52]).

compositions of the sample. Mainly, by increasing the temperature at a constant rate, the decrease in the mass of the sample is recorded. The sample is located on a balance with a platinum melting pot, which is placed inside a furnace, and the procedure is generally carried on under air gas. As a result of this analysis, mass against temperature or time plot is obtained to measure the changes in the physical and chemical properties of the sample. The obtained data provide information about thermal stability of the remained sample, dehydration, pyrolysis, solid/gas interactions, etc. [54]

3.2 Mechanical characterization

Mechanical characterization of the scaffolds plays a critical role in tissue engineering applications. The designed scaffold's mechanical properties have to match the mechanical properties of the desired tissue. The mechanical strength of a scaffolds is crucial especially for in vivo applications, where the scaffold exposed mechanical loading repeatedly.

Most common characterization technique for nanofibrous scaffolds is tensile testing or nano-tensile testing. The theory is based on the attachment of the scaffold from both sides to the grips of the tensile testing machine and then pulling the scaffold with a constant rate until the rupture occurs (**Figure 12a**). The results give information about the stress-strain values, modulus, and strength of the scaffolds. But there are two limitations, which need to be overcome. Firstly, sample gripping is a problem because fibers tend to slip from the grips or break at the grips. These machines generally are not equipped to perform under micro sizes, so the small

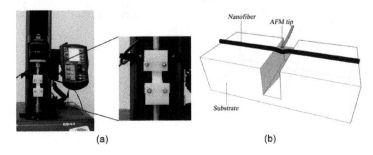

(a)   (b)

Figure 12.
(a) Tensile testing of nanofiber scaffolds, and (b) nanoscale bending test schematic. (a was author's unpublished thesis images, b was reproduced with the permission from Zhou et al. [55]).

size of the specimen is a major limitation for this process. Second, alignment of the scaffolds is needed because randomly oriented fibers may lead to premature sample failure due to unwanted bending caused by misalignment [56].

Another characterization technique that is widely used is bending test for nanofibers (nanoscale three-point bending test). The capability of an AFM system to apply forces in the nano/pico-Newton range and measure the deformation in the range of Angstroms has made this characterization method very useful. The nanofiber sample is produced or deposited on a substrate with holes in it. Then the nanofiber is positioned such that a part of it is suspended over a hole. The adhesion between the sample and the substrate is enough for the test to be performed without a failure (**Figure 12b**). With three-point bending test, Young modulus and fracture strength can be obtained. The downside of this method is that it is only limited to samples that can be produced using AFM anodization [45, 56].

3.3 Chemical characterization

3.3.1 X-ray photoelectron spectroscopy (XPS)

X-ray photoelectron spectroscopy (XPS) technique is one of the most powerful characterization techniques because of the ability of giving chemical information about the surface of the material, both elemental and molecular composition (**Figure 13a**). It can also differentiate chemical states of the same element to determine their depth distribution at a thickness between 5 and 10 nm. Useful electron signal is obtained only from a depth around 10–100 A° on the surface. Basically, the surface is irradiated by hitting the core electrons of the atoms. X-ray absorption causes the removal of an electron from one of the innermost atomic orbitals, and the kinetic energy of the emitted electron is recorded. The recorded kinetic energy is then converted into a spectrum by a computer. Binding energies of the elements from the sample will be determined according to the peaks from the spectrum. In literature, the kinetic energy and binding energy values assigned to each element can be found. This method often requires an argon ion bombardment step to eliminate surface impurities [45, 59].

3.3.2 Fourier's transform infrared spectroscopy (FTIR)

Fourier's transform infrared spectroscopy (FTIR) is a technique used to collect an infrared spectrum of an emission or an absorption of a solid, liquid, or gas. It is used to identify organic, inorganic, and polymeric materials utilizing IR light to

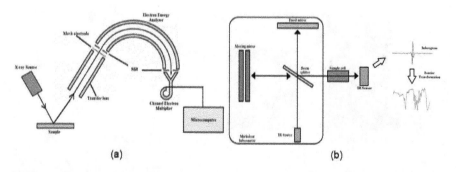

(a) (b)

Figure 13.
(a) Working principle of X-ray photoelectron spectroscopy, and (b) schematic of Fourier's transform infrared spectroscopy. (a was adapted from Seyama et al. [57], b was adapted from Lee et al. [58]).

scan samples. Standard FTIR setup is composed of a source, sample cell, detector, amplifier, A/D converter, and a computer (**Figure 13b**).

IR radiation is passed through the sample, and the emitted radiation could be absorbed and/or transmitted from the sample. Changes in the patterns of the absorption bands pinpoint a change in the composition of the material. So, the obtained signals are amplified, changes got detected by the detector, converted by the A/D converter, and as a result, a spectrum will be obtained. The obtained spectrum provides information about chemical composition of the material because the wavelength of absorbed light indicates characteristics of the chemical bonds. Just like fingerprints, two individual molecular structures cannot generate same IR spectrum. Every molecule has a specific fingerprint, which makes this technique a valuable tool for chemical identification. Also, this feature makes FTIR very preferable for many analyses such as determining the components in a mixture, identifying unknown materials, detecting contaminants in a material, finding additives, or determining the quality of a sample [60].

3.3.3 Raman spectroscopy (RS)

Raman spectroscopy (RS) method is based on irradiating a sample with a powerful laser source consisting of a monochromatic beam and measuring the scattered beam from a specific angle. During light scattering, the energy of most of the scattered light becomes equal to the energy of light interacting with the specimen. This type of elastic scattering is called Rayleigh scattering. In addition to elastic scattering, if a small part of the scattered light includes inelastic scattering, it is called Raman scattering. In Raman scattering, the excess or decrease in the energy of the scattered light relative to the energy of the light interacting with the molecule is as much as the energy difference between the energy levels of the molecule interacting with the light. This excess or scarcity at the energy levels is called the Raman shift. These shifts are measured in Raman spectroscopy (**Figure 14**) [62].

This method is used to evaluate vibrational and rotational frequency modes in physical and chemical systems. The intensity of Raman scattering depends on the change in polarizability. RS is suitable for the qualitative and quantitative analysis of organic, inorganic, and biological systems. With the obtained spectrum, unknown material identification, material differences, crystallinity, and material amount can be determined [60].

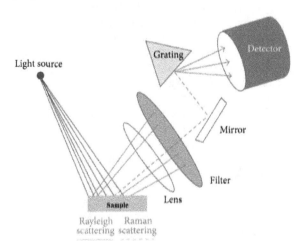

Figure 14.
Working principle of Raman scattering. (Figure was adapted from Kim [61]).

4. Tissue engineering applications

As mentioned in the introduction section, natural and synthetic fibers from different sources have been widely used in many areas for hundreds of years, and tissue engineering area happens to be one of them. TE maintains an alternative way to the restoration and regeneration of the injured tissues. TE is an interdisciplinary field that requires knowledge of biological, chemical, and engineering sciences toward the objective of tissue regeneration using cells, factors, and biomaterials alone or in combination with each other. With the light of this brief information, widely used TE applications of nanofibers are discussed below.

4.1 Muscoskeletal tissue engineering applications

4.1.1 Bone tissue engineering (BTE)

Bone tissue is mainly composed of organic bone matrix, which mostly includes collagen fibers (95% of these fibers are collagen type I) and inorganic compounds such as hydroxyapatite crystals [63]. There is a global need for bone grafts because of the high incidence of bone defects, which are caused by bone tumors, infections, and bone loss by traumas. Main treatment approaches for these injuries are auto-grafts, allografts, or xenografts. But there are some challenges to these approaches such as inflammation, scarring, infection, immunological graft rejection, hema-tomas, high-cost procedures, etc. [64, 65] At this point, bone tissue engineering (BTE) approaches present an alternative treatment way for these injuries. BTE field aims to form materials that can outperform bone autografts and allografts. The ultimate goal is to manufacture a scaffold that can be implanted to the defect area and then remodeled by patient's own cells. The key is to fulfill the role of the extra-cellular matrix (ECM) in the defect area. The design of the scaffolds for BTE is also modeled by the structure and function of healthy bone tissue, which is crucial to its function, for example, highly porous trabecular bone or highly dense cortical bone, which surrounds the trabecular bone. But still, regardless of recent advancements in TE and RM, reconstruction of critical-size bone injuries is still challenging [66].

For bone tissue regeneration, wide range of biomaterials can be used to mimic the function, structure, and composition of bone ECM with proper osteogenic activity. First studies for stimulating bone regeneration were done by ES of widely used polymers such as polycaprolactone (PCL), polylactic acid (PLA), gelatin, silk, and chitosan. The common feature of all these polymers was biodegradability because if the scaffolds are not biodegradable, a second surgery is necessary to remove the scaffold, which can result in infection, patient discomfort, or additional costs. According to the study of Cai et al., a 3D PCL/PLA scaffold was produced, and its bone regeneration efficiency was investigated in a rabbit tibia bone defect model by using human embryonic stem cell–derived mesenchymal stem cells (hESC-MSCs) [67]. They reported that the attachment of the hESC-MSCs to the 3D scaffold was successful due to the differentiation of the cells from round-like shape to a spindle-like form. Additionally, the histology and radiography studies resulted in 3D bone tissue formation after 6 weeks. Another study conducted by Nedjari, et al. is based on the development of a novel 3D honeycomb-shaped scaffold made by electrospun hybrid nanofibers, which includes poly(l-lactide-ε-caprolactone) and bone ECM protein fibrinogen (FBG) (**Figure 15**) [68].

Results of this study indicate that PLCL/FBG scaffolds support osteogenic differentiation of human adipose-derived mesenchymal stem cells (hADMSCs). Besides ES, melt ES writing is a promising method to design scaffolds with controlled structure. Abdalhay et al. manufactured PCL/HAp composite 3D

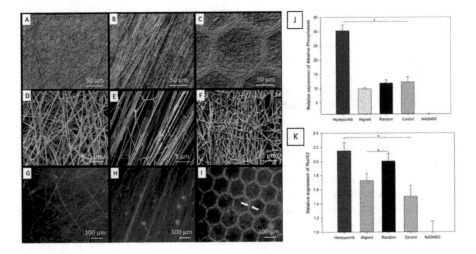

Figure 15.
SEM images of random (A, D), aligned (B, E) and honeycomb (C, F) PLGL-FBG nanofibers. Bottom row represents the immunofluorescent imaging of FBG within the fibers (red) on random (G), aligned (H), and honeycomb (I)-shaped scaffolds (the arrows indicate the higher accumulation of FBG at the walls of the honeycomb shapes.). Relative expression of alkaline phosphatase gene (J) and RunX2 gene (K) of ADMS cultured for 21 days in osteogenic medium on different shaped scaffolds. (Figure was reproduced with the permission of Nedjari et al. [68]).

scaffolds with high porosity (96–98%) by using melt ES writing method [69]. According to the results, infiltration and proliferation of seeded osteoblasts were achieved, which supports high interconnectivity and porosity of the PCL/HAp scaffolds. Velioglu et al. fabricated 3D-printed PLA scaffolds with different pore sizes for trabecular bone repair and regeneration. Their findings showed that the resemblance between 3D-printed scaffolds and native trabecular bone in terms of pore size, porosity, and mechanical properties of the scaffolds, the 3D-printed PLA scaffolds printed in this study can be considered as candidates for bone substitutes in bone repair [70]. In 2019, Lukasova et al. produced 3D and 2D nanofibrous scaffolds by using centrifugal ES and needleless ES methods, respectively. Cyclone device was used as a spinneret for centrifugal ES, and the spin guidance was sideways. Needleless ES on the other hand was performed by using Nanospider® technology. Scaffolds were then tested for metabolic activity, cell differentiation, and proliferation by using hMSCs. Scaffolds obtained with centrifugal ES showed higher cell proliferation due to their 3D, porous, and interconnective architecture [71].

4.1.2 Tendon/ligament tissue engineering

Tendon/ligament injuries, which are caused by tears, ruptures, traumas, and inflammation, result in severe pain and are generally seen in physically active young patients. Natural healing of these tissues is challenging due to their poor healing capacity and scar tissue formation, which then result in poor mechanical properties. Standard treatment approaches for these injuries are grafts or artificial prosthesis. Autografts are considered "gold standard" because of their lack of immune response, but they are limited by donor site availability and morbidity. Allografts hold the same concerns as in BTE, which are rejection, risk of disease transmission, and high re-rupture rates caused by mismatches between the donor and the recipient. To overcome these challenges, TE approaches are widely used for tendon/ligament tissue repair [72].

These soft tissues are mainly composed of dense and aligned collagen fibers, so the mechanic load of tendons and ligaments is restricted to one direction. As a result, scaffolds composed of aligned nanofibers are highly promising for tendon/ligament tissue repair studies because they can mimic the anisotropic nature of the native tissues. In the light of these information, a novel, multilayer scaffold was proposed by Yang et al., which was composed of fibrous PCL and methacrylated gelatin produced by dual ES [73]. The scaffold was formed by five sheets, which were cross-linked. The scaffold was then reinforced with gelatin layer bearing the stem cells, which were treated with TGF-β3 for 7 days to stimulate differentiation. Results showed an increase of tendon markers tenascin-C and scleraxis, which implies the scaffolds were porous enough for the diffusion of bioactive molecules. Another study conducted by Perikamana et al. reported that immobilization of platelet-derived growth factor (PDGF) in a gradient scaffold, which is also composed of aligned nanofibers, enhanced the expression of tenomodulin compared with a non-modified nanofiber scaffold [74]. Rinoldi et al., fabricated a bead-on-string fibrous

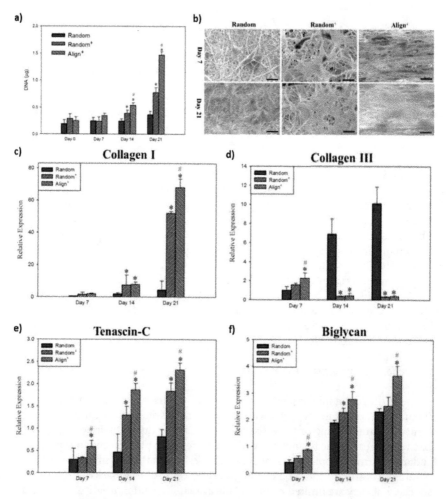

Figure 16.
The cell proliferation rates from DNA assays (a) and SEM images (b) of tenocytes cultured on different core-shell nanofiber scaffolds. The relative mRNA expression of type-I collagen (c), type III collagen (d), tenascin-C (e), and biglycan (f) by tenocytes after cultured on different core-shell nanofiber scaffolds for 7, 14, and 21 days. (Figures are reproduced with the permission from Chen et al. [76]).

scaffold and incorporated with silica particles to enhance the biological activities and modify the properties of the scaffolds such as wettability, degradation rate, etc. The results imply that their bead-to-string fibrous scaffold is a significant candidate for guided tissue regeneration [75]. In a recent study, Chen et al. proposed a three types of core-shell nanofibrous scaffolds [76]. For one group, HA (hyaluronic acid) is the core and PCL (random) is the shell while the other groups are HA/PRP (platelet-rich plasma) core–PCL shell (Random+) and HA/PRP core–PCL shell (Align+) (**Figure 16**). Tenocytes were used in in vitro studies, and the cells in Align+ showed the highest cell proliferation rate while Random+ is also significantly higher than Random study group. According to the expression studies, by day 14, Random+ and Align+ showed significant downregulation of collagen III gene expression when shift of collagen III to collagen I occurs during the tendon maturation.

It is safer to study tendon/ligament tissue regeneration compared with tendon/ligament-bone interface regeneration, which is also called the enthesis. Regeneration enthesis is exceptionally challenging due to its complex and gradient structure. The enthesis possesses location-dependent changes such as gradients, in terms of composition of ingredients and structural properties.

Still, there are many studies and research groups trying to fabricate scaffolds to repair tissue-tissue interfaces by incorporating bone-like biomaterials such as hydroxyapatite, hyaluronan, etc. (**Figure 17**) [74, 77]. Yet still, it remains a challenge in the field, which requires much more time and effort.

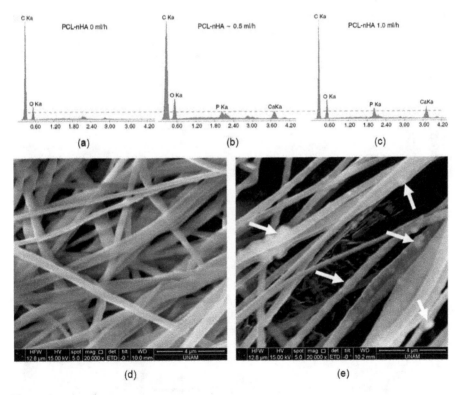

Figure 17.
Energy-dispersive spectroscopy (EDS) spectrum of polycaprolactone (PCL)-only and PCL-nano-hydroxyapatite (nHA) meshes at different flow rates corresponding to different nHA concentrations (a–c). Representative scanning electron microscopy (SEM) micrographs taken from (d) polycaprolactone (PCL)-rich and (e) nano-hydroxyapatite (nHA)-rich surfaces of the spatially graded meshes. White arrows in E indicate nHA particulates embedded into nanofibers. (Figure was reproduced with the permission from Bayrak et al. [77]).

4.1.3 Cartilage tissue engineering

Other than tendons and ligaments, cartilage tissue is another class of connective tissues that presents elastic behavior and protects the end of bones at joints. Nose, ears, knees, and many other parts of the body contain cartilage tissue. The main ECM components of dense cartilage tissue are collagen and proteoglycans, which are produced by a low number of chondrocytes. After an articular cartilage injury such as rupture, trauma, aging, etc., remodeling and regeneration of the native tissue are challenging due to the low availability of chondrocytes and the complex structure of the tissue. Current approaches are mostly grafts, decellularized structures, microfracture, etc.; however, these approaches pose significant risk to the patient such as inflammation, rejection, implant loosening, or failure.

To repair the damaged articular cartilage, Tuli et al., prepared nanofibrous PCL scaffolds by ES method. The fetal bovine chondrocytes (FBCs) were seeded onto these scaffolds and examined in terms of their ability to maintain chondrocytes in a functional state. According to their results, PCL scaffold seeded with FBC is able to preserve the chondrocyte phenotype by expressing cartilage-specific ECM components such as aggrecan and collagen [78]. In another study, electrospun gelatin/PLA nanofibers were fabricated, and one group was modified by cross-linking with hyaluronic acid to examine the ability to repair cartilage damage. These scaffolds were then subjected to an in vivo study on rabbits using an articular cartilage injury model. Results of the in vivo studies demonstrated that the hyaluronic-acid-modified scaffolds could increase the repair of cartilage along with their super-absorbent properties and cytocompatibility (**Figure 18**) [79].

Another research group fabricated a scaffold by using coaxial ES with poly (L-lactide-co-caprolactone) and collagen as the shell and kartogenin solution as the

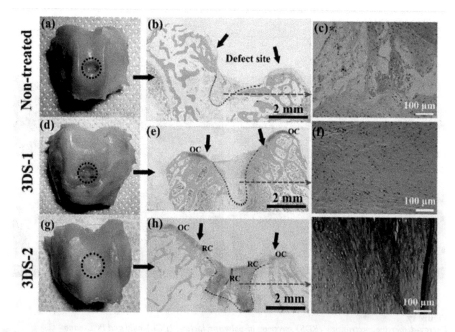

Figure 18.
Macroscopic images (a, d, and g) of the cartilage joints from three groups at 12 weeks after surgery. Histological analysis of cartilage defect area from three groups at 12 weeks after surgery, stained with safranin O-fast green (b, e, and h) and H&E (c, f, and i). Arrows and dotted lines indicated the defect sites. (OC: Original cartilage tissue. RC: Repaired cartilage tissue.) (Figure was reproduced with the permission from Chen et al. [79]).

core. Kartogenin's release behavior was monitored over 2 months, and it is shown that the proliferation and chondrogenic differentiation of rabbit bone-marrow-derived MSCs are increased due to the chondrogenesis inducement properties of kartogenin [80]. Furthermore, incorporation of cartilage-derived ECM into nanofibrous scaffolds is another novel way for stimulation of chondrogenic bioactivity [81].

4.2 Skin tissue engineering

The skin is the largest organ in mammals and acts as a physical barrier between the human body and the external environment, which means it is directly exposed to harmful microbial, thermal, mechanical, and chemical damage. Skin tissue, mainly composed of epidermis, dermis, and subcutaneous layer, suffers from integral skin loss with every injury, which can cause functional imbalance in case of large full-thickness skin defects or loss of large skin areas. Skin loss can occur for many reasons, such as disorders, burns, and chronic wounds. For years, autografts and allografts have been used to treat burns or other skin defects, yet the inability of damaged skin tissue to fully heal has opened up the field of tissue engineering for repair broadly to resolve skin-related defects. The basic prerequisite for a material

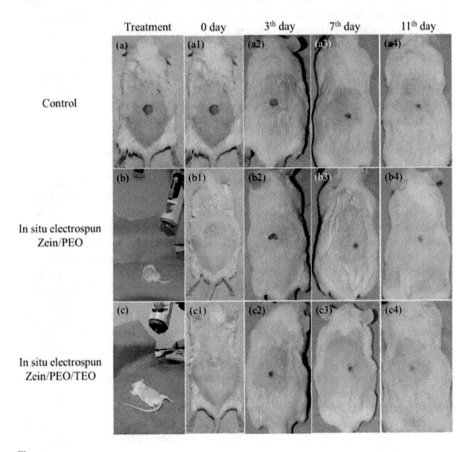

Figure 19.
In situ deposition of electrospun zein/PEO and zein/PEO/TEO fibrous meshes onto wounds of Kunming mice: (a) no treatment, (b) zein/PEO (control group), and (c) zein/PEO/TEO (study group); gross observation of wounds healing at 0, 3, 7, and 11 days after injury for no treatment (a1–4), zein/PEO (b1–4), and zein/PEO/TEO (c1–4), respectively. (Figure is reproduced with the permission from Liu et al. [87]).

to qualify as a biomaterial is biocompatibility, which is the ability of a material to perform with an appropriate host response [82].

Nanofibrous scaffolds with high porosity can enable cell respiration, infiltration, and absorb exudates. Natural polymers such as chitosan, collagen, and elastin are widely used biomaterials for wound dressing according to their biocompatible and biodegradable properties [83]. Ghosal et al. extracted the silk sericin protein (SS) and blended it with PCL, fabricating a scaffold by using emulsion ES method to examine the effect of the silk sericin protein in the scaffold morphology and proliferation of human primary skin fibroblasts. Results showed an increase in proliferation of the cells on PCL/SS scaffolds [84]. Nanoparticles due to their antioxidant and antibacterial properties are also widely used in this field. Augustine et al. incorporated cerium oxide (CeO_2) nanoparticles into electrospun poly (3-hydroxybutyrate-co-3-hydroxyvalerate) scaffolds and analyzed the wound healing properties. The results showed increased cell proliferation, angiogenesis, and wound healing [85]. Chantre et al., prepared a scaffold by centrifugal ES composed of hyaluronic acid to repair cutaneous tissue. In vitro test showed that due to the high porosity, the infiltration of seeded dermal fibroblasts was successful, and scaffolds present biocompatible and bioactive properties. In vivo studies supported their research as well by the acceleration of the tissue formation, neovascularization, and re-epithelialization [86]. Recently, portable electrospinning devices have been widely used to understand in situ deposition of fibers for wound coverage. This technology allows fibrous scaffolds to form directly on the wound site in a matter of minutes.

For example, Liu et al. fabricated electrospun zein/poly (ethylene oxide) nanofibrous scaffolds modified with thyme essential oil (TEO) by using portable handheld ES device directly onto partial thickness wounds on mice dermal tissue defect (**Figure 19**). It is found that electrospun nanofibers improved the wound healing process within 11 days [87].

4.3 Cardiovascular tissue engineering

Cardiovascular diseases such as coronary artery, cardiomyopathy, hypertension, valve disorders, heart failure, etc., are the leading cause of death globally, and the incidence rates are drastically increasing day by day. Common approach is vascular graft transplantation, but it has some limitations such as lack of organ donors, mismatches, preexisting vascular diseases. These limitations cause a need for more stable, flexible grafts with low toxicity and immunity. Since cardiac tissue ECM causes cardiomyocytes to form into fiber-like cell bundles and these bundles elongate and align themselves, a polymeric scaffold that could mimic this specific feature of cardiac tissue could be a potential candidate for cardiovascular tissue engineering.

To stimulate myocardial regeneration, a 3D PCL-based scaffold with hexagonal structure was fabricated using melt electrowriting method. The aim of the study was to create functional cardiac patches, human induced pluripotent stem cell-derived cardiomyocytes (iPSC-CM) were seeded onto these scaffolds. Results of the in vitro studies showed increased cell alignment, cardiac-maturation-related markers, and sarcomere content. Furthermore, in vivo studies, which are conducted on a contracting porcine heart with a minimally invasive approach, showed that the scaffolds express successful biaxial deformation and also supported high tensile stress [88]. Recently, many studies focus on the development of conductive nanoporous scaffolds for cardiovascular tissue engineering approaches. Bertuoli et al. developed an electrospun conducting and biocompatible uniaxial and core-shell fibers having PLA, PEG, and polyaniline (PAni) for cardiac tissue engineering (**Figure 20**) [89]. They produced PLA, PLA/PAni, and PLA/PEG/PAni fibers and

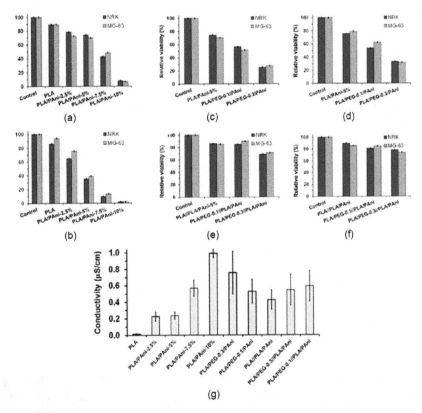

Figure 20.
Biocompatibility of PLA/PAni uniaxial fibers expressed as relative viability of normal rat fibroblasts (NRK) and osteosarcoma (MG-63) culture cells onto the fibrous mats after (a) 24 h (cell adhesion) and (b) 96 h (cell proliferation). Biocompatibility of (c, d) PLA/PEG/PAni uniaxial and (e, f) PLA/PEG//PLA/PAni core-shell fibers expressed as relative viability of NRK and MG-63 cells onto the fibrous mats after (c, e) 24 h (cell adhesion) and (d, f) 96 h (cell proliferation). (g) Electrical conductivity of PLA, PLA/PAni, PLA/ PEG/PAni, PLA//PLA/PAni, and PLA/PEG//PLA/PAni fibrous mats. (Figures were reproduced with the permission from Bertuoli et al. [89]).

core-shell PLA/PLA/PAni and PLA/PEG//PLA/PAni fibers successfully via uniaxial and coaxial electrospinning, respectively. The proposed PLA/PAni-5% uniaxial and PLA//PLA/PAni coaxial fibers offer very good adhesion for cardiac cells, also being able to modulate cell shape and orientation, something important for the characteristic anisotropy of the cardiac tissue.

Liang et al. fabricated a conductive nanofibrous scaffold by encapsulating polypyrole (PPY), which is a conductive polymer, in silk fibroin electrospun fibers. Neonatal rat cardiomyocytes (NRCM) and iPSC-CM cells were used to evaluate cardiomyocyte contraction studies. Results showed that both cell lines attached and proliferated onto these scaffolds successfully. Contraction study indicated that scaffolds with different amount of PPY exhibit contraction behavior starting from day 5 [90]. Another group developed a lab-on-a-chip system integrated with PLLA and PU nanofiber mats for cardiovascular diseases. The aim of the study was to create a model of hypoxic myocardial tissue. The microfluidic system allows simultaneously conducting cell cultures under different circumstances. Cardiac cell lines were used for this study, and results showed that cell viability was high, and cells were positioned parallelly on the scaffolds. The hypoxia study indicates that the amount of ATP molecules decreased during biochemical simulation [91].

4.4 Neural tissue engineering

Injuries affecting peripheral or central nervous systems can cause long-lasting loss of neurological functions due to the severity of the injuries. The usual path of an injury is the inhibition of nerve regeneration, which triggers the formation of compact scar tissue at the defect site. The scar tissue inhibits the connection of the axons across the gap, which will result in disruption of the native tissue and signaling pathways. Short nerve injuries, which have less than 20 mm transaction gap between nerves, are usually repaired surgically; however, long-distance nerve defects require nerve healing/regeneration. For long-distance nerve defects, allografts are usually the gold standard, but there are some disadvantages such as limited donor source, nerve size mismatch, neuroma, etc. To overcome these

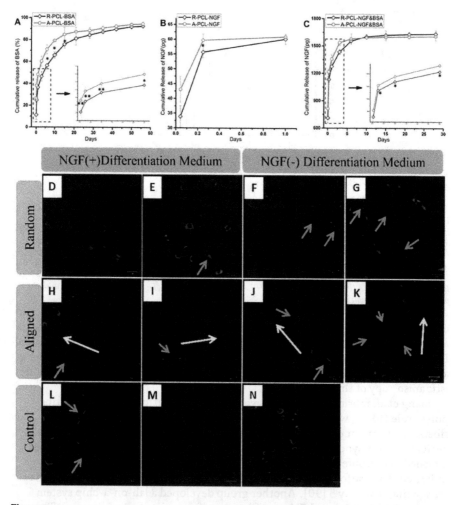

Figure 21.
In vitro release profiles of BSA and NGF. (A) Release curve of BSA from (R/A)-PCL-BSA scaffolds, (B) release properties of NGF from (R/A)-PCL-NGF scaffolds, (C) NGF release from (R/A)-PCL-NGF&BSA scaffolds. Fluorescent images of PC12 cells cultured for 8 days on the surface of different samples with labeling of cytoplasm (red) and nuclei (blue). (D) R-PCL, (E) R-PCL-BSA, (F) R-PCL-NGF, (G) R-PCL-NGF&BSA, (H) A-PCL, (I) A-PCL-BSA, (J) A-PCL-NGF, (K) A-PCL-NGF&BSA, (L) CS-positive, (M) CS-blank for immunofluorescent staining, (N) CS-negative. (yellow arrow indicates the alignment direction for the underlying nanofibers, and blue arrow shows the neurite-bearing PC12 cells.) (Figure was reproduced with the permission from Hu et al. [92]).

challenges, design and fabrication of nerve grafts composed of synthetic or natural polymers are a promising approach for neural tissue engineering.

Nanoporous scaffolds for neural tissue engineering purposes should provide enough surface area for Schwann cells growth and migration, which will direct axons to elongate. Since the orientation structure of axons is axial, some researchers recommended the use of aligned scaffolds, which can provide better contact guidance for cells. Hu et al. fabricated aligned and random PCL scaffolds via ES, and PC-12 (pheochromocytoma of the rat adrenal medulla) neural-like cells were seeded (**Figure 21**).

The results showed that aligned scaffold increased the length of the neurites and directed the extension parallely to the fiber axis. The study also showed that NGF and bovine serum albumin (BSA) incorporated PCL core-shell nanofibrous scaffolds provide sustained release of NGF and neuronal marker expressions and differentiation of PC-12 cells, which indicates that cells were responded to released NGF [92]. Zhang et al. fabricated a conductive scaffold composed of polyaniline (PAN) and poly (L-lactic-co-ε-caprolactone)/silk fibroin nanofibers with incorporation of nerve growth factor (NGF) by using coaxial ES method. The scaffolds successfully support the neurite outgrowth of PC-12 cells, and under electrical stimulation, the amount of neurite-bearing cells and median neurite length were increased [93]. Oxidative stress has a negative impact on nerve cells, so novel approaches, which include antioxidant agents, were investigated. Wang et.al fabricated an antioxidant scaffold composed of lignin/PCL copolymer, and results showed increased mechanical properties of the scaffold and antioxidant activity on cells.

5. Conclusions

The need for less invasive treatment approaches, biocompatible, tailorable, and biodegradable tissue constructs, which can properly mimic native tissues, is still a major challenge for tissue engineering field, and the surface is barely scratched. But nanofibers, due to their tailorable structure, variety of biomaterial options, fabrication routes, and application areas became a popular class of nanomaterials for tissue engineering field. Besides tissue engineering, are other biomedical applications such as drug delivery, biosensor technology, etc. Applications described in this chapter are only a minor proportion of all the results proving the great potential and usefulness of nanofibers. Within this chapter, different fabrication routes, characterization methods, and tissue engineering applications are explained briefly.

Conflict of interest

The authors declare no conflict of interest.

Author details

Ece Bayrak
TOBB University of Economics and Technology, Ankara, Turkey

*Address all correspondence to: ebayrak@etu.edu.tr

IntechOpen

References

[1] Zeleny J. The electrical discharge from liquid points and a hydrostatic method of measuring the electric intensity at their surfaces. Physical Review. 1914;**3**:69-91

[2] Reneker DH, Yarin AL, Fong H, Koombhongse S. Bending instability of electrically charged liquid jets of polymer solutions in electrospinning. Journal of Applied Physics. 2000;**87**: 4531-4547

[3] Baumgarten PK. Electrostatic spinning of acrylic microfibers. Journal of Colloid and Interface Science. 1971;**36**:71-79

[4] Bharwaj N, Kundu SC. Electrospinning: A fascinating fiber fabrication techniques. Biotechnology Advances. 2010;**28**(3):325-347

[5] Gugulothu D, Barhoum A, Nerella R, Ajmer R, Bechlany M. Fabrication of nanofibers: Electrospinning and non-electrospinning techniques. In: Handbook of Nanofibers. Swtizerland: Springer Cham; 2018. pp. 1-34

[6] Huang ZM, Zhang YZ, Kotaki M, Ramakrishna S. A review on polymer nanofibers by electrospinning and their applications in nanocomposites. Composites Science and Technology. 2003;**63**(15):2223-2253

[7] Gonçalves AM, Moreira A, Weber A, Williams GR, Costa PF. Osteochondral tissue engineering: The potential of electrospinning and additive manufacturing. Pharmaceutics. 2021;**13**(7):983

[8] Ura DP, Rosell-Llompart J, Zaszczyńska A, Vasilyev G, Gradys A, Szewczyk PK, et al. The role of electrical polarity in electrospinning and on the mechanical and structural properties of as-spun fibers. Materials. 2020;**13**(18):4169

[9] Wu YK, Wang L, Fan J, Shou W, Zhou BM, Liu Y. Multi-jet electrospinning with auxiliary electrode: The influence of solution properties. Polymers. 2018;**10**(6):572

[10] Li WJ, Shanti RM, Tuan RS. Electrospinning Technology for Nanofibrous Scaffolds in tissue engineering. In: Kumar CSSR, editor. Nanotechnologies for the Life Sciences. 1st ed. Weinheim: Wiley; 2007. pp. 135-187. DOI: 10.1002/9783527610419.ntls0097

[11] Haider A, Haider S, Kang IK. A comprehensive review summarizing the effect of electrospinning parameters and potential applications of nanofibers in biomedical and biotechnology. Arabian Journal of Chemistry. 2018;**8**:1165-1188

[12] Migliaresi C, Ruffo GA, Volpato FZ, Zeni D. Advanced electrospinning setups and special fibre and mesh morphologies. In: Neves NM, editor. Electrospinning for Advanced Biomedical Applications and Therapies. 1st ed. United Kingdom: Smithers Rapra; 2012. pp. 22-68

[13] Yoon J, Yang HS, Lee BS, Yu WR. Recent progress in coaxial electrospinning: New parameters, various structures and wide applications. Advanced Materials. 2018;**30**(42):1704765

[14] Brown TD, Dalton PD, Hutmacher DW. Melt electrospinning today: An opportune time for an emerging polymer process. Progress in Polymer Science. 2016;**56**:116-166

[15] Taghavi SM, Larson RG. Regularized thin-fiber model for nanofiber formation by centrifugal spinning. Physical Review E. 2014;**89**(2):023011

[16] Blachowicz T, Ehrmann A. Most recent developments in electrospun

magnetic nanofibers: A review. Journal of Engineered Fibers and Fabrics. 2020;**15**:1558925019900843

[17] Yoon YI, Park KE, Lee SJ, Park WH. Fabrication of microfibrous and nanofibrous scaffolds: Melt and hybrid electrospinning and surface modification of poly(L-lactic acid) with plasticizer. BioMed Research International. 2013;**2013**:309048. DOI: 10.1155/2013/309048

[18] Zhang LH, Duan XP, Yan X, Yu M, Ning X, Zhao Y, et al. Recent advances in melt electrospinning. RSC Advances. 2016;**6**(58):53400-53414

[19] Zhang X, Lua Y. Centrifugal spinning: An alternative approach to fabricate nanofibers at high speed and low cost. Polymer Reviews. 2014;**54**:677-107

[20] Loordhuswamy AM, Krishnaswamy VR, Korrapati PS, Thinakaran S, Rengaswami GDV. Fabrication of highly aligned fibrous scaffolds for tissue regeneration by centrifugal spinning technology. Materials Science and Engineering: C. 2014;**42**:799-807

[21] Xu J, Liu X, Zhang Z, Wang L, Tan R, Zhang D. Controllable generation of nanofibers through a magnetic-field-assisted electrospinning design. Material Letters. 2019;**247**: 19-24

[22] Liu Y, Zhang X, Xia Y, Yang H. Magnetic-field-assisted electrospinning of aligned straight and wavy polymeric nanofibers. Advanced Materials. 2010;**22**:2454-2457

[23] Li TT, Yan M, Xu W, Shiu BC, Lou CW, Lin JH. Mass-production and characterizations of polyvinyl alcohol/ sodium alginate/graphene porous nanofiber membranes using needleless dynamic linear electrospinning. Polymers. 2018;**10**(10):1167

[24] Nikmaram N, Roohinejad S, Hashemi S, Koubaa M, Barba FJ, Abbaspourrad A, et al. Emulsion-based systems for fabrication of electrospun nanofibers: Food, pharmaceutical and biomedical applications. RSC Advances. 2017;**7**(46):28951-28964

[25] Niu H, Lin T. Fiber generators in needleless electrospinning. Journal of Nanomaterials. 2012;**12**:1-13

[26] Partheniadis I, Nikolakakis O, Laidmæ I, Heinamaki J. A mini-review: Needleless electrospinning of nanofibers for pharmaceutical and biomedical applications. PRO. 2020;**8**(6):673

[27] Buzgo M, Mickova A, Rampickova M, Doupnik M. Blend electrospinning, coaxial electrospinning, and emulsion electrospinning techniques. In: Tampieri A, Focarete ML, editors. Core-Shell Nanostructures for Drug Delivery and Theranostis. 1st ed. Sawston, UK: Woodhead Publishing, Elsevier; 2018. pp. 325-347

[28] Zhang X, Shi X, Gautrot JE, Peijs T. Nanoengineered electrospun fibers and their biomedical applications: A review. Nano. 2021;**7**(1):1-34

[29] Arafat MT, Tronci G, Yin J, Wood DJ, Russell SJ. Biomimetic wet-stable fibres via wet spinning and diacid-based crosslinking of collagen triple helices. PRO. 2015;**77**:102-112

[30] Wang L, Lundahl MJ, Greca LG, Papageorgiou AC, Borghei M, Rojas OJ. Effects of non-solvents and electrolytes on the formation and properties of cellulose I filaments. Scientific Reports. 2019;**9**(1):1-11

[31] Puppi D, Chiellini F. Wet-spinning of biomedical polymers: From single fibre production to additive manufacturing of three-dimensional scaffolds. Polymer International. 2017;**66**(12):1690-1696

[32] Ramakrishna S, Fujihara K, Teo WE, Lim TK, Ma Z. An Introduction to Electrospinning and Nanofibers. 1st ed. Singapore: World Scientific Press; 2005. p. 10

[33] Ondarcuhu T, Joachim C. Drawing a single nanofibre over hundreds of microns. Europhysics Letters. 1998;**42**(2):215

[34] Wang J, Nain AS. Suspended micro/nanofiber hierarchical biological scaffolds fabricated using non-electrospinning STEP technique. Langmuir. 2014;**30**:13641-13649

[35] Wng Y, Zheng M, Lu H, Feng S, Ji G, Cao J. Template synthesis of carbon nanofibers containing linear mesocage arrays. Nanoscale Research Letters. 2010;**5**(6):913-916

[36] Liu S, Shan H, Xia S, Yan J, Ding B. Polymer template synthesis of flexible SiO$_2$ nanofibers to upgrade composite electrolytes. ACS Applied Materials & Interfaces. 2020;**12**(28):31439-31447

[37] Ma PX, Zhang R. Synthetic nano-scale fibrous extracellular matrix. Journal of Biomedical Materials Research Part A. 1999;**46**:60-72

[38] Zhang R, Ma PX. Processing polymer scaffolds: Phase separation. In: Atala A, Lanza R, Lanza RP, editors. Methods of Tissue Engineering. 1st ed. San Diego: Academic Press; 2002. pp. 715-724

[39] Sharma J, Lizu M, Stewart M, Zygula K, Lu Y, Chauhan R, et al. Multifunctional nanofibers towards active biomedical therapeutics. Polymers. 2015;**7**(2):186-219

[40] Xu XD, Jin Y, Liu Y, Zhang XZ, Zhuo RX. Self-assembly behavior of peptide amphiphiles (PAs) with different length of hydrophobic alkyl tails. Colloids and Surfaces B: Biointerfaces. 2010;**81**(1):329-335

[41] Wang L, Gong C, Yuan X, Wei G. Controlling the self-assembly of biomolecules into functional nanomaterials through internal interactions and external stimulations: A review. Nanomaterials. 2019;**9**(2):285

[42] Raaijmakers MJ, Benes NE. Current trends in interfacial polymerization. Prograss in Polymer Science. 2016;**63**:86-142

[43] Huang J, Kaner RB. A general chemical route to polyaniline nanofibers. Journal of the American Chemical Society. 2004;**126**(3):851-855

[44] Freitas TV, Sousa EA, Fuzari GC Jr, Arlindo EP. Different morphologies of polyaniline nanostructures synthesized by interfacial polymerization. Materials Letters. 2018;**224**:42-45

[45] Sirc J, Hobzova R, Kostina N, Munzarova M, Juklickova M, Lhotka M, et al. Morphological characterization of nanofibers: Methods and application in practice. Journal of Nanomaterials. 2012;**2012**:327369. DOI: 10.1155/2012/327369

[46] Oliviera JE, Mattoso LH, Orts WJ, Medeiros ES. Structural and morphological characterization of micro and nanofibers produced by electrospinning and solution blow spinning: A comparative study. Advances in Materials Science and Engineering. 2013. DOI: 10.1155/2013/409572

[47] Megelski S, Stephens JS, Chase DB, Rabolt JF. Micro-and nanostructured surface morphology on electrospun polymer fibers. Macromolecules. 2002;**35**(22):8456-8466

[48] Deng X, Xiong F, Li X, Xiang B, Li Z, Wu X, et al. Application of atomic force microscopy in cancer research. Journal of Nanobiotechnology. 2018;**16**(1):1-15

[49] Birdi KS. Scanning Probe Microscopes: Applications in Science

and Technology. 1st ed. London: CRC Press, Taylor & Francis Group; 2003. p. 16. DOI: 10.1201/9780203011072

[50] Kaur H, Bhagwat SR, Sharma TK, Kumar A. Analytical techniques for characterization of biological molecules–proteins and aptamers/ oligonucleotides. Bioanalysis. 2019;**11**(02):103-117

[51] Bunaciu AA, Udristioiu EG, Aboul-Enein HY. X-ray diffraction: Instrumentation and applications. Critical Reviews in Analytical Chemistry. 2015;**45**(4):289-299

[52] Loganathan S, Valapa RB, Mishra RK, Pugazhenthi G, Thomas S. Thermogravimetric analysis for characterization of nanomaterials. In: Thomas S, Thomas R, Zachariah A, Mishra R, editors. Thermal and Rheological Measurement Techniques for Nanomaterials Characterization. 1st ed. Elsevier; 2017. pp. 67-108

[53] Coats AW, Redfern JP. Thermogravimetric analysis. A review. Analyst. 1963;**88**(1053):906-924

[54] Saadatkhah N, Carillo Garcia A, Ackermann S, Leclerc P, Latifi M, Samih S, et al. Experimental methods in chemical engineering: Thermogravimetric analysis-TGA. The Canadian Journal of Chemical Engineering. 2020;**98**(1):34-43

[55] Zhou J, Cai Q, Liu X, Ding Y, Xu F. Temperature effect on the mechanical properties of electrospun PU nanofibers. Nanoscale Research Letters. 2018;**13**(1):1-5

[56] Tan EPS, Lim CT. Mechanical characterization of nanofibers. Composites. 2006;**66**(9):1102-1111

[57] Seyama H, Soma M, Theng BKG. X-ray photoelectron spectroscopy. In: Bergaya F, Lagaly G, editors. Developments in Clay Science. 1st ed.

Amsterdam, Netherlands: Elsevier; 2013. pp. 161-176

[58] Lee SH, Park SM, Lee LP. Optical methods in studies of olfactory system. In: Park T, editor. Bioelectronic Nose. Dordrecht: Springer; 2014. pp. 191-220

[59] Seah MP. Quantitative AES and XPS: Convergence between theory and experimental databases. Journal of Electron Spectroscopy and Related Phenomena. 1999;**100**:55-73

[60] Polini A, Yang F. Physicochemical characterization of nanofiber composites. In: Ramalingam M, Ramakrishna S, editors. Nanofiber Composites for Biomedical Applications. 1st ed. Sawston, UK: Woodhead Publishing, Elsevier; 2017. pp. 97-115

[61] Kim HH. Endoscopic Raman spectroscopy for molecular fingerprinting of gastric cancer: Principle to implementation. BioMed Research International. 2015; **2015**:670121. DOI: 10.1155/2015/ 670121

[62] Titus D, Samuel EJJ, Roopan SM. Nanoparticle characterization techniques. In: Kumar Shukla A, Iravani S, editors. Green Synthesis, Characterization and Applications of Nanoparticles. 1st ed. Amsterdam: Elsevier; 2019. pp. 303-319

[63] de Melo Pereira D, Habibovic P. Biomineralization-inspired material Design for Bone Regeneration. Advanced Healthcare Materials. 2018;**7**(22):1800700

[64] Silber J, Anderson DG, Daffner SD, Brislin BT, Leland JM, Hilibrand AS, et al. Donor site morbidity after anterior iliac crest bone harvest for single-level anterior cervical discectomy and fusion. Spine. 2003;**28**(2):134-139

[65] Mankin HJ, Hornicek FJ, Raskin KA. Infection in massive bone allografts.

Clinical Orthopaedics and Related Research. 2005;**432**:210-216

[66] Du Y, Guo JL, Wang J, Mikos AG, Zhang S. Hierarchically designed bone scaffolds: From internal cues to external stimuli. Biomaterials. 2019;**218**:119334

[67] Cai YZ, Zhang GR, Wang LL, Jiang YZ, Ouyang HW, Zou XH. Novel biodegradable three-dimensional macroporous scaffold using aligned electrospun nanofibrous yarns for bone tissue engineering. Journal of Biomedical Materials Research Part A. 2012;**100**(5):1187-1194

[68] Nedjari S, Awaja F, Altankov G. Three dimensional honeycomb patterned fibrinogen based nanofibers induce substantial osteogenic response of mesenchymal stem cells. Scientific Reports. 2017;**7**(1):1-11

[69] Abdal-hay A, Abbasi N, Gwiazda M, Hamlet S, Ivanovski S. Novel polycaprolactone/hydroxyapatite nanocomposite fibrous scaffolds by direct melt-electrospinning writing. European Polymer Journal. 2018;**105**: 257-264

[70] Velioglu ZB, Pulat D, Demirbakan B, Ozcan B, Bayrak E, Erisken C. 3D-printed poly (lactic acid) scaffolds for trabecular bone repair and regeneration: Scaffold and native bone characterization. Connective Tissue Research. 2019;**60**(3):274-282

[71] Lukášová V, Buzgo M, Vocetková K, Sovková V, Doupník M, Himawan E, et al. Needleless electrospun and centrifugal spun poly-ε-caprolactone scaffolds as a carrier for platelets in tissue engineering applications: A comparative study with hMSCs. Materials Science and Engineering: C. 2019;**97**:567-575

[72] Olender E, Uhrynowska-Tyszkiewicz I, Kaminski A. Revitalization of biostatic tissue

allografts: New perspectives in tissue transplantology. In Transplantation Proceedings. 2011;**43**(8):3137-3141

[73] Yang G, Lin H, Rothrauff BB, Yu S, Tuan RS. Multilayered polycaprolactone/gelatin fiber-hydrogel composite for tendon tissue engineering. Acta Biomaterialia. 2016;**35**:68-76

[74] Perikamana SKM, Lee J, Ahmad T, Kim EM, Byun H, Lee S, et al. Harnessing biochemical and structural cues for tenogenic differentiation of adipose derived stem cells (ADSCs) and development of an in vitro tissue interface mimicking tendon-bone insertion graft. Biomaterials. 2018;**165**:79-93

[75] Rinoldi C, Kijeńska E, Chlanda A, Choinska E, Khenoussi N, Tamayol A, et al. Nanobead-on-string composites for tendon tissue engineering. Journal of Materials Chemistry B. 2018;**6**(19): 3116-3127

[76] Chen CH, Li DL, Chuang ADC, Dash BS, Chen JP. Tension stimulation of tenocytes in aligned hyaluronic acid/platelet-rich plasma-Polycaprolactone Core-sheath nanofiber membrane scaffold for tendon tissue engineering. International Journal of Molecular Sciences. 2021;**22**:11215

[77] Bayrak E, Ozcan B, Erisken C. Processing of polycaprolactone and hydroxyapatite to fabricate graded electrospun composites for tendon-bone interface regeneration. Journal of Polymer Engineering. 2017;**37**(1):99-106

[78] Tuli R, Li WJ, Tuan RS. Current state of cartilage tissue engineering. Arthritis Research & Therapy. 2003;**5**(5):1-4

[79] Chen W, Chen S, Morsi Y, El-Hamshary H, El-Newhy M, Fan C, et al. Superabsorbent 3D scaffold based on electrospun nanofibers for cartilage tissue engineering. ACS Applied Materials & Interfaces. 2016;**8**(37):24415-24425

[80] Yin H, Wang J, Gu Z, Feng W, Gao M, Wu Y, et al. Evaluation of the potential of kartogenin encapsulated poly (L-lactic acid-co-caprolactone)/collagen nanofibers for tracheal cartilage regeneration. Journal of Biomaterials Applications. 2017;**32**(3): 331-341

[81] Garrigues NW, Little D, Sanchez-Adams J, Ruch DS, Guilak F. Electrospun cartilage-derived matrix scaffolds for cartilage tissue engineering. Journal of Biomedical Materials Research Part A. 2014;**102**(11):3998-4008

[82] Borena BM, Martens A, Broeckx SY, Meyer E, Chiers K, Duchateau L. Regenerative skin wound healing in mammals: State-of-the-art on growth factor and stem cell based treatments. Cellular Physiology and Biochemistry. 2015;**36**:1-23

[83] Demir M, Büyükserin F, Bayrak E, Türker NS. Development and characterization of metronidazole-loaded and chitosan-coated PCL electrospun Fibres for potential applications in guided tissue regeneration. Trends in Biomaterials & Artificial Organs. 2021;**35**(3):255-263

[84] Ghosal K, Manakhov A, Zajíčková L, Thomas S. Structural and surface compatibility study of modified electrospun poly (ε-caprolactone) (PCL) composites for skin tissue engineering. AAPS PharmSciTech. 2017;**18**(1):72-81

[85] Augustine R, Hasan A, Patan NK, Dalvi YB, Varghese R, Antony A, et al. Cerium oxide nanoparticle incorporated electrospun poly (3-hydroxybutyrate-co-3-hydroxyvalerate) membranes for diabetic wound healing applications. ACS Biomaterials Science & Engineering. 2019;**6**(1):58-70

[86] Chantre CO, Gonzalez GM, Ahn S, Cera L, Campbell PH, Hoerstrup SP, et al. Porous biomimetic hyaluronic acid and extracellular matrix protein nanofiber scaffolds for accelerated cutaneous tissue repair. ACS Applied Materials & Interfaces. 2019;**11**(49):45498-45510

[87] Liu JX, Dong WH, Mou XJ, Liu GS, Huang XW, Yan X, et al. In situ electrospun zein/thyme essential oil-based membranes as an effective antibacterial wound dressing. ACS Applied Biomaterials. 2019;**3**(1):302-307

[88] Castilho M, van Mil A, Maher M, Metz CH, Hochleitner G, Groll J, et al. Melt electrowriting allows tailored microstructural and mechanical design of scaffolds to advance functional human myocardial tissue formation. Advanced Functional Materials. 2018;**28**(40):1803151

[89] Bertuoli PT, Ordono J, Armelin E, Perez-Amodio S, Baldissera AF, Ferreira CA, et al. Electrospun conducting and biocompatible uniaxial and Core–Shell fibers having poly(lactic acid), poly(ethylene glycol), and polyaniline for cardiac tissue engineering. ACS Omega. 2019;**4**: 3660-3672

[90] Liang Y, Mitriashkin A, Lim TT, Goh JCH. Conductive polypyrrole-encapsulated silk fibroin fibers for cardiac tissue engineering. Biomaterials. 2021;**276**:121008

[91] Kobuszewska A, Kolodziejek D, Wojasinski M, Jastrzebska E, Ciach T, Brzozka Z. Lab-on-a-chip system integrated with nanofiber mats used as a potential tool to study cardiovascular diseases (CVDs). Sensors and Actuators B: Chemical. 2021;**330**: 129291

[92] Hu J, Tian L, Prabhakaran MP, Ding X, Ramakrishna S. Fabrication of nerve growth factor encapsulated aligned poly (ε-caprolactone) nanofibers and their assessment as a

potential neural tissue engineering
scaffold. Polymers. 2016;**8**(2):54

[93] Zhang J, Qiu K, Sun B, Fang J,
Zhang K, Hany EH, et al. The aligned
core–sheath nanofibers with electrical
conductivity for neural tissue
engineering. Journal of Materials
Chemistry B. 2014;**2**(45):7945-7954

Chapter 14

Composite Metamaterials: Classification, Design, Laws and Future Applications

Tarek Fawzi and Ammar A.M. Al-Talib

Abstract

The development of science and applications have reached a stage where the naturally existed materials are not meeting the required properties. Metamaterials (MMs) are artificial materials that obtain their properties from their accurately engineered meta-atoms rather than the characteristics of their constituents. The size of the meta-atom is small compared to light's wavelength. A metamaterial (MM) is a term means beyond material which has been engineered in order to possess properties that does not exist in naturally-found materials. Currently, they are made of multiple elements such as plastics and metals. They are being organized in iterating patterns at a scale that is smaller than wavelengths of the phenomena it influences. The properties of the MMs are not derived from the forming materials but their delicate size, geometry, shape, orientation, and arrangement. These properties maintain MMs to manipulate the electromagnetic waves via promoting, hindering, absorbing waves to attain an interest that goes beyond the natural materials' potency. The apt design of MMs maintains them of influencing the electromagnetic radiation or sound in a distinctive technique never found in natural materials. The potential applications of MMs are wide, starting from medical, aerospace, sensors, solar-power management, crowd control, antennas, army equipment and reaching earthquakes shielding and seismic materials.

Keywords: metamaterials, permittivity, permeability, electromagnetic wave, hyperbolic, chiral, transmission matrix, reflection matrix, anisotropic

1. Introduction

Metamaterials are synthetic composite structures with peculiar material characteristics. They have protruded as a promising material for several science disciplines comprising physics, chemistry, engineering, and material science. MM has been described as structures designed according to an imposed geometry to be exploited in a definite application. The characteristics of MMs are derived from the microstructure, inherent properties, architecture, and the features' size within it. The 3D architecture can substantially substitute some material characters such as photonic band-gaps, negative thermal expansion, negative Poisson ratio and negative refractive indices [1].

The fabrication of composite metamaterials (CMMs) enables the manipulation of the microstructure and the novel geometry resulted in new or developed properties which were never found in bulk materials. Also, CMM expands the design space occupied by MM [1]. The industrial applications incorporated MM and CMM in lightweight materials, micro-electromechanical systems, sensors, energy storage and photovoltaic.

2. Types of metamaterials

Negative index materials (NIMs) propagate wave in a way where the wave does not parallel the Poynting vector or (energy and phase velocities are anti-parallel). The right-handed vectors of wave vector, magnetic and electric fields (k, H &E) respectively, in positive index material (PIM) transforms into left-handed triplet in negative index material, where negative refraction occurs to the propagating light beams [2]. Such a phenomenon was perceived in three systems: left-handed (or double negative) MMs [3], hyperbolic MMs [4] and in photonic crystals near band-gap edge [5].

2.1 Double negative MMs

Negative index materials (NIMs) are one of the most researched MMs. They can only be structured via meta-atoms or artificial structure that defeats the boundaries imposed on matter-light interactions due to the need for negative index of refraction which cannot be found unless both of permittivity and permeability are negative, **Figure 1** illustrates the first NIM in the world [6].

The light beam does not parallel Poynting vector because this vector is being determined via the equation:

$$S = E \times H*$$ (1)

Where E is electric field and H is magnetic field H. This means that the direction of energy velocity counteracts phase velocity direction. In general, the refractive index relation is: $n = \pm\sqrt{\varepsilon\mu}$ (Where ε is permittivity and μ is permeability), in this case the negative value must be taken because $\varepsilon_r < 0$, $\varepsilon_i > 0$ and $\mu_r < 0$, $\mu_i > 0$. Hence, $\varepsilon_r\mu_i + \varepsilon_i\mu_r$ is negative. Also, the refraction angle is negative depending on Snell's law:

$$\sin\theta_r = \sin\theta_i/n$$ (2)

Figure 1.
The first NIM in the world [6].

(θ is the refraction angle of the wave vector).Which means that both incident and refracted beams are on the same side of the normal. To discover or design a NIM, it is by the negative values of both permittivity and permeability. Natural materials' magnetic susceptibility is quite small compared to dielectric one, this restricts the atoms interaction to the electric constituent of the electromagnetic (EM) wave and neglects the magnetic constituent undeveloped, for that reason most of the natural materials possess a (μ) value close to 1 [6].

2.2 Zero-index materials

The most important property that these materials possess is a zero-phase delay. Some phenomena like infinite wavelength and quasi-infinite phase velocity can be implied from this delay. In this material, EM wave is static in the spatial field. While it allows the energy transportation due to the time domain flexibility. One of the pioneer applications of these materials is to show how it can enable the user for controlling the emission of an internal source to gather the power in a small angular field around the normal [7]. The critical angles' reflection is (0) depending on Snell's law. The external source (outside of the material) beam cannot be transmitted due to the total reflection, while the internal source results in a perpendicular beam to the surface of the beam [6].

It is being complicated to perceive a zero-index material with nullified permittivity and permeability, while many realizations of materials with one nullified parameter (permittivity or permeability) have been achieved. Zero or near zero index MMs have another fascinating property in graded-index structures known as anomalous absorption near zero index transition. In which magnetic permeability and dielectric permittivity gradually moves among negative, zero and positive values [8].

2.3 Mechanical MMs

The distinctive and tunable characteristics of MM have attracted the researchers in the mechanics discipline. Materials with negative Poisson ratio (auxetic) and near zero Poisson ratio (anepirretic) showed compression within load applying [9], such a phenomenon led to optimized shock absorbing, shear and indentation resistance. For those properties, they have been often used in medical devices (stents), protective gear and gaskets. Reports have illustrated that the geometrical design is the main reason behind auxetic property [10, 11]. The phase distribution and combination can be organized in order to produce an auxetic material, via the reaction between the host and fiber enforcement [12].

Lightweight - strength materials are another focal interest for researchers. Materials that combines both of light weight and strength are expected to be filled up in the Ashby's plot, here density reduces significantly [13]. The strength can be conserved while the weight decreased via applying 3D structuring. The mechanical properties in the micro-scale are changing according to the intrinsic and extrinsic size impact. Such properties are well known in metals, but they are also exhibited in CMM micro-lattice [14]. The researchers nowadays, are working on MMs that possess mechanical anisotropy [15, 16], programmable materials [15, 17], and nil-thermal expansion CMM [18].

2.4 Photonic MMs

MMs can dominate the path of light, and imposing a definite geometry, enables MMs to produce polarization filtering [19], negative refractive indices [20] and photonic bandgaps [21]. Negative refractive index (left-handed) materials can be applied to create a reversed Doppler effect, optical tunneling, super lenses and electromagnetic camouflage [22]. Split ring resonators are the most used structures to discover these impacts, exposing it to exterior magnetic field makes the current in the ring evolve electromagnetic field, **Figure 2** shows the most common designs of photonic MMs [1]. Gaps possess big capacitance that impacts the frequency of the resonance. The disposition is giving the structure a great quality factor, also, cloaking often related to Split-ring resonators (SRR) due to the induction of the opposite flux to the occurring one [1].

Presenting a recurring layered structure in a 3D material can result in a photonic bandgap [23, 24]. The matching between wavelength and distance between two layers hinders the light passing and reflects it. The common name of this phenomenon is band-stops, and it attracted the researchers to be applied in optical networks approaching the semiconductors' electrical networks. The woodpile design is the most eminent structure due to its simple design [25–27]. Also, the simple fabrication of colloidal crystals via self-assembly has attracted researchers, as this method has resulted in a structure that can be used as templates because its voids can be eliminated and inverse opals can be originated. The light polarization can be controlled via chiral structure as a substitute to noble materials (gold and silver). Such a polarizer has been manufactured with template-assisted electro-deposition from pure electro-deposited gold [28], while others covered two-photon lithography pints with electroless deposited silver [29].

Also, MMs can control sound wave propagation depending on the geometrical design, the similar behavior of sound and electromagnetic waves makes the same concept applicable for photonic MMs. Conceptually, phononic and photonic crystals are the same, plenty 2D phononic crystals have been manufactured of silicone with array of holes, they are able to filter definite phonons' wavelength because of the stimulated bandgap [30, 31]. Stereo-lithography succeeded in producing 3D

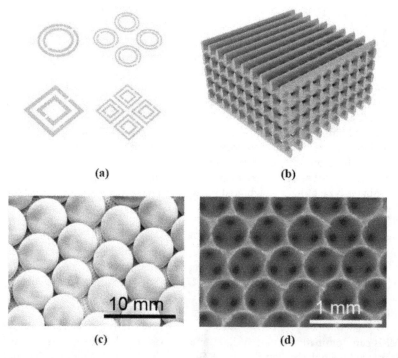

(a) (b)

(c) (d)

Figure 2.
The most common designs for photonic MMs: (a) SRRs, (b) wood pile structures, (c) colloidal crystals, and (d) inverse opals [1].

phononic crystals out of acrylic polymer and metallic constituents, which have increased the ratio of the broad band vibration propagation. Polymer has been used in creating fano-like dampening [32]. Whilst silencers with inertial local resonant have been created for acoustic lensing [33].

2.5 Chiral MMs

Chiral means lacking of mirror symmetry, chiral medium is subcategory of bianisotropic, where magnetic and electrical fields are conjugated together. The optical response of the general chiral media has been described by these two fundamental equations:

$$\overline{D} = \varepsilon_0 \overline{\overline{\varepsilon}} \overline{E} + \frac{i}{c_0} \overline{\overline{\chi}} \overline{H} \tag{3}$$

$$\overline{B} = -\frac{i}{c_0} \overline{\overline{\chi}} \overline{E} + \mu_0 \overline{\overline{\mu}} \overline{H} \tag{4}$$

Where $\overline{\overline{\varepsilon}}, \overline{\overline{\mu}}$ *and* $\overline{\overline{\chi}}$ are permittivity, permeability, and chirality tensors, respectively [34]. A zero in the subscript position (X_0) for any of the variables illustrates the vacuum. $\overline{H}$ magnetic field, $\overline{B}$ magnetic induction, $\overline{E}$ electric field and $\overline{D}$ is the electric displacement. If the chiral material is isotropic, ε, μ, and χ scalars are used to simplify and facilitate the fundamental parameters. The left and right handed circular polarization (LCP &RCP) of the refractive indices are given via a different equation:

$$n_{\pm} = \sqrt{\varepsilon\mu} \pm \chi \tag{5}$$

Dissimilar phase accumulation will result via the waves according to the handedness, while bothare having corresponding impedance:

$$Z = Z_0 \sqrt{\varepsilon/\mu} \pm \chi \tag{6}$$

LCP and RCP overlapping with identical amplitude will illustrate a linearly polarized wave. The refractive index difference between two circularly polarized waves gives a rotation in the value of an angle

$$\Theta = (n_+ - n_-) \frac{\pi d}{\lambda_0} \tag{7}$$

or

$$\Theta = [\arg(T_+) - \arg(T_-)]/2 \tag{8}$$

Where d is the thickness of the medium, λ_0 is the wavelength in the vacuum, T is the transmission coefficient in different spin conditions. This is the optical rotation's mechanisms and physical consequences [34].

It can be inferred that if chirality χ is strong enough, the occurring refraction may be negative in the case of one circularly polarized light albeit both ε and μ are positive [35].

The retrieval method can be applied to acquire effective parameters from scattering ones [36] to address the characters of the MM. The optical response forms an apt way to describe the properties of the MM. John matrices can be used in defining optical response of planar MM, which relate the scattered fields and the complex amplitudes of the incident [37].

$$\begin{pmatrix} E_r^x \\ E_r^y \end{pmatrix} = \begin{pmatrix} r_{xx} & r_{xy} \\ r_{yx} & r_{yy} \end{pmatrix} \begin{pmatrix} E_i^x \\ E_i^y \end{pmatrix} = R \begin{pmatrix} E_i^x \\ E_i^y \end{pmatrix} \tag{9}$$

$$\begin{pmatrix} E_t^x \\ E_t^y \end{pmatrix} = \begin{pmatrix} t_{xx} & t_{xy} \\ t_{yx} & t_{yy} \end{pmatrix} \begin{pmatrix} E_i^x \\ E_i^y \end{pmatrix} = T \begin{pmatrix} E_i^x \\ E_i^y \end{pmatrix} \tag{10}$$

T & R, are the transmission and reflection matrices for linear polarization. E_r^x, E_i^x, & E_t^x are the reflected, incident, and the transmitted electric fields polarized along the x axis, respectively. Y direction has similar notations with an apt superscript (Y). Applying the Cartesian base to a circular base result in Jones' matrices for the two conditions of circular polarization.

$$\begin{aligned} R_{\text{circ}} &= \begin{pmatrix} r_{++} & r_{+-} \\ r_{-+} & r_{--} \end{pmatrix} = \Lambda^{-1} R \Lambda \\ &= \frac{1}{2} X \begin{pmatrix} r_{xx} + r_{yy} + i(r_{xy} - r_{yx}) & r_{xx} - r_{yy} - i(r_{xy} - r_{yx}) \\ r_{xx} + r_{yy} + i(r_{xy} + r_{yx}) & r_{xx} + r_{yy} - i(r_{xy} - r_{yx}) \end{pmatrix} \end{aligned} \tag{11}$$

$$T_{\text{circ}} = \begin{pmatrix} t_{++} & t_{+-} \\ t_{-+} & t_{--} \end{pmatrix} = \Lambda^{-1} T \Lambda = \frac{1}{2} X \begin{pmatrix} t_{xx} + t_{yy} + i(t_{xy} - t_{yx}) & t_{xx} - t_{yy} - i(t_{xy} - t_{yx}) \\ t_{xx} + t_{yy} + i(t_{xy} + t_{yx}) & t_{xx} + t_{yy} - i(t_{xy} - t_{yx}) \end{pmatrix} \tag{12}$$

Where $\Lambda = \frac{1}{\sqrt{2}} \begin{pmatrix} 1 & 1 \\ i & -i \end{pmatrix}$ is the variation of the fundamental matrix, subscript +/− indicates the circularly polarized waves along +z direction whether it is clockwise or counterclockwise.

Symmetrical distribution has been an effective method to anticipate to which extent the symmetrical structure influences the properties of the structure and Jones matrices' characters. If the MM represent a certain symmetry group, the original and the Jones matrices must be identical. In Jones matrices of the linear polarizations, the mirror symmetry results in an absent to all the off-diagonal elements ($r_{xy} = r_{yx} = t_{xy} = t_{yx} = 0$) with respect to the incident plane. Hence, the Jones matrices for circular polarizations become symmetric ($r_{--} = r_{++}, r_{-+} = r_{+-}$, $t_{+-} = t_{-+}, t_{--} = t_{++}$) [37].

Furthermore, chirality nullifies polarization rotation as the optical activity has been always described by $\theta = [\arg(t_{++}) - \arg(t_{--})]/2$. The correspondence of the two circularly polarized waves results in similar efficiency for the polarization conversion due to the correspondence of the off-diagonal elements. Inevitably, mirror symmetric structures have neither CD nor optical activity, and this result explains why chirality exist in the structures that suffers mirrors symmetries deficiency -from Jones matrix perspective-. **Figure 3** shows various types of 3D chiral mechanical metamaterials [38].

Other rotational symmetries (threefold & fourfold) with respect to symmetry guides to $r_{xx} = r_{yy}$, and $r_{xy} = -r_{yx}$ [37]. For circular polarization, $r_{+-} = r_{-+} = 0$ and the transmission matrix possess the same reasoning so that $t_{+-} = t_{-+} = 0$. The reciprocity of De Hoop declares that the matrix of reflection follows the general identity $R = R^T$, where t performs the transpose operation in the specific case of normal incidence [39]. Combining the aforementioned restrictions will guide to a linear polarization conversions where ($r_{xy} = r_{yx} = 0$). Thus, Jones' matrices can be formed as

$$R_{\text{circ}} = \begin{pmatrix} r_{xx} & 0 \\ 0 & r_{xx} \end{pmatrix} = r_{xx} I \ (I, \text{ is the identity matrix}) \tag{13}$$

$$T_{\text{circ}} = \begin{pmatrix} t_{xx} + it_{xy} & 0 \\ 0 & t_{xx} + it_{xy} \end{pmatrix} \tag{14}$$

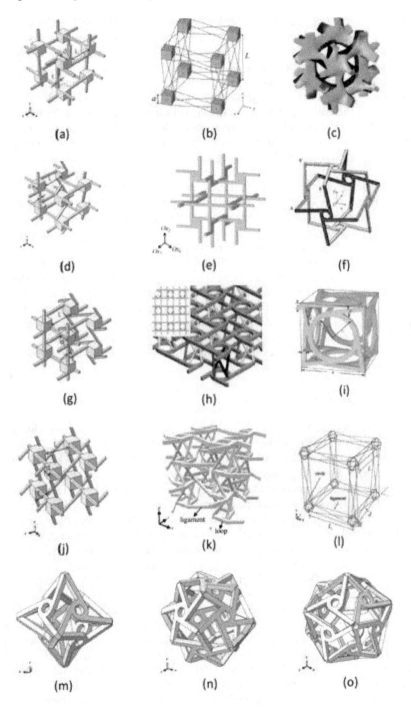

Figure 3.
Various types of 3D chiral mechanical metamaterials: (a) 3D anti-tetrachiral mechanical metamaterials; (b) 3D chiral metastructures; (c) computational optimized auxetic lattice with 3D anti-tetrachiral configuration; (d) 3D chiral-antichiral-antichiral mechanical metamaterials; (e) alternative anti-tetrachiral lattices; (f) 3D chiral lattice with negative Poisson ratio; (g) 3D chiral-antichiral-antichiral mechanical metamaterials; (h) 3D cellular metamaterials with planar anti-tetrachiral topology; (i), compression-twist chiral mechanical metamaterials; (j) 3D chiral-chiral-chiral mechanical metamaterials; (k) 3D cellular metamaterials with planar tetrachiral topology; (l) alternative 3D chiral unit cell; (m) 3D chiral pyramid lattice; (n) 3D chiral dodecahedron lattice; (o) 3D chiral regular icosahedrons lattice prepared by the author [38].

Hence, all the polarizations' reflection coefficients are identical. Scalar χ expresses the chirality parameter that nullifies the polarization conversions which leads to symmetric structure of threefold/fourfold [40]. The identical illumination reflection may occur inthe LCP or RCP material due to being isotropic material, in other words, it can only be determined by the ration between permeability and permittivity of the material and never related to the parameters of the chirality [41]. Moreover, if there are no losses at all the energy conversion, also, reciprocity controls the coefficients of the transmission and leads to identical spin states [39]. Conversion between two spin states of the polarization would occur to chiral MMs such as in twofold (C2) MMs becoming anisotropic [34]. It is notable that the 2D planar structures' disability results in a structural chirality in 3D space, caused by the off-diagonal elements equality in Jones matrices due to the in-plane mirror symmetry output ($t_{++} = t_{--}$). Still, the dichroism phenomenon being supported in the transmission of two spin states via a devised structure supported with unequal sufficiency in polarization conversions ($t_{+-} \neq t_{-+}$) [42] and this phenomenon is known as asymmetric transmission in reciprocal materials [43].

Metasurfaces, is an innovative method for circularly polarized light manipulation [44]. The circularly polarized light optical response spatially adjusting the achiral elements phase response than designing chiral metamolecules are manipulated via metasurfaces being diverse from the common MMs. The aforementioned approach, justifies several optical phenomena like high efficiency holograms [45], optical vortex generation [46], all-dielectric focusing [47], optical angular momentum achromatic generation [48] and dispersionless irregular refraction and reflection [49].

2.5.1 Optical chiral MMs fabrication

The optical chiral MMs depends on structures consisted of nanoscaled blocks. The two main fabricating techniques (top-down & bottom-up) can be used to fabricate optical chiral materials. The conventional example of chiral MMs is the 3D helix structure. It is being fabricated by direct laser writing completed with electrochemical precipitation of gold. The array of helical pores is being fabricated via positive-tone photo-resist, accompanied with spinning onto a glass substrate. This method is not applicable for bichiral-structures where the left- and right-handed spirals arranged in three orthogonal spatial axes [50]. Whilst using electroless silver plating with direct laser writing results in bichiral plasmonic crystals [51]. Also, 3D chiral plasmonic nano-structure can be fabricated via glancing angle deposition [52]. Moreover, the On-edge lithography is a pioneer method to manufacture a 3D chiral material [53].

Top-down manufacturing technologies are expensive, requiring a lot of time, non-scalable but capable of manufacturing structures below 100 nm. Unlike, self-assembled technology (bottom-up approach) it is tunable, cost effective and a fast process. It exploits the basic forces of nature in converting the blocks of the building into multi-atom systems. The equilibrium state of the structure depends on the accurate balance among the distinct forces [54]. One can take advantage of the delicate positioning and manipulate the metal's nano-particles with chemical compositions, sizes and geometries.

2.5.2 Chiral MMs applications

In the second harmonic generation, chiroptical influences are naturally bigger than their linear equivalents [55]. In this process, two photons are being converted into a single photon in a double frequency [56]. The response can be described via nonlinear polarization, as same as the electric dipole approximation, it can be presented via this equation:

$$P_i^{NL}(2\omega) = \chi_{ijk}^{(2)} E_j(\omega) E_k(\omega) \tag{15}$$

Where ω is the angular frequency, $\chi_{ijk}^{(2)}$ is the second order susceptibility tensor, E is the electric field and the Cartesian indices are i, j & k.

The weak chiroptical in nature are languid, the unique mechanism that overcome this obstacle is the superchiral fields. The chiroptical signals are being enhanced by several nano-structures such as dielectric nanoparticles [56], plasmonic structure [57], and negative index MMs [58]. It has been suggested that the chiral Purcell factor can help in characterizing the optical resonator capability and enhance the chiroptical signals [59]. Future optical systems like polarization sensitive interactive and imaging display, can be developed by achieving an active manipulation over the MMs' chirality. The full control includes the reconfiguration of the molecule from right- handed enantiomer to left- handed counterpart, vice versa. **Figure 4** illustrates a design and fabricated chiral-airfoil with water-jet cutting technique [60].

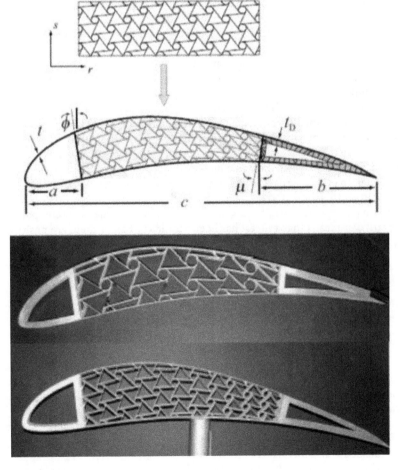

Figure 4.
Chiral airfoil and aerodynamic performance simulation [60].

2.6 Hyperbolic MMs (HMMs)

This term has been inspired from the topology of the isofrequency surface. They are an extreme anisotropy, as they act like metal for polarization or direct light propagation while they act like dielectric for others due to positive and negative permittivity tensor constituents. Its extreme anisotropy results in propagation of light on the surface and within the material [61]. Thus, its applications are promising in the field of controlling optical signals, sensors, imaging, and enhanced plasmons resonance effect [62]. For instance, this equation: $k_x^2+k_y^2+k_z^2=\omega^2/c^2$, expresses the spherical isofrequency surface that implicated by the linear dispersion and isotropic behavior of waves propagating in vacuum, where $\vec{k}$ is the propagating wave vector, ω is the radiation frequency and c is the light velocity in space.

For an extraordinary wave in a uniaxial medium, the relation becomes:

$$\frac{k_x^2 + k_y^2}{\varepsilon_{zz}} + \frac{k_z^2}{\varepsilon_{xx}} = \frac{\omega^2}{c^2} \tag{16}$$

If the case is about anisotropic the spherical isofrequency surface transforms into elliptical one in the vacuum. While in extreme anisotropic like $\varepsilon_\parallel \varepsilon_\perp < 0$, the isofrequency surface become an open hyperboloid. Here, the scrutinized material has two different behaviors, metal and dielectric (insulator), according to the direction.

The fundamental property of such material is the large magnitude, while the properties of large wave vector waves like evanescence and exponential decay are the most important ones in vacuum. However, waves propagation with infinite wave vectors in the typical limit are permitted via the isofrequency surface open form in hyperbolic media [63]. Thus, the vanishing waves don't occur in such medium which is a promising property for plethora of devices using hyperbolic media [64].

In order to classify the hyperbolic media, it will be enough to know the components' signal. In other words, the first type (I HMMs) has one negative component of the dielectric tensor ($\varepsilon_{zz} < 0$; ε_{xx}, $\varepsilon_{yy} > 0$) while the other type (II HMM) have two negative components (ε_{xx}, $\varepsilon_{yy} < 0$; $\varepsilon_{zz} > 0$). Whilst, if the three components are negative, it means that a metal has been acquired. If all of them are positive, then, it means that the medium is dielectric **Figure 5** [66].

2.6.1 HMMs designing

HMMs requires metal and dielectric to be used in the structure to act like metal and insulator at once. The exalted-k propagating wave, derived from the metallic content of the structure to develop the dispersion behavior of the material. The light-matter coupling is necessary because of the metallic polaritonic properties as it creates a hurdle to induce high-k waves. Absolutely, HMMs structure should have optically active phonons known as phonon-polaritonic or free electron metal known as Plasmon- polaritonic. The high-k modes caused via the surface Plasmon polariton as near-field coupling at bothof the interfaces metal and dielectric. While the dielectric-metal lattices provides the bloch modes which in turn creates a fertile base for the high-k modes [34].

The dielectric and metal layers forms superlattice (multilayer) resulting in the excessive anisotropy [67]. The thickness of the layer must be shorter than the operating wavelength in order to be valid. The most important factors that

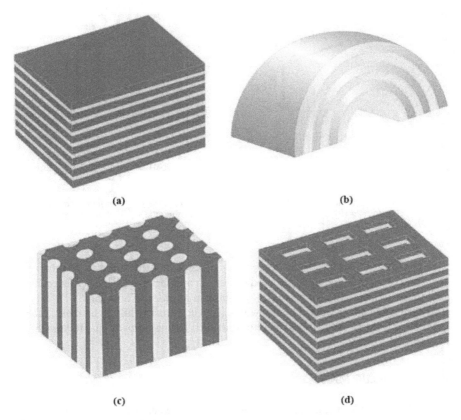

Figure 5.
(a) Metal/dielectric multilayered structures, (b) cylindrical metal/dielectric multilayered structures, (c) metal wires array in a dielectric matrix, and (d) fishnet structure [65].

influence the absorption and the impedance matching of the MMs are the metal plasma frequency and the loss [68]. The behavior of the hyperbolic refers to the dielectric's high index and the plasmonic metals' wide spectra. For instance, silver, gold, and alumina forms a fertile base for MMs at the frequencies of the ultraviolet. Adding high index dielectrics like SiN or titania to expand the design ratio to comprise visible wavelengths [69].

Concerning the near IR wavelengths, substituting plasmonic metals such as silver and gold for their reflective behavior requires materials provided with low plasma frequency. These materials are consisted of transparent conducting oxides or transition metal nitrides and applicable for HMMs [68]. The hyperbolic behavior can be achieved via a dielectric host with metallic nanowires. It possesses great advantages like high transmission, low loss and wide broadband. **Figure 6** is showing a good illustration of the two types [34].

2.6.2 HMMs fabrication

The multilayer design depends on precipitating smooth ultrathin layers of dielectric and metal, but surface roughness causes light scattering and material loss. But the minor aberration of the layer thickness does not effectively influence the medium response [70]. This strict stoichiometry is achievable by using pulsed laser deposition or reactive sputtering [68].

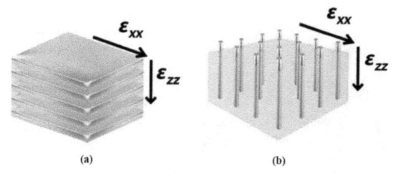

Figure 6.
(a) Illustrates multilayer structure consisting of alternating metallic and dielectric layers forming a metal-dielectric superlattice, and (b) shows nano-wire structure consisting of metallic nanorods embedded in a dielectric host [34].

Fabricating HMMs has been successfully achieved by using anodizing aluminum [71], another method depends on anodic alumina membranes [72]. The dielectric medium structure shall be a periodic nano-porous in order to host the electrodeposited silver (or gold) nanowires [73]. It is worthwhile to be mentioned that the behavior of the material can be controlled by the porosity due to the fill fraction of the metal.

2.7 Semiconductor MMs

Usually, a special surface exists at the interface between a dielectric and a noble metal, is known as surface plasmonic polariton (SPP) [74]. This surface has opened new applications such as high order harmonic generation [75], MM design [76], sensors [73], microelectronics [77], lasers [78], photovoltaics [79] and photonics [80]. Moreover, research has shown that the dispersion of SPP can be manipulated or excited in a prescribed manner via nano structuring the metal surface [81–83].

HMMs' ability to prop large wave vectors has enabled the researchers to use it in several intriguing applications, such as, hyper-lens [84] and sub-wavelength imaging [85], which were impracticable with natural materials. Between two anisotropic MMs a type of surface wave has been scrutinized other than SPPs and Dyakonov waves [34]. These waves promoted via the nanostructured MM, cross the light track and fundamental share that has low frequencies stabilize above the light line in free space, which divide radiative and nonradiative areas.

The semiconductor's effective permittivity can be calculated via:

$$\varepsilon_{1,3}(\omega) = \varepsilon_\infty - \frac{\omega_P^2}{\omega^2 + i\delta\omega},\tag{17}$$

where ω_P is the frequency of the plasma and ε_∞ is the background permittivity. The effective medium approach, helps in describing the optical response, it describes the size of the wavelength of the radiation to be compared to the thickness of the studied layer [86]. At the interface, the tangential constituents' matching of the magnetic and electrical fields, implicates the relation of the dispersion for the surface states located at the boundaries splitting two anisotropic media [87]. For example, doping silicon heavily, resulted in a metal-like properties at terahertz frequencies, where it is promising to be used in applications instead of metals [88]. The frequency range of the surface can be manipulated via adjusting layers' thickness and permittivity [89]. The resonant behavior of $\varepsilon_\perp$ can be achieved by manipulating the doping ratio, this manipulation results in tuning of resonant frequencies over the thresholds of the frequency ranges. Also, fill factions of the semiconductor sheet and dielectric, will influence the resonant behavior of $\varepsilon_\perp$.

2.8 Quantum and atomistic MMs

The domains of quantum MMs studies in near IR or optical region are still shallow but promising as the quantum degrees of freedom are incorporated [90]. In the photonic structure, the quantum wells have been used to describe the permittivity influence over the structure behavior electromagnetically. Studying layered MMs supplied with two quantum wells of GaAs, showed an effective permittivity tensor resulting in a negative refraction [91]. Many proposals have been done to extend the quantum magnetism of the MMs via organic synthesis or molecular engineering. Theory showed that Cu-CoPc $_2$ (copper phthalocyanine and cobalt phthalocyanine chains) provided a relatively robust ferromagnetism [92].

A chain of studies has been implemented that can be described as a development of work. It has been discovered that a full quantum process happens between two level atoms and a quantized electromagnetic field [93]. Then Cavity Array MMs (CAM), where 2D network of coupled atom-optical cavities were scrutinized to analyze the model via 2D photonic crystal membrane [34]. For a reasonable hypothesis, Jaynes-Cummings-Hubbard Hamiltonian method can be used to depict a system that exhibits a quantum phase transition [94]. So, it is possible to work as a quantum simulator [95]. Also, negative refraction and cloaking phenomena were elucidated. Moreover, the polaritons hybrids that formed of atomic and photonic states are an exciting system.

The dielectric function ε_{QD} expresses the quantum dots [96]:

$$\varepsilon_{QD}(\omega) = \varepsilon_b + (f_c(E_h) - f_v(E_h)) \frac{a}{\omega^2 - \omega_0^2 + 2i\omega\gamma} \qquad (18)$$

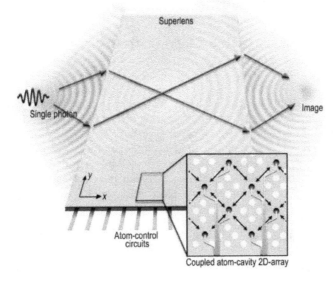

Figure 7.
A cavity array metamaterials [34].

Where, $(f_c(E_h) - f_v(E_h))$, is the difference between population levels.
Figure 7, is showing cavity array metamaterials [34].

3. MMs applications

Besides all the aforementioned applications, MMs are promising candidates for micro-robotics and micro-electromechanical systems (MEMS). Nowadays, the CMMs researchers focus is on the microscale, where applying them in micro-robotics and MEMS is spatially constrained. Moreover, literature reports endorsed those substantial applications of non-mechanical MM depend on the architecture microstructure.

Furthermore, MMs have been used in sensors and in micro-robotics as Negative permeability and permittivity of SRR arrays introduced them as promising sensors for molecule detection, deposit sensing, mechanical strain, temperature, gas detection and concentration [97].

The change in resonance frequency stimulates the detection, SRR is similar to IC-oscillator and the frequency is being calculated by:

$$f = x = \frac{1}{2\pi\sqrt{LC}} \tag{19}$$

Where L is inductance, C is capacitance of the narrow gap section.

Mechanical contortion changes the capacitance because of the geometrical change of the gap region. SRR was built with conductive polymer to sense the gas via the changes of the dielectric of the adsorption of the gas in the polymer [98]. (Publications of sensing SRR).

Also, CMMs are considered fundamental candidate for micro-robotics application due to the existence of anisotropy in the design which can be defined through the shape and the change in the material. A plenty of magnetic micro-robotics and chemical propulsion have been scrutinized [99–101]. These micro-robotics, have been applied in environmental cleaning devices [99, 100], drug delivery system and cargo transport [102, 103].

4. Composite metamaterials (CMMs) synthesis

4.1 Additives manufacturing

Plethora of MM applications require microscale structure which complicates the manufacturing process. Additive manufacturing offers many benefits, such as decreasing the resource effort which develop sustainability, expedites design, prowess step from macroscale into nanoscale, and manufacturing on demand. This method is the most apt method in manufacturing CMMs, taking in consideration its simple batch manufacturing, modeling, research and development [104]. Thus, it is already applied in medical products, electronics, machining, aerospace and automobile who are possible users for the CMMs [105].

4.2 Methods and size

Additive and subtractive methods have been used in MM & CMMs production. Selective laser melting and sintering (SLM & SLS) are additive manufacturing methods that can be used in micromachining and established mesoscale. In microscale, subtractive methods with the necessary resolution have been scarce, besides the available methods like focused ion beam milling, has limited degree of freedom. Anyway, this method is critical to evolve a metamaterial with the desired characters. Whilst, negative Poisson ratio, mechanical linearity, zero thermal expansion and programmable mechanical MMs are size independent, while the other MM applications are size-dependent. Applications that depend on effect, such as artificial bandgap in acoustic and optics, SRR's resonance frequency require manufacturing in the appropriate size range. Moreover, microscale manufacturing is required in micro robotics and MEMS devices.

5. Conclusion

The development of MMs and Metasurfaces is clarifying their exotic behavior. Detail work is still required to understand, analyze, fabricate and finding the effect of the design on these materials engineering. Although the recent studies have shown changes in the common principles and laws of EM waves, photonics and optics for the future. NIMs and Zero-index materials need to be studied and their law must be exposed because manipulating such materials may help in conquering new fields of applications were been impossible to be anticipated.

Acknowledgements

The authors would like to thank Dr. WAEL FAWZI for his ineffable contribution in completing this work.

Conflict of interest

The authors declare no conflict of interest.

Nomenclature

MM	Metamaterial
HMM	Hyperbolic metamaterials
CMM	Composite metamaterial
SPP	Surface plasmonic polariton
PIM	Positive index material
CAM	Cavity Array metamaterial
EM	Electromagnetic
MEMS	Micro-electromechanical systems
SRR	Split-ring resonators
SLM	Selective laser melting
LCP	Left handed circular polarization
SLS	Selective laser sintering
RCP	Right handed circular polarization

Author details

Tarek Fawzi[1*] and Ammar A.M. Al-Talib[2]

1 National Sun Yat-sen University (NSYSU), Kaohsiung City, Taiwan

2 Faculty of Engineering, UCSI University Kuala Lumpur, Malaysia

*Address all correspondence to: tarekfawzi8@gmail.com

IntechOpen

References

[1] Schiirch, P. and Philippe, L. Swiss federal laboratories for materials and science and technology, Switzerland. 2021.

[2] Veselago, V.G. The electrodynamics substances with simultaneously negative values of ε and m, Soviet Physics Uspekhi. 10 (1968) 509e514. DOI: 10.1070/PU1968v010n04ABEH003699

[3] Smith, D.R. Padilla, W.J. Vier, D.C. et al., Composite medium with simultaneously negative permeability and permittivity, Phys. Rev. Lett. 84 (2000) 4184e4187. DOI: 10.1103/PhysRevLett.84.4184

[4] Smith, D.R. Schurig, D. Electromagnetic wave propagation in media with indefinite permittivity and permeability tensors, Phys. Rev. Lett. 90 (2003) 077405e077409.DOI: 10.1103/PhysRevLett.90.077405

[5] Luo, C.Y. Johnson, S.G. Joannopoulos, J.D. et al., All-angle negative refraction without negative effective index, Phys. Rev. B 65 (2002) 201104e201107.

[6] Sun, J. Litchinister, N.M. Metamaterials, University at Buffalo, The State University of New York, NY, USA. Fundamentals and Applications of Nanophotonics. http://dx.doi.org/10.1016/B978-1-78242-464-2.00009-9, 2016.

[7] Enoch, S. Tayeb, G. Sabouroux, P. Guérin, N. Vincent, P. A metamaterial for directive emission, Phys. Rev. Lett. 89 (2002) 213902. DOI: 10.1103/PhysRevLett.89.213902

[8] Litchinitser, N.M. Maimistov, A.I. Gabitov, I.R. Sagdeev, R.Z. Shalaev, V. M. Metamaterials: electromagnetic enhancement at zero-index transition, Opt. Lett. 33 (2008) 2350e2352.

[9] Yuan, S.Q., Chua, C.K., Zhou, K., 2019. Advanced Materials Technologies 4 (3), 9. DOI: doi.org/10.1002/admt.201800419

[10] Stavric, M., Wiltsche, A., 2019. Nexus Network Journal21(1),79–90.

[11] Hengsbach, S., Lantada, A.D., 2014. Smart Materials and Structures 23 (8), 10.

[12] Alderson, K.L., Simkins, V.R., Coenen, V.L., et al., 2005.Physica Status SolidiB-Basic Solid StatePhysics242 (3),509–518.

[13] Ashby, M., 2013.Scripta Materialia. V68(1),4–7.

[14] Bauer, J. Hengsbach, S. Tesari, I. Schwaiger, R. Kraft, O. 2014. Proceedings of the National Academy of Sciencesofthe United StatesofAmerica111 (7),2453–2458.

[15] Florijn, B., Coulais, C., van Hecke, M., 2014. Physical Review Letters 113 (17), 5.DOI:doi:10.1103/PhysRevLett.113.175503

[16] Coulais C, Teomy E, de Reus K, Shokef Y, van Hecke M. Combinatorial design of textured mechanical metamaterials. Nature. 2016 Jul 28;535 (7613):529-532. doi: 10.1038/nature18960. PMID: 27466125.

[17] Silverberg JL, Evans AA, McLeod L, Hayward RC, Hull T, Santangelo CD, Cohen I. Applied origami. Using origami design principles to fold reprogrammable mechanical metamaterials. Science. 2014 Aug 8;345 (6197):647-650. doi: 10.1126/science.1252876. PMID: 25104381.

[18] Qu, J. Y., M. Kadic, A. Naber, and M. Wegener. 2017."Micro-Structured Two-Component 3D Metamaterials with Negative Thermal-Expansion Coefficient from Positive

Constituents."Scientific Reports 7. DOI: 10.1038/srep40643

[19] Kaschke, J., Wegener, M., 2015. Optics Letters Vol. 40, issue 17, pp. 3986-3989(2015) , DOI: 10.1364/OL.40.003986

[20] Rill, M.S. Plet, C. Thiel, M, 2008. Nature Materials 7(7),543–546.

[21] Teyssier, J., Saenko, S., van der Marel, D. *et al.* Photonic crystals cause active colour change in chameleons. Nat Commun **6**, 6368 (2015). DOI: 10.1038/ncomms7368

[22] Enoch, S., Tayeb, G., Sabouroux, P., Guerin, N., Vincent, P., 2002. Physical Review Letters 89 (21), 4. DOI: 10.1103/PhysRevLett.89.213902

[23] Purcell, E.M., Torrey, H.C. and Pound, R.V. (1946) Physical Review, 69, 681. DOI: 10.1103/PhysRev.69.37

[24] Yablonovitch, E., 1987.Physical Review Letters58 (20),2059–2062. DOI: 10.1103/PhysRevLett.58.2059

[25] LaFratta CN, Fourkas JT, Baldacchini T, Farrer RA. Multiphoton fabrication. Angew Chem Int Ed Engl. 2007;46(33):6238-6258. DOI: 10.1002/anie.200603995.

[26] Mizeikis, V., Juodkazis, S., Tarozaite, R., et al., 2007.Optics Express 15(13),8454–8464. DOI: 10.1364/OE.15.008454

[27] Nagpal, P., Han, S.E., Stein, A., Norris, D.J., 2008.Nano Letters 8 (10),3238–3243. DOI: 10.1021/nl801571z

[28] Gansel, J.K., Latzel, M., Frolich, A., et al., 2012.Applied Physics Letters100 (10),3. DOI: 10.1063/1.3693181

[29] Yan, Y.J., Rashad, M.I., Teo, E.J., et al., 2011.Optical Materials Express 1(8),1548–1554.

[30] Sledzinska, M. Graczykowski, B. Alzina, F. Lopez, J.S. Torres, C.M.S., 2016. Microelectronic Engineering 149, 41–45.

[31] Wu, T.T., Wu, L.C., Huang, Z.G., 2005. Journal of Applied Physics 97 (9), 7. DOI: 10.1063/1.1893209

[32] Ghaffarivardavagh, R., Nikolajczyk, J., Anderson, S., Zhang, X., 2019. Physical Review B 99 (2), 10. DOI: 10.1103/PhysRevB.99.024302.

[33] Bigoni, D., Guenneau, S., Movchan, A.B., Brun, M., 2013. Elastic metamaterials with inertial locally resonant structures: Application to lensing and localization, Physical Review B 87 (17), 6. DOI: 10.1103/PhysRevB.87.174303

[34] Gric, T. Hess, O. Chapter 1 - Types of Metamaterials, Phenomenon of MMs, Micro and Nano Technologies, 2019, Pages 1-39.

[35] Zhang S, Park YS, Li J, Lu X, Zhang W, Zhang X. Negative refractive index in chiral metamaterials. Physical review letters. 2009 Jan 12;102(2): 023901.

[36] Zhao, R. Koschny, T. Soukoulis, C. M. Opt. Express 18 (2010) 14553–14567. DOI: 10.1364/OE.18.014553

[37] Menzel, C. Rockstuhl, C. Lederer, F. Phys. Rev. A 82 (2010) 053811. DOI: 10.1103/PhysRevA.82.053811

[38] Wu, W. Hu, W. Qian, G. Liao, H. Xu, X. and Betro, F. Mechanical design and multifunctional applications of chiral mechanical metamaterials: A review, Materials and Design 180 (2019) 107950, https://doi.org/10.1016/j.matdes.2019.107950)

[39] Kaschke, J. Blome, M. Burger, S. Wegener, M. Tapered N-helical metamaterials with three-fold rotational symmetry as improved circular

polarizers. Opt. Express 22 (2014) 19936-19946. DOI: 10.1364/OE.22.019936

[40] Saba, M. Turner, M.D. Mecke, K. Gu, M. Schroder-Turk, G.E. Phys. Rev. B 88 (2013) 245116. DOI: 10.1103/PhysRevB.88.245116

[41] Bingnan, W. Jiangfeng, Z. Thomas, K. Maria, K. Costas, M.S. Journal Opt. A Pure Appl. Opt. 11 (2009) 114003.

[42] Plum E, Fedotov VA, Zheludev NI. Planar metamaterial with transmission and reflection that depend on the direction of incidence. Applied Physics Letters. 2009 Mar 30;94(13):131901.

[43] Kenanakis G, Xomalis A, Selimis A, Vamvakaki M, Farsari M, Kafesaki M, Soukoulis CM, Economou EN. Three-dimensional infrared metamaterial with asymmetric transmission. ACS Photonics. 2015 Feb 18;2(2):287-294.

[44] Yu N, Capasso F. Flat optics with designer metasurfaces. Nature materials. 2014 Feb;13(2):139-150.

[45] Zheng G, Mühlenbernd H, Kenney M, Li G, Zentgraf T, Zhang S. Metasurface holograms reaching 80% efficiency. Nature nanotechnology. 2015 Apr;10(4):308-312.

[46] Ma X, Pu M, Li X, Huang C, Wang Y, Pan W, Zhao B, Cui J, Wang C, Zhao Z, Luo X. A planar chiral meta-surface for optical vortex generation and focusing. Scientific reports. 2015 May 19;5(1):1-7.

[47] Lin D, Fan P, Hasman E, Brongersma ML. Dielectric gradient metasurface optical elements. science. 2014 Jul 18;345 (6194):298-302.

[48] Pu M, Li X, Ma X, Wang Y, Zhao Z, Wang C, Hu C, Gao P, Huang C, Ren H, Li X. Catenary optics for achromatic generation of perfect optical angular

momentum. Science Advances. 2015 Oct 1;1(9):e1500396.

[49] Huang L, Chen X, Muhlenbernd H, Li G, Bai B, Tan Q, Jin G, Zentgraf T, Zhang S. Dispersionless phase discontinuities for controlling light propagation. Nano letters. 2012 Nov 14; 12(11):5750-5755.

[50] Thiel M, Rill MS, von Freymann G, Wegener M. Three-dimensional bi-chiral photonic crystals. Advanced Materials. 2009 Dec 11;21(46): 4680-4682.

[51] Radke A, Gissibl T, Klotzbücher T, Braun PV, Giessen H. Three-dimensional bichiral plasmonic crystals fabricated by direct laser writing and electroless silver plating. Advanced Materials. 2011 Jul 19;23(27):3018-3021.

[52] Singh JH, Nair G, Ghosh A, Ghosh A. Wafer scale fabrication of porous three-dimensional plasmonic metamaterials for the visible region: chiral and beyond. Nanoscale. 2013;5 (16):7224-7228.

[53] Dietrich K, Lehr D, Helgert C, Tünnermann A, Kley EB. Circular dichroism from chiral nanomaterial fabricated by on-edge lithography. Advanced Materials. 2012 Nov 20;24 (44):OP321-OP325.

[54] Min Y, Akbulut M, Kristiansen K, Golan Y, Israelachvili J. The role of interparticle and external forces in nanoparticle assembly. Nanoscience And Technology: A collection of reviews from Nature journals. 2010:38-49.

[55] Petralli-Mallow T, Wong TM, Byers JD, Yee HI, Hicks JM. Circular dichroism spectroscopy at interfaces: a surface second harmonic generation study. The Journal of Physical Chemistry. 1993 Feb;97(7):1383-1388.

[56] Boyd, R.W. Nonlinear Optics, Academic Press, San Diego, CA, 2003.

[57] García-Etxarri A, Dionne JA. Surface-enhanced circular dichroism spectroscopy mediated by nonchiral nanoantennas. Physical Review B. 2013 Jun 10;87(23):235409.

[58] Yoo S, Cho M, Park QH. Globally enhanced chiral field generation by negative-index metamaterials. Physical Review B. 2014 Apr 23;89(16): 161405.

[59] Yoo S, Park QH. Chiral light-matter interaction in optical resonators. Physical review letters. 2015 May 21;114 (20):203003.

[60] Spadoni, A. Application of chiral cellular materials for the design of innovative components, Dissertations & Theses-Gradworks, 2008

[61] High AA, Devlin RC, Dibos A, Polking M, Wild DS, Perczel J, De Leon NP, Lukin MD, Park H. Visible-frequency hyperbolic metasurface. Nature. 2015 Jun;522(7555):192-196.

[62] Takayama O, Lavrinenko AV. Optics with hyperbolic materials. JOSA B. 2019 Aug 1;36(8):F38-F48.

[63] Jacob Z, Smolyaninov II, Narimanov EE. Broadband Purcell effect: Radiative decay engineering with metamaterials. Applied Physics Letters. 2012 Apr 30;100(18):181105.

[64] Guo Y, Newman W, Cortes CL, Jacob Z. Applications of Hyperbolic Metamaterial Substrates. Advances in Opto Electronics. 2012 Jan 1.

[65] Sun, J. Litchinitser, N.M. and Zhou, J. Indefinite by nature: from ultraviolet to terahertz, ACS Photonics 1 (2014) 293-303.

[66] Korobkin D, Neuner B, Fietz C, Jegenyes N, Ferro G, Shvets G. Measurements of the negative refractive index of sub-diffraction waves propagating in an indefinite permittivity medium. Optics express. 2010 Oct 25;18(22):22734-22746.

[67] Xiong Y, Liu Z, Sun C, Zhang X. Two-dimensional imaging by far-field superlens at visible wavelengths. Nano letters. 2007 Nov 14;7(11):3360-3365.

[68] Naik GV, Kim J, Boltasseva A. Oxides and nitrides as alternative plasmonic materials in the optical range. Optical materials express. 2011 Oct 1;1 (6):1090-1099.

[69] Lu D, Liu Z. Hyperlenses and metalenses for far-field super-resolution imaging. Nature communications. 2012 Nov 13;3(1):1-9.

[70] Liu, H. Wang, B. Leong, E.S. Yang, P. Zong, Y. Si, G. Teng, J. Maier, S.A. ACS Nano 4 (6) (2010) 3139–3146.

[71] Pollard RJ, Murphy A, Hendren WR, Evans PR, Atkinson R, Wurtz GA, Zayats AV, Podolskiy VA. Optical nonlocalities and additional waves in epsilon-near-zero metamaterials. Physical review letters. 2009 Mar 27;102(12):127405.

[72] Noginov MA, Barnakov YA, Zhu G, Tumkur T, Li H, Narimanov EE. Bulk photonic metamaterial with hyperbolic dispersion. Applied Physics Letters. 2009 Apr 13;94(15):151105.

[73] Kabashin AV, Evans P, Pastkovsky S, Hendren W, Wurtz GA, Atkinson R, Pollard R, Podolskiy VA, Zayats AV. Plasmonic nanorod metamaterials for biosensing. Nature materials. 2009 Nov;8(11):867-871.

[74] Tsakmakidis KL, Hermann C, Klaedtke A, Jamois C, Hess O. Surface plasmon polaritons in generalized slab heterostructures with negative permittivity and permeability. Physical Review B. 2006 Feb 9;73(8):085104.

[75] Kim, S.-W. Nat. Photonics 5 (11) (2011) 677–681.

[76] Shalaev VM. Optical negative-index metamaterials. Nature photonics. 2007 Jan;1(1):41-48.

[77] MacDonald KF, Sámson ZL, Stockman MI, Zheludev NI. Ultrafast active plasmonics. Nature Photonics. 2009 Jan;3(1):55-58.

[78] Park IY, Kim S, Choi J, Lee DH, Kim YJ, Kling MF, Stockman MI, Kim SW. Plasmonic generation of ultrashort extreme-ultraviolet light pulses. Nature Photonics. 2011 Nov;5 (11):677-681.

[79] Atwater HA, Polman A. Plasmonics for improved photovoltaic devices. Materials for sustainable energy: a collection of peer-reviewed research and review articles from Nature Publishing Group. 2011:1-1.

[80] Liu H, Lalanne P. Microscopic theory of the extraordinary optical transmission. Nature. 2008 Apr;452 (7188):728-731.

[81] Williams CR, Andrews SR, Maier SA, Fernández-Domínguez AI, Martín-Moreno L, García-Vidal FJ. Highly confined guiding of terahertz surface plasmon polaritons on structured metal surfaces. Nature Photonics. 2008 Mar;2(3):175-179.

[82] Huang L, Chen X, Bai B, Tan Q, Jin G, Zentgraf T, Zhang S. Helicity dependent directional surface plasmon polariton excitation using a metasurface with interfacial phase discontinuity. Light: Science & Applications. 2013 Mar;2(3):e70-.

[83] Lin J, Mueller JB, Wang Q, Yuan G, Antoniou N, Yuan XC, Capasso F. Polarization-controlled tunable directional coupling of surface plasmon polaritons. Science. 2013 Apr 19;340 (6130):331-334.

[84] Rho J, Ye Z, Xiong Y, Yin X, Liu Z, Choi H, Bartal G, Zhang X. Spherical hyperlens for two-dimensional sub-diffractional imaging at visible frequencies. Nature communications. 2010 Dec 21;1(1):1-5.

[85] Shvets G, Trendafilov S, Pendry JB, Sarychev A. Guiding, focusing, and sensing on the subwavelength scale using metallic wire arrays. Physical review letters. 2007 Aug 2;99(5): 053903.

[86] Gric T, Hess O. Tunable surface waves at the interface separating different graphene-dielectric composite hyperbolic metamaterials. Optics express. 2017 May 15;25(10): 11466-11476.

[87] Iorsh I, Orlov A, Belov P, Kivshar Y. Interface modes in nanostructured metal-dielectric metamaterials. Applied Physics Letters. 2011 Oct 10;99(15): 151914.

[88] Li S, Jadidi MM, Murphy TE, Kumar G. Terahertz surface plasmon polaritons on a semiconductor surface structured with periodic V-grooves. Optics express. 2013 Mar 25;21(6): 7041-7049.

[89] Gric, T. Prog. Electromagn. Res. 46 (2016) 165–172.

[90] Plumridge J, Phillips C. Ultralong-range plasmonic waveguides using quasi-two-dimensional metallic layers. Physical Review B. 2007 Aug 15;76(7): 075326.

[91] Plumridge JR, Steed RJ, Phillips CC. Negative refraction in anisotropic waveguides made from quantum metamaterials. Physical Review B. 2008 May 22;77(20):205428..

[92] Wu W. Modelling copper-phthalocyanine/cobalt-phthalocyanine chains: towards magnetic quantum metamaterials. Journal of Physics: Condensed Matter. 2014 Jul 3;26(29): 296002.

[93] Quach JQ, Su CH, Martin AM, Greentree AD, Hollenberg LC. Reconfigurable quantum metamaterials. Optics express. 2011 Jun 6;19(12): 11018-11033.

[94] Henry RA, Quach JQ, Su CH, Greentree AD, Martin AM. Negative refraction of excitations in the Bose-Hubbard model. Physical Review A. 2014 Oct 31;90(4):043639.

[95] Greentree AD, Tahan C, Cole JH, Hollenberg LC. Quantum phase transitions of light. Nature Physics. 2006 Dec;2(12):856-861.

[96] Holmström P, Thylén L, Bratkovsky A. Dielectric function of quantum dots in the strong confinement regime. Journal of Applied Physics. 2010 Mar 15;107(6):064307.

[97] Chen, T., Li, S.Y., Sun, H., 2012. Sensors 12(3),2742–2765.

[98] Vena, A., Sydanheimo, L., Tentzeris, M. M., Ukkonen, L., 2015. IEEESensorsJournal 15 (1), 89–99.

[99] Serra, A., Gomez, E., Valles, E., 2015.Electrochimica Acta 174,630–639.

[100] Garcia-Torres, J., Serra, A., Tierno, P., Alcobe, X., Valles, E., 2017. ACS Applied Materials and Interfaces 9 (28),23859–23868.

[101] Paxton WF, Baker PT, Kline TR, Wang Y, Mallouk TE, Sen A. Catalytically induced electrokinetics for motors and micropumps. Journal of the American Chemical Society. 2006 Nov 22;128(46):14881-14888.

[102] Nelson BJ, Kaliakatsos IK, Abbott JJ. Microrobots for minimally invasive medicine. Annual review of biomedical engineering. 2010 Aug 15;12: 55-85.

[103] Barbot, A., Decanini, D., Hwang, G., 2016.Scientific Reports 6, 8.

[104] Thomas D. Costs, benefits, and adoption of additive manufacturing: a supply chain perspective. The International Journal of Advanced Manufacturing Technology. 2016 Jul; 85 (5):1857-1876.

[105] Ford, S., Despeisse, M., 2016. Journalof Cleaner Production 137,1573–1587.

Chapter 15

Nanocomposite Material Synthesized Via Horizontal Vapor Phase Growth Technique: Evaluation and Application Perspective

Muhammad Akhsin Muflikhun, Rahmad Kuncoro Adi and Gil Nonato C. Santos

Abstract

The synthesis of nanomaterials has been reported by many researchers using different methods. One of the methods that can be used with perfect pureness and have less pollution in the synthesized materials results is the vapor phase growth technique (VPGT). Several types of nano shapes materials were reported such as nanoparticles, nanorods, nano triangular, nanosphere, and nanocrystal. The synthesis method has a fundamental process where the nanomaterials evaporated and condensed based on the temperature difference. There are three important variables, i.e., stochiometric ratio of source materials, temperature and baking time. The synthesis was occured in the quartz tube and sealed in the vacuum condition. This create the material was synthesis in pure and isolated conditions. The application of the nanomaterials synthesized via Horizontal Vapor Phase Growth (HVPG) can be implemented in anti-pathogen, anti-bacterial, gas sensing and coating applications.

Keywords: HVPG, synthesis nanomaterials, phase transition, anti-bacterial, coating, sensing applications

1. Introduction

Recently, nanotechnologies and nanoscience have raised high hopes for a new potential industry revolution [1]. They produced materials of various types at nanoscales [2]. Nanotechnology is commonly used in many applications such as in industrial, medical, agriculture, aerospace, energy, automotive, and food. Many researchers have conducted to explore the field of nanotechnology. They focus to obtain to get the best nanomaterials with optimal mechanical and physical properties [3–5].

Nanomaterials are a wide class of materials that have a range of the dimension 1 nm-100 nm at least and they are made from nanoparticles [2, 6, 7]. There are several methods to synthesis a nanomaterial that was developed by many researchers.

There is chemical reduction [8, 9], chemical vapor deposition [10], photochemical [11], electrochemical [12], green synthesis [13], photochemical [11], Horizontal Vapor Phase Growth (HPVG) [14, 15], photochemical [11], microwave [16], sol–gel [17], and sonochemical [18]. One of the methods that successfully synthesized nanocomposite materials is the HPVG technique. HVPG was proven to be used to develop and able to produce material with various dimensional of nanostructures like Fe_2O_3 [19], Ln_2O_3 [20], $Ag-TiO_2$ and SnO_2 [21]. This technique offers some advantages like economical, reliable method, and less source material with high purity [22, 23]. HVPG technique is also capable to create nanomaterials with various shapes such as nanoparticles, nanotubes, nanorods, and triangular nanomaterials [14, 24–26]. They were evaluated to investigate the structure, chemical composition, hardness, and morphological behavior of the nanocomposite material [27]. Previous research about evaluated nanocomposite material synthesized with the HVPG technique is present in **Table 1**.

The applications of HVPG techniques on synthesized nanomaterials were varies among the different sectors. Tibayan et al. [23, 29] used HVPG to synthesized Ag/SnO2 nanocomposite materials that can be used in coating applications. The characterization process used UV filtering analysis to evaluate the UV blocking. The results showed that the UV can be blocked efficiently. Moreover, DFT analysis using An and Ag as the sample material also showed that the entire spectrum of UV light can be absorbed with the model. The summary of the previous research about synthesized nanomaterials using the HVPG techniques can be seen in **Table 2**.

Based on the explanation above, nanomaterials were synthesized by the HVPG technique and evaluated using several tests such as Scanning Electron Microscope (SEM), Energy-Dispersive X-ray Spectroscopy (EDX), X-Ray Diffraction Analysis (XRD), Density-Functional Theory (DFT), Applied Behavior Analysis (ABA),

Source	Method to evaluate	Function
Muflikhun et al. [28]	SEM	To determine the image and measure the synthesized nanomaterial
	EDX	To determine the composition of the material
	AFM	To determine the 3D surface roughness of nanocomposite material
Tibayan et al. [29]	SEM	To determine the image and measure the zone in synthesized nanomaterial
	DFT	To determine the atomic geometry and electronic properties the nanocomposite material
	ABA	To determine the metabolic activity in the material
Motlagh et al. [30]	SEM	To investigate the dispersion of nanoparticles
	AFM	To investigate the surface roughness of the coating
	Elcometer Motorized Scratcher	To determine the scratch resistance
	Sheen Pendulum Hardness (707 KP)	To determine the surface hardness
	Micro-Tri Gloss Meter (BYK-Gardner)	To determine the gloss coating
Reyes and Santos [31]	SEM	To determine the image of nanomaterial
	EDX	To determine the composition of the material

Table 1.
Nanocomposite materials evaluation on previous research.

Source	Years	Nanomaterials name
Muflikhun et al. [32]	2019	Synthetic Silver (Ag/TiO$_2$)
Shimizu et al. [24]	2016	Skewered phthalocyanine [Fe^{III}(Pc)(SCN)]$_n$
Quitaneg and Santos [33]	2011	Cadnium Selenide (CdSe)
Reyes and Santos [31]	2011	Tin Dioxide (SnO$_2$)
Buot and Reyes [34]	2012	Carbon-silver (C/Ag)
Muflikhun et al. [15]	2017	Silver-graphene (Ag/Ge)

Table 2.
Previous research of synthesized nanomaterials using HVPG technique.

and Atomic Force Microscope (AFM), representing advances in the development of nanomaterials. The above discussion can also be used for further research in developing large-scale nanomaterials that can be applied in the industrial world.

2. The HVPG method

Detailed of the HVPG method are described in the study of Muflikhun et al. [21]. The step-by-step synthesis of nanomaterials is shown in **Figure 1**. It can be explained in the sub chapter 2.1 to 2.3 where all process was followed.

2.1 Preparation

Before the source material is inserted into the tube, the materials should be measured in the high precision weight scale to get the exact amount of material with a total weight of 35 mg. If the sample was composed of more than 1 sample, it should be divided into the individual sample. For example, the target synthesis was Ag/TiO$_2$ nanocomposite materials. The source materials that need to be prepared were Ag powder 17.5 mg and TiO$_2$ powder 17.5 mg. the combination should be equal to reach optimum results.

After material measurement, the next step was tube preparation. The tube was purchased from the general market nearby the university. The tube was originally used as the heater tube used in the toaster or heating unit. The tube before used as the place to grow nanomaterials, it should be washed in the ultrasonicator to remove the pollutant and impurity. After washed, the tube then dried until the water was evaporated. The one-side seal was applied to ensure the material that poured into the tube is trapped in the base. The one edge seal was also the final point of preparation where the specimen can be poured into the tube and then placed in the Thermionic High Vacuum System (THVS).

2.2 Sealing

Sample sealing is another important aspect to ensure the material is sealed properly. The tube before final sealing was placed vertically where the material was at the bottom. The upper tube is then joined with the pipe for the vacuum process. The process was occurred at a very low pressure (10 Torr^{-6} Torr). The low pressure has purposes pushing the melting point of the materials. Sealing is also a critical process when the sealing is not proper, the leak will allow outer air to go inside the tube and the vacuum process failed. The sealing process also needs special treatment due to the used blow torch that used LPG and O$_2$ gas.

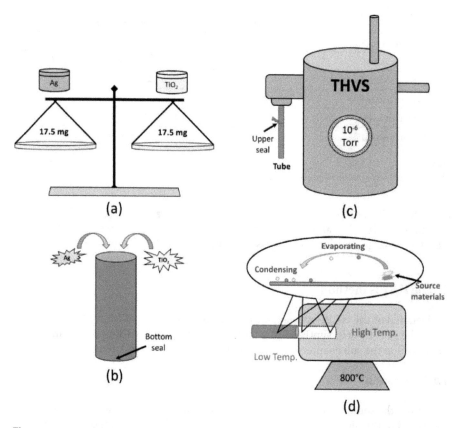

Figure 1.
(a) Material measurement, (b) pouring the materials into the tube, (c) final sealing process in vacuum condition, (d) baking process.

2.3 Baking

The HVPG method to synthesis nanomaterial is based on the temperature difference between 2 points. It is shown that the material that received heat in a continuous-time, the phase of material will change as follow:

From **Figure 2**, it is shown that the phase of the material will be melt and then evaporate after the critical point was overlapped. This phase transition occurred due to high temperatures in the furnace. The material then condenses and then become solid after moving from high temperature (inside the tube) to low temperature (outside the tube).

3. The study results and discussion

After cured with designated temperature and time, the nanomaterial inside the tube can be characterized and evaluated using the electron microscope. Since the unique phase transition of the materials that flow inside the tube, the nanomaterial growth inside the tube is also fascinating in terms of shape and size. The study from Muflikhun et al. [15] showed that flower-like and jellyfish-like nano-silver was successfully grown from the combination of Ag/Ge materials. The silver was grown on the graphene multilayer as shown in **Figure 3**. The growth variable in HVPG for the Ag/Ge was 1200°C and 6 hours baking time.

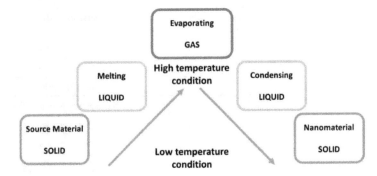

Figure 2.
Phase transition of materials in HVPG [21].

Figure 3.
(a) Jellyfish-like silver nanomaterial, (b) flower-like silver nanomaterial.

Another study conducted by Bernardino and Santos [35] shows that HVPG can be used to synthesis Gallium Oxide/Tin Oxide Nanostructures. The synthesis was used variable temperature 1200°C and 6 hours baking time. The results showed that different nano shape was grown such as nanowire, nano particles, and nano crystal.

There are 3 main variables to grow nanomaterial using the HVPG technique: time, temperature, and the weight ratio of the source material. During the baking process in the furnace, time and temperature play an important role to develop the shape of nanomaterials in the results. These two variables have been reported by Muflikhun et al. [28, 32]. The time was set to 4 hours, 6 hours, and 8 hours and the temperature was set to 800°C, 1000°C, and 1200°C. By using these 6 parameters, the 27 combinations of time and temperature can be achieved. The results of these results showed that the more different nanomaterials shape was successfully synthesized with different shapes of nanomaterials as reported below (**Table 3**).

No.	Temperature	Baking time	Zone	Material shape and diameter
1	800°C	4 Hours	1	Nanoparticles
2			2	Micro particles
3			3	Micro particles
4		6 Hours	1	Nanospheres, Nanoparticles
5			2	Nanoparticles
6			3	Nanoparticles
7		8 Hours	1	Nanoparticles
8			2	Nanotubes, Nanospheres
9			3	Nanoparticles
10	1000°C	4 Hours	1	Nanoparticles
11			2	Nanospheres, Nano-triangular, Nanorods
12			3	Nanospheres, Nano-triangular
13		6 Hours	1	Nanoparticles
14			2	Nanoparticles, Nanospheres
15			3	Nanoparticles
16		8 Hours	1	Nanoparticles, Nanorods
17			2	Nanorods, Nanoparticles
18			3	Nanorods, Nanoparticles
19	1200°C	4 Hours	1	Nanoparticles
20			2	Nanospheres, Nanoparticles
21			3	Nanoparticles
22		6 Hours	1	Nanoparticles
23			2	Nanocrystal, Nano-triangular
24			3	Nanoparticles
25		8 Hours	1	Nanoparticles
26			2	Nanospheres, Nanorods, Nanocrystal
27			3	Nanorods

Table 3.
The different shapes of nanomaterials with different temperatures and baking times.

The third variable that play the important role was the ratio of the source material. For that variable, the combination of two different materials was reported by Tibayan et al. [23, 29]. The study was used Ag/SnO$_2$ and HVPG as the method to synthesis nanomaterial. The variable study based on the ratio between Ag and SnO$_2$ where stoichiometric ratio mixtures of 0:5, 1:4, 2:3, 3:2, 4:1, and 5:0 were used. Since the different of material mixture were added, the results showed that different nano shape have been reported as seen in the **Figure 4**. The time for baking is 8 hours in the 800°C temperature condition.

Based on the previous work done by many researchers used HVPG to synthesis nanomaterials, it is shown that the HVPG method can ensure the pureness and the high quality of the nano shape due to the excellent sealing process and occurred in the vacuum condition.

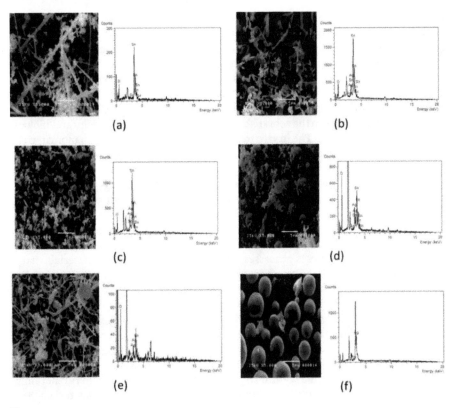

Figure 4.
Representative sample of Ag/SnO2 with different ratio. (Ag:SnO$_2$) (a) 0:5, (b) 1:4, (c) 2:3, (d) 3:2, (e) 4:1, (f) 5:0. Pictures were retrieved from [23].

4. Future trend

The pandemic that occurred in 2019 as known as COVID-19 has demonstrated the need for rapid, excellent, and robust technology that can prevent the virus and future diseases that may occur in the world. Researchers have been searching the new technology and they found that one of the best fits of the future technology that can be applied in almost all aspects of human life was founded in nanomaterials and nanotechnology [36, 37].

Nanotechnology become the most relevant solution for the problem of human life in the present day and the future due to the fact that nanotechnology has been proven to be applied in a different application. The various metal oxide was summarized by Shkodenko et al. [38] that useful to be applied as anti-bacterial technology. Pasquale et al. [37] were given an alternative perspective to disinfect the virus using a combination of TiO$_2$ nanomaterials in the photocatalytic process. It is shown that future applications can be nearly applied that photocatalysts for air, surface and water were available. Moreover, in medical based especially in dentistry, the graphene-based nanomaterial can be applied to eradicate microbial with good results [39]. Future manufacturing technology can potentially be combined with nanotechnology, as in additive manufacturing. Different nanomaterials, such as Carbon Nano Tubes, Carbon Nano Fibers, Graphene Oxide, Metal Nanoparticles, and Metal Oxide Nanoparticles, can be combined in the various polymer matrices [40]. It is shown that by adding nanomaterials, the properties of the matrices can be significantly improved. The field of automotive also become concerned by the scientist. Kotia et al. [41] reported that nanomaterials were used in automotive engine to improve the efficiency and performance of the machine. The future application in the field of optical sensing, biological imaging and photodynamic therapy was reported by Chen et al. [42]. The studies from many researchers were summarized and they reported that chemiluminescence resonance energy transfer platform based on nanomaterial successfully fabricated. The study related to drug delivery and toxicity have been evaluated by Jia et al. [43]. The nanomaterial was tested on the zebrafish to determine the effect of toxicity and biological related safety concerns. The in vivo toxicological profiles of different nanomaterials, including Ag nanoparticles (NPs), CuO NPs, silica NPs, polymeric NPs, quantum dots, nanoscale metal–organic frameworks, etc., that appear in zebrafish have been evaluated. Furthermore, mechanical testing related to strain sensing using graphene nanomaterial has been reported by Mehmood et al. [44]. Graphene nanomaterial has been chosen because its excellent properties in thermal, electrical and mechanical strength.

There are several aspect that related to nanomaterials. One of the most important aspects related to the synthesis of nanomaterials were about the environmental aspect where many syntheses processes were used catalysts or other materials that can harm the environment. To prevent the issue related to the environment, green synthesis was launched by many researchers as an alternative to producing nanomaterials. Different materials were introduced such as: biocompatible reagents, synthesis process by microorganisms, using plant mediated synthesis, improve the waster treatment system using nano filtering process, etc. [45]. It can be summarized that in the future, nanomaterials and nanotechnology still became the alternative and major material that used in various applications. In that point of view, HVPG technique that used to synthesis nanomaterial can be applied in further high scale synthesis process to fulfill the community needs related to nanomaterials [46].

5. Conclusion

The characteristics of synthesis nanomaterials using HVPG has been described and reported in the present study. The details aspect of the synthesis and the sample of the results of the synthesis nanomaterial also presented. It is shown that the HVPG can be used to synthesis various type of nanomaterials with the following advantages: excellent pureness ratio, free of external impurity during synthesis process, simple procedure and setup, environmentally friendly, and used recycle material (quartz tube) that previously used as the heating components. Since the application of nanomaterials can be found in very wide aspects, the synthesis process of nanomaterials using HVPG can be an alternative method. The future trend as shown in the present study ensure the sustainability of the synthesis nanomaterial without compromising with the environment and related to human healthy aspect.

Acknowledgements

The authors would like to thank Mechanical and Industrial Engineering Department, Gadjah Mada University and De La Salle University, Philippines, for the support and funding.

Conflict of interest

"The authors declare no conflict of interest."

Author details

Muhammad Akhsin Muflikhun[1*], Rahmad Kuncoro Adi[2] and Gil Nonato C. Santos[3]

1 Mechanical and Industrial Engineering Department, Universitas Gadjah Mada, Yogyakarta, Indonesia

2 Mechanical Engineering Department, Universitas Muhammadiyah Yogyakarta, Indonesia

3 Physics Department, De La Salle University, Manila, Philippines

*Address all correspondence to: akhsin.muflikhun@ugm.ac.id

IntechOpen

References

[1] Kreyling WG, Semmler-Behnke M, Chaudhry Q. A complementary definition of nanomaterial. Nano Today. 2010;5:165-168. DOI: 10.1016/j.nantod. 2010.03.004

[2] Khan I, Saeed K, Khan I. Nanoparticles: Properties, applications and toxicities. Arabian Journal of Chemistry. 2019;12:908-931. DOI: 10.1016/j.arabjc.2017.05.011

[3] Mitter N, Hussey K. Moving policy and regulation forward for nanotechnology applications in agriculture. Nature Nanotechnology. 2019;14:508-510. DOI: 10.1038/s41565-019-0464-4

[4] Keskinbora KH, Jameel MA. Nanotechnology applications and approaches in medicine: A review. Journal of Nanoscience & Nanotechnology Research. 2018;2:6

[5] Baris K. Availibility of renewable energy sources in Turkey: Current situation, potential, government policies and the EU perspective. Energy Policy. 2012;42:377-391. DOI: 10.1016/j.enpol. 2011.12.002

[6] Goyal RK. Nanomaterials and Nanocomposites: Synthesis, Properties, Characterization Techniques, and Applications. CRC Press. 2017. DOI: 10.1201/9781315153285

[7] Bratovcic A. Different applications of nanomaterials and their impact on the environment. International Journal of Materials Science and Engineering. 2019;5:1-7. DOI: 10.14445/23948884/ ijmse-v5i1p101

[8] Khan Z, Al-Thabaiti SA, Obaid AY, Al-Youbi AO. Preparation and characterization of silver nanoparticles by chemical reduction method. Colloids and Surfaces. B, Biointerfaces. 2011;82:513-517. DOI: 10.1016/j.colsurfb. 2010.10.008

[9] Liu D, Pan J, Tang J, Liu W, Bai S, Luo R. Ag decorated SnO2 nanoparticles to enhance formaldehyde sensing properties. Journal of Physics and Chemistry of Solids. 2019;124:36-43. DOI: 10.1016/j.jpcs.2018.08.028

[10] Beier O, Pfuch A, Horn K, Weisser J, Schnabelrauch M, Schimanski A. Low temperature deposition of antibacterially active silicon oxide layers containing silver nanoparticles, prepared by atmospheric pressure plasma chemical vapor deposition. Plasma Processes and Polymers. 2013;10:77-87. DOI: 10.1002/ppap. 201200059

[11] Gabriel JS, Gonzaga VAM, Poli AL, Schmitt CC. Photochemical synthesis of silver nanoparticles on chitosans/ montmorillonite nanocomposite films and antibacterial activity. Carbohydrate Polymers. 2017;171:202-210. DOI: 10.1016/j.carbpol.2017.05.021

[12] Khaydarov RA, Khaydarov RR, Gapurova O, Estrin Y, Scheper T. Electrochemical method for the synthesis of silver nanoparticles. Journal of Nanoparticle Research. 2009;11:1193-1200. DOI: 10.1007/s11051-008-9513-x

[13] Patra S, Mukherjee S, Barui AK, Ganguly A, Sreedhar B, Patra CR. Green synthesis, characterization of gold and silver nanoparticles and their potential application for cancer therapeutics. Materials Science & Engineering. C, Materials for Biological Applications. 2015;53:298-309. DOI: 10.1016/j.msec. 2015.04.048

[14] Muflikhun MA, Chua AY, Santos GNC. Statistical design analysis of silver-titanium dioxide nanocomposite materials synthesized via horizontal vapor phase growth (HVPG). Key Engineering Materials. 2017;735:210-214. DOI: 10.4028/www. scientific.net/KEM.735.210

[15] Muflikhun MA, Castillon GB, Santos GNC, Chua AY. Micro and nano silver-graphene composite manufacturing via horizontal vapor phase growth (HVPG) technique. Materials Science Forum. 2017;**901**:3-7. DOI: 10.4028/www.scientific.net/msf.901.3

[16] Hong Y. Van der Waals epitaxy of InAs nanowires vertically aligned on single-layer graphene. Nano Letters. 2012;**12**:1431-1436. DOI: 10.1021/nl204109t

[17] Nair KK, Kumar P, Kumar V, Harris RA, Kroon RE, Viljoen B, et al. Synthesis and evaluation of optical and antimicrobial properties of Ag-SnO2 nanocomposites. Physica B: Condensed Matter. 2018;**535**:338-343. DOI: 10.1016/j.physb.2017.08.028

[18] Kumar B, Smita K, Cumbal L, Debut A, Pathak RN. Sonochemical synthesis of silver nanoparticles using starch: A comparison. Bioinorganic Chemistry and Applications. 2014;**2014**:1-8. DOI: 10.1155/2014/784268

[19] De Mesa DMB, Santos GNC, Quiroga RV. Synthesis-and-characterization-of-Fe2O3-nanomaterials-using-HVPC-growth-technique. International Journal of Scientific & Engineering Research. 2012;**3**(8):1-12

[20] Ayeshamariam A, Kashif M, Muthu Raja S, Sivaranjani S, Sanjeeviraja C, Bououdina M. Synthesis and characterization of In2O3nanoparticles. Journal of the Korean Physical Society. 2014;**64**:254-262. DOI: 10.3938/jkps.64.254

[21] Muflikhun MA, Santos GNC. A standard method to synthesize Ag, Ag/Ge, Ag/TiO2, SnO2, and Ag/SnO2 nanomaterials using the HVPG technique. MethodsX. 2019;**6**:2861-2872. DOI: 10.1016/j.mex.2019.11.025

[22] Uon L, Santos GN, Chua A. Synthesis and characterization of titanium dioxide nanomaterials via horizontal vapor phase growth (hvpg) technique. ASEAN Engineering Journal. 2020;**10**:93-100. DOI: 10.11113/aej.v10.16699

[23] Tibayan EB, Muflikhun MA, Villagracia ARC, Kumar V, Santos GNC. Structures and UV resistance of Ag/SnO2 nanocomposite materials synthesized by horizontal vapor phase growth for coating applications. Journal of Materials Research and Technology. 2020;**9**(3):4806-4816. DOI: 10.1016/j.jmrt.2020.03.001

[24] Shimizu E, Santos GN, Yu DE. Nanocrystalline axially bridged iron phthalocyanine polymeric conductor: (μ-Thiocyanato)(phthalocyaninato) iron(III). Journal of Nanomaterials. 2016;**2016**:1-7

[25] Espulgar WV, Santos GNC. Antimicrobial silver nanomaterials synthesized by HVPCG Technique. International Journal of Scientific and Engineering Research. 2011;**2**:8-11. http://www.ijser.org

[26] Briones JC, Castillon G, Delmo MP, Santos GNC. Magnetic-field-enhanced morphology of tin oxide nanomaterials for gas sensing applications. Journal of Nanomaterials. 2017;**2017**:1-11. DOI: 10.1155/2017/4396723

[27] Qadir A, Le TK, Malik M, Amedome Min-Dianey KA, Saeed I, Yu Y, et al. Representative 2D-material-based nanocomposites and their emerging applications: A review. RSC Advances. 2021;**11**:23860-23880. DOI: 10.1039/d1ra03425a

[28] Muflikhun MA, Chua AY, Santos GNC. Structures, morphological control, and antibacterial performance of ag/tio 2 nanocomposite materials. Advances in Materials Science and Engineering. 2019;**2019**:1-12. DOI: 10.1155/2019/9821535

[29] Tibayan EB, Muflikhun MA, Kumar V, Fisher C, Villagracia ARC, Santos GNC. Performance evaluation of Ag/SnO2 nanocomposite materials as coating material with high capability on antibacterial activity. Ain Shams Engineering Journal. 2020;**11**(3):767-776. DOI: 10.1016/j.asej.2019.11.009

[30] Labbani Motlagh A, Bastani S, Hashemi MM. Investigation of synergistic effect of nano sized Ag/TiO2 particles on antibacterial, physical and mechanical properties of UV-curable clear coatings by experimental design. Progress in Organic Coatings. 2014;**77**:502-511. DOI: 10.1016/j.porgcoat.2013.11.014

[31] Reyes RDL, Santos GNC. Growth mechanism of SnO2 nanomaterials De- rived from backscattered electron image and EDX observations. International Journal of Scientific and Engineering Research. 2011;**2**:1-4

[32] Muflikhun MA, Frommelt MC, Farman M, Chua AY, Santos GNC. Structures, mechanical properties and antibacterial activity of Ag/TiO2 nanocomposite materials synthesized via HVPG technique for coating application. Heliyon. 2019;**5**:1-21. DOI: 10.1016/j.heliyon.2019.e01475

[33] Quitaneg AC, Santos GNC. Cadmium selenide quantum dots synthesized by HVPC growth technique for sensing copper ion concentrations. International Journal of Scientific and Engineering Research. 2011;**2**:1-4

[34] Buot RMN, Santos GNC. Synthesis and characterization of C-Ag nanomaterials for battery electrode application. International Journal of Scientific and Engineering Research. 2012;**3**:1-3

[35] Bernardino LD, Santos GNC. Synthesis and characterization of gallium oxide/tin oxide nanostructures via horizontal vapor phase growth technique for potential power electronics application. Advances in Materials Science and Engineering. 2020;**2020**:1-14. DOI: 10.1155/2020/8984697

[36] Hussain CM. Handbook of Functionalized Nanomaterials for Industrial Applications. Amsterdam: Elsevier; 2020

[37] De Pasquale I, Lo Porto C, Edera MD, Curri L, Comparelli R. TiO 2 -based nanomaterials assisted photocatalytic treatment for virus inactivation : perspectives and applications. Current Opinion in Chemical Engineering. 2021;**34**:1-10.100716. DOI: 10.1016/j.coche.2021.100716

[38] Shkodenko L, Kassirov I. Metal oxide nanoparticles against bacterial biofilms : Perspectives and limitations. Microorganisms. 2020;**2**:1-21

[39] Radhi A, Mohamad D, Suhaily F, Rahman A, Manaf A. Mechanism and factors influence of graphene- based nanomaterials antimicrobial activities and application in dentistry. Journal of Materials Research and Technology. 2020;**11**:1290-1307. DOI: 10.1016/j.jmrt.2021.01.093

[40] N.V. Challagulla, V. Rohatgi, D. Sharma, R. Kumar, ScienceDirect recent developments of nanomaterial applications in additive manufacturing : A brief review, Current Opinion in Chemical Engineering. 2020;**28**:75-82. https://doi.org/10.1016/j.coche.2020.03.003

[41] Kotia A, Chowdary K, Srivastava I, Kumar S, Kamal M, Ali A. Carbon nanomaterials as friction modi fi ers in automotive engines : Recent progress and perspectives. Journal of Molecular Liquids. 2020;**310**:113200. DOI: 10.1016/j.molliq.2020.113200

[42] Chen J, Qiu H, Zhao S. Trends in Analytical Chemistry Fabrication of

chemiluminescence resonance energy transfer platform based on nanomaterial and its application in optical sensing , biological imaging and photodynamic therapy. Trends in Analytical Chemistry. 2020;**122**:115747. DOI: 10.1016/j. trac.2019.115747

[43] Jia H, Zhu Y, Duan Q, Chen Z, Wu F. Nanomaterials meet zebrafish : Toxicity evaluation and drug delivery applications. Journal of Controlled Release. 2019;**311-312**:301-318. DOI: 10.1016/j.jconrel.2019.08.022

[44] Mehmood A, Mubarak NM, Khalid M, Walvekar R, Abdullah EC, Siddiqui MTH, et al. Journal of environmental chemical engineering graphene based nanomaterials for strain sensor application — a review. Journal of Environmental Chemical Engineering. 2020;**8**:103743. DOI: 10.1016/j.jece.2020.103743

[45] Mondal P, Anweshan A, Purkait MK. Chemosphere green synthesis and environmental application of iron-based nanomaterials and nanocomposite : A review. Chemosphere. 2020;**259**:127509. DOI: 10.1016/j.chemosphere.2020. 127509

[46] Muflikhun MA, Chua AY, Santos GNC. Dataset of material measurement based on SEM images of Ag/TiO2 nanocomposite material synthesized via Horizontal Vapor Phase Growth (HVPG) technique. Data Br. 2020;**28**:1-9. DOI: 10.1016/j. dib.2019.105018

Section 4

Nanofertilizers and Its Applications in Agriculture and Environment

Recent Advances in Nano-Enabled Fertilizers towards Sustainable Agriculture and Environment: A Mini Review

Challa Gangu Naidu, Yarraguntla Srinivasa Rao,
Dadi Vasudha and Kollabathula Vara Prasada Rao

Abstract

Food creation be directed expand uniquely to take care of the developing human populace; however, this should be accomplished while at the same time decreasing unfriendly natural effects. In such manner, there is expanding interest in the utilization of nanomaterials as composts for further developing plant mineral sustenances that are crippling Indian agriculture. To address these problems, there is a need to explore one of the frontier technologies like nano-technology to precisely detect and deliver correct quantity of nutrients that promote the productivity. Nano-technology uses synthesized materials that are 10–9 nm in size to improve the productivity, yield and crop quality. Research has proved beyond doubt that the nano-fertilizers that contain readily available nutrients in nano-scale have increased uptake, absorption and improved bioavailability in the plant body compared to the conventional bulk equivalents. This audit assesses the current writing on ENMs utilized as pesticides and manures, and features basic information holes that should be addressed to guarantee maintainable use of nanotechnology in horticulture to accomplish worldwide food security. Designing nanoparticles-based nanofertilizers offer advantages in crop nourishment of the board by upgrading abiotic stress resilience and improving farming efficiency towards the advancement of brilliant and supportable future horticulture.

Keywords: fertilizers, nanotechnology, nanomaterials, nanofertilizers, plant nutrition

1. Introduction

The ascent in worldwide populace, joined with further developed pay also dietary changes, is driving a consistently expanding food request that is relied upon to ascend by 70% in 2050 [1]. Agriculture is the significant wellspring of food and feed for people and homegrown creatures. In any case, rural yield bugs, environmental change occasions like dry spell, and low supplement use productivity are critical deterrents to accomplishing worldwide food security [2]. More than 22,000 types of plant microbes, weeds, bugs and vermin are assaulting ranch produce globally [3]. Annually, China also the United States use around 1806 and

386 a huge number of kilograms of pesticides, individually. However, financial misfortunes brought about by crop infections and vermin in the United States are assessed at a few billions of dollars every year. In the United States, endeavors to battle contagious microbes alone surpass $600 million for every annum [4, 5]. This degree of financial misfortune also shortcoming in food creation keep on frustrating endeavors pointed towards accomplishing and keeping up with food security [4]. The board of plant infections and nuisances is especially difficult, both as far as ideal recognizable proof of infection and because of the predetermined number of administration choices.

The best methodology among the regular techniques for illness the board procedures is the turn of events of host opposition crop varieties [6]. However, not all crops intrinsically have obstruction qualities against pathogenic infections what's more there keeps on being huge cultural disquiet over hereditarily changed food sources. These sources contain like copper, manganese along with zinc as a micro elemental metals in farm safeguard. Nonetheless, based on viability of regular compost having specific revisions obstructed, due to existence of minimum supplement required in which would be impartial solubility [7, 8]. There has been interest in the utilization of nanotechnology in horticulture for almost 15 years, in spite of the fact that effective application has been to some degree tricky. By the by, the utilization of designed nanomaterials (ENMs) in plant infection the executives and soil treatment has earned expanded interest as of late, with different reports illustrating critical potential. Various ENMs have been accounted for to further develop development, improve supplement use proficiency, and stifle sicknesses in plants in nursery tests and a modest number of field trials [9, 10] also, the utilization of ENMs as an expected option in the insurance of plants against bugs and weeds is acquiring interest, albeit hardly any investigations have been led in this area [11]. This audit assesses current potential open doors for the utilization of ENMs in horticulture, zeroing in on nanotechnology-empowered manures and pesticides (counting microbicides, bug sprays and herbicides), from now on alluded to as nanofertilizers and nanopesticides. A number of the detailed articles were fundamentally assessed in view of the adequacy of ENMs utilized in the examination, the trial plan, possible ecological effects, and relative correlation with traditional business items. Notwithstanding reviewing the current writing, a conversation of

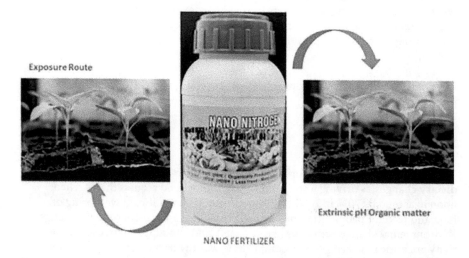

Exposure Route

Extrinsic pH Organic matter

NANO FERTILIZER

Figure 1.
Schematic representation approach for distribution and accumulation of nanofertilizers in crops.

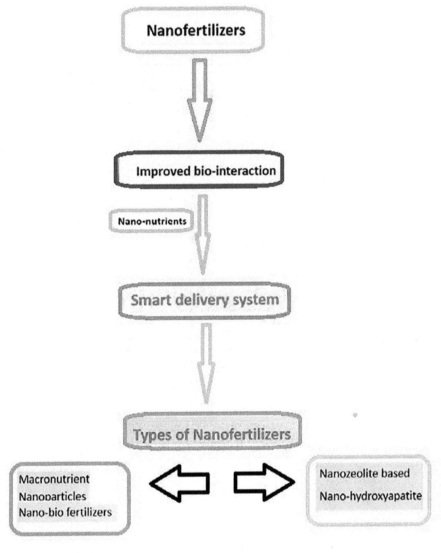

Figure 2.
Schematic portrayal of the conceivable take-up system, aggregation, and circulation of nanofertilizers in crops.

potential instruments of activity is incorporated (**Figures 1-5**), just as points of view on information holes to be filled, before the effective what's more economical use of nanotechnology in farming.

The agriculture area today is confronting an exceptional strain for accomplishing extensive effectiveness in food security utilizing options to synthetic manures [12]. New methodologies and advancements are required in the event that worldwide agricultural creation and request are to be satisfied in a financially and naturally supportable way.

Materials that are of up to 100 nm molecule size in somewhere around one aspect are for the most part delegated nanomaterials [13–15] and are the reason for nanotechnology [15]. Different sorts of various metals are consisting of their nanotechnological applications [16–19].

Given the novel properties of nanomaterials for example, high surface-to-volume proportion, controlled-discharge energy to designated destinations and

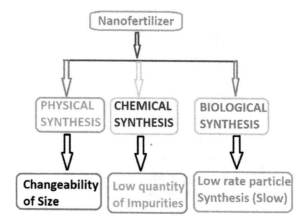

Figure 3.
Techniques for the union of nanofertilizers. Physical (hierarchical), synthetic (base up) and natural drew closer for the development of nanofertilizers.

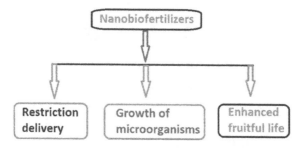

Figure 4.
Systems of activity of nanobiofertilizers in crops. A graph showing the major useful impacts of nanobiofertilizers.

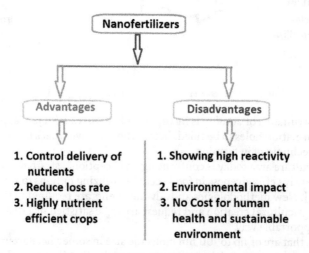

Figure 5.
Benefits and constraints of nanofertilizers. A right utilization of nanofertilizers to further develop crop yields needs before market execution a cautious assessment not just of their benefits for the physiology of plants, yet in addition of their expected impediments for the climate and human wellbeing its societical impact.

sorption limit, nanotechnology has a high significance for the plan and utilization of new manures [8].

Nanofertilizers are supplements exemplified/covered with nanomaterial for the control and slow conveyance of at least one supplements to fulfill the objective supplement prerequisites of plants [20]. These "brilliant composts" are right now being viewed as a promising option [21], to the limit that they are in a few cases viewed as the favored type of manures over the customary ones [22, 23].

2. Nanofertilizer in plant development and improvement

Plants need a few fundamental supplements (macronutrients and micronutrients) for solid development and yield. Assuming there is lacking one of those fundamental parts, then, at that point, the plant will be unable to develop from the seed appropriately. Be that as it may, the presence of an overabundance measure of supplements can likewise hurt plants. A satisfactory inventory of these supplements to fulfill the needs of the essential cell process is a difficult matter. Consequently, a legitimate and explicit conveyance of supplements is severely needed for plants to finish their life cycle. At the point when the mineral supplements were consumed by the plant have many capacities to play in the plant's body. They can assist with making and arrange plant tissue. They are the constituents of different proteins, colors, catalysts, and engaged with cell flagging and digestion. Till now 17 components (N, P, K, H, B, C, O, Mg, S, Cl, Ca, Mn, Fe, Ni, Cu, Zn, Mo) have been recognized as fundamental supplements for plant development and advancement. Among them, nitrate, phosphorus, potassium, magnesium are significant fundamental components required by the plant. They cannot be assimilated straightforwardly from the climate, yet plant retains them through their roots. In this manner, the nanoscale-aspect of nanofertilizers has turned into an innovative arrangement for supplement lack issues. Nanofertilizers are made out of a few nanoparticles including metal oxides, carbon based, also other nanoporous materials changing upon their synthesis and mix properties. It tends to be integrated by hierarchical (physical), base up (compound), and organic methodologies [14].

Nanofertilizers have turned into an incredible option for soil the executives to decrease the over-use of ordinary composts. Furthermore, the sluggish delivery component is offering the chance for utilizes as per development and natural status. Besides, nanofertilizers have shown incredible reaction to provoke plant development and usefulness yet take-up, movement, and gathering of utilizing nanoparticles is as yet not yet distinct.

2.1 Benefits of nanofertilizers

There is a developing strain on the agribusiness area to satisfy the persistently expanding requests of the reliably developing human populace. Synthetic manures are believed to be key for further developing yield efficiency and are widely applied through various techniques. Notwithstanding, crop use is by and large not exactly 50% of the applied measure of manure, and the excess measure of minerals expected to arrive at the designated site might filter down, so that they become fixed in soil or add to water contamination. It has been accounted for that key macronutrient components, including N, P, and K, applied to the dirt are lost by 40–70%, 80–90% also 50–90%, separately [18], causing a significant loss of assets Besides, producers will quite often utilize rehashed uses of these manures to accomplish wanted more significant returns, which conversely can prompt a lessening in soil ripeness and increment salt focuses in this way causing future yield misfortunes. Besides, lopsided use of treatment without control on supplement delivery can weaken item quality. Consequently, it is urgent to grow slow/control discharge composts not exclusively to build crop creation and quality, yet in addition to improve the supportability in green creation. Nanofertilizer applications in farming might fill in as an open door to accomplish maintainability towards worldwide food creation. There is a huge food creation strain on the area as healthful lacks in human populaces are essentially a direct result of utilizing less nutritious food and a low dietary admission of products of the soil [18]. Significant advantages of nanofertilizers over traditional synthetic composts depend on their supplement conveyance framework [12]. They control the accessibility of supplements in crops through sluggish/control discharge instruments. Such a sluggish conveyance of supplements is related with the covering or solidifying of supplements with nanomaterials [10]. By taking benefit of this sluggish supplement conveyance, cultivators can build their crop development due to reliably long haul conveyance of supplements to plants. For instance, supplements can be delivered north of 40–50 days in a sluggish discharge style rather than the 4–10 days by the traditional manures [11]. In traditional supplement the board frameworks, a big part of the applied compost is lost in draining or becomes inaccessible for the plant on the grounds that of extreme accessibility preventing the roots to take-up or in some cases causing poisonous impacts on the plant. Besides, nanofertilizers diminish the requirement for transportation and application costs. Another benefit of utilizing little amounts is that the dirt does not get stacked with salts that typically are inclined to over-application utilizing regular composts on a short-or long haul premise. One more benefit for utilizing nanofertilizers is that they can be blended by the supplement necessities of expected yields. In such manner, biosensors can be appended to another inventive compost that controls the conveyance of the supplements as per soil supplement status, development time of a crop or natural conditions. Plants are touchy towards micronutrient accessibility during crop development and adverse consequences result as leafy foods with helpless nourishment. In regular supplement the executives framework, it is undeniably challenging to control the micronutrient conveyance to a particular yield, yet nanofertilizers give the open door to the producers for providing satisfactory measures of supplements [8].

For example, the majority of the plant developing regions overall are insufficient in specific micronutrients (for example Zn and Fe), so nanofertilizers can go about as powerful and proficient stronghold items for crop and new food items. Nanofertilzers increment the bioavailability of supplements through their high explicit surface region, smaller than normal size and high reactivity [12]. Then again, by giving adjusted sustenance, nanofertilizers empower the plant to battle different biotic and abiotic stresses, with by and large clear benefits [20, 21]. Nonetheless, the broad utilization of nanofertilizers in farming might have a few significant constraints, which should likewise be thought of and will be talked about later in detail.

2.2 Micronutrient nanofertilizers

Micronutrients are those components that are needed by the plant in follow/low amounts, however are fundamental to keep up with essential metabolic processes in plants [22–26]. Plant development is profoundly subject to zinc (Zn) since it is a primary part or administrative co-factor for different chemicals furthermore proteins [27]. This micronutrient is likewise engaged with the blend of carbs, protein digestion, and the guideline of auxins, and gives guard to plants against hurtful microbes [28]. Then again, boron (B) is not just engaged with the biosynthesis of plant cell divider and its lignification, yet additionally assumes a significant part in plant development and different other physiological cycles [29]. Subsequently, those are having basic nature legitimate measures the metallic (B and Zn) properties of accomplishing greatest recoveries with great assessment of their productivity [30]. Agriculture applications—i.e. additionally announced with respect to filled seeding in supplement arrangement in addition to elastic sort expanded through natural product recovery contrasted [31, 32]. Additionally, it was noticed an impressive yield increment utilizing Zn nanoparticles as a supplement source in rice, maize, wheat, potato, sugarcane and sunflower [15]. In addition, settled maghemite nanoparticles applied through water system in arrangement structure in soil as a nanofertilizer further developed the development rate and chlorophyll substance contrasted with the control (chelated iron) in *Brassica napus* [33]. Iron (Fe) is additionally a significant supplement needed by plants in minute amounts for keeping up with appropriate development and advancement. In spite of the fact that it is needed in follow sums, its lack or abundance prompts impedance in key plant metabolic and physiological cycles, accordingly prompting diminished yield [34–37].

2.3 Limitation of nanofertilizers

With regards to practical agribusiness, late advancement is without a doubt seeing the effective utilization of some nanofertilizers for accomplishing upgraded crop efficiency. Be that as it may, the intentional presentation of this innovation in rural exercises could result in numerous accidental non-reversible results [13]. In this situation, new natural and accidental wellbeing security issues can restrict the utilization of this innovation in plant harvests' efficiency. Nanomaterial phytotoxicity is likewise an issue in such manner since various plants react distinctively to different nanomaterials in a portion subordinate way [38–42]. Consequently, it is urgent to think about the benefits of nanofertilizers, yet additionally their restrictions before market execution. Bioavailability just as accidental ecological effects upon openness to natural frameworks, limit their acknowledgment to reception in practical farming and the agriculture areas [8]. Hazard evaluation furthermore risk recognizable proof of the nanomaterials including nanomaterial or then again compost life cycle evaluation are basic just as building up needs for toxicological exploration. This is especially evident considering the amassing of nanoparticles in plants and potential wellbeing concerns. For sure, the utilization of nanofertilizers got from nanomaterials have raised genuine worries connected with sanitation, human and food security.

3. Conclusions and future prospects

Nanofertilizers have a huge effect in the horticulture area for accomplishing improved usefulness and protection from abiotic stresses. Consequently, encouraging uses of nanofertilizers in the agrifood biotechnology what's more cultivation areas cannot be ignored. Besides, the possible advantages of nanofertilizers have invigorated a incredible interest to expand the creation capability of rural harvests under the current environmental change situation. The fundamental financial advantages of the utilization of nanofertilizers are decreased draining and volatilization related with the utilization of customary manures. At the same time, the notable positive effect on yield and item quality has a gigantic potential to expand producers' net revenue through the use of this innovation. Notwithstanding, regardless of the intriguing results of nanofertilizers in the field of farming, up until this point, their significance has not however been engaged towards attractiveness. Vulnerability connected with the cooperation of nanomaterials with the climate and expected impacts on human wellbeing should be investigated exhaustively prior to spreading nanofertilizers at a business scale. Future investigations should be centered around creating extensive information in these underexplored regions in request to present this clever wilderness in maintainable agribusiness. Thusly, nanofertilizer application wellbeing and the investigation of the harmfulness of various nanoparticles utilized for nanofertilizer creation should be an examination need. Besides, a top to bottom assessment of the impact of nanofertilizers in the dirts with various physio-substance properties is fundamental to suggest a particular nanofertilizer for a particular yield and soil type. Biosynthesized nanoparticles-based composts and nanobiofertilizers ought to be investigated further as a promising innovation to further develop yields while accomplishing supportability.

Acknowledgements

The researchers express their gratitude to Vignan's Institute of Information Technology (VIIT) and Vignan's Institute of Pharmaceutical Technology (VIPT) for their inspiration and for sharing the article for publication.

Conflict of interest

There are no conflicts of interest declared by the authors.

Author details

Challa Gangu Naidu[1]*, Yarraguntla Srinivasa Rao[2], Dadi Vasudha[3] and Kollabathula Vara Prasada Rao[4]

1 Department of Basic Sciences and Humanities (BS & H), Vignan's Institute of Information Technology (VIIT), Visakhapatnam, Andhra Pradesh, India

2 Department of Pharmaceutics, Vignan Institute of Pharmaceutical Technology, Visakhapatnam, India

3 Department of Pharmaceutical Chemistry, Vignan Institute of Pharmaceutical Technology, Visakhapatnam, India

4 Department of Pharmaceutical Analysis, Vignan Institute of Pharmaceutical Technology, Visakhapatnam, India

*Address all correspondence to: naiduiict@gmail.com; naidu064@gmail.com

IntechOpen

References

[1] Malhotra SK. Water soluble fertilizers in horticultural crops: An appraisal. Indian Journal of Agricultural Sciences. 2016;**86**:1245-1256

[2] Zhang X et al. Managing nitrogen for sustainable development. Nature. 2015;**528**:51-59

[3] Congreves KA, Van Eerd LL. Nitrogen cycling and management in intensive horticultural systems. Nutrient Cycling in Agroecosystems. 2015;**102**:299-318

[4] Chhipa H. Nanofertilizers and nanopesticides for agriculture. Environmental Chemistry Letters. 2017;**15**:15-22

[5] Van Eerd LL et al. Comparing soluble to controlled-release nitrogen fertilizers: Storage cabbage yield, profit margins, and N use efficiency. Canadian Journal of Plant Science. 2018;**98**:815-829

[6] Lü S et al. Multifunctional environmental smart fertilizer based on L-aspartic acid for sustained nutrient release. Journal of Agricultural and Food Chemistry. 2016;**64**:4965-4974

[7] Raliya R, Saharan V, Dimkpa C, Biswas P. Nanofertilizer for precision and sustainable agriculture: Current state and future perspectives. Journal of Agricultural and Food Chemistry. 2017;**66**:6487-6503

[8] Feregrino-Pérez AA, Magaña-López E, Guzmán C, Esquivel K. A general overview of the benefits and possible negative effects of the nanotechnology in horticulture. Scientia Horticulturae. 2018;**238**:126-137

[9] Chhipa H, Joshi P. Nanofertilisers, nanopesticides and nanosensors, Agriculture. Nanoscience in Food and Agriculture Reviews. 2016;**20**:247-282

[10] Solanki P, Bhargava A, Chhipa H, Jain N, Panwar J. Nano-fertilizers and their smart delivery system, Agriculture. Nanoscience in Food and Agriculture. 2016;**20**:81-101

[11] Chen X. Controlled-release Fertilizers As a Means to Reduce Nitrogen Leaching and Runoff in Container-grown Plant Production, Nitrogen in Agriculture-Updates. London, UK: IntechOpen; 2018. pp. 33-52

[12] Liu R, Lal R. Potentials of engineered nanoparticles as fertilizers for increasing agronomic productions. Science Total Environment. 2015;**514**:131-139

[13] Kah M. Nanopesticides and nanofertilizers: Emerging contaminants or opportunities for risk mitigation? Frontiers in Chemistry. 2015;**3**:64

[14] Kim DY et al. Recent developments in nanotechnology transforming the agricultural sector: A transition replete with opportunities. Journal of the Science of Food and Agriculture. 2018;**98**:849-864

[15] Monreal CM, Derosa M, Mallubhotla SC, Bindraban PS, Dimkpa C. Nanotechnologies for increasing the crop use efficiency of fertilizer-micronutrients. Biology and Fertility of Soils. 2016;**52**:423-437

[16] Raliya R, Tarafdar JC. ZnO nanoparticle biosynthesis and its effect on phosphorous- mobilizing enzyme secretion and gum contents in clusterbean (*Cyamopsis tetragonoloba* L.). Agricultural Research. 2013;**2**:48-57

[17] Raliya R, Biswas P, Tarafdar JC. TiO$_2$ nanoparticle biosynthesis and its physiological effect on mung bean (*Vigna radiata* L.). Biotechnology Reports. 2015;**5**:22-26

[18] Raliya R, Tarafdar JC, Biswas P. Enhancing the mobilization of native phosphorus in the mung bean rhizosphere using ZnO nanoparticles synthesized by soil fungi. Journal of Agricultural and Food Chemistry. 2016;**64**:3111-3118

[19] Tan W et al. Surface coating changes the physiological and biochemical impacts of nano-TiO$_2$ in basil (*Ocimum basilicum*) plants. Environmental Pollution. 2017;**222**:64-72

[20] Zuverza-Mena N et al. Exposure of engineered nanomaterials to plants: Insights into the physiological and biochemical responses: A review. Plant Physiology and Biochemistry. 2017;**110**:236-264

[21] Mahdieh M, Sangi MR, Bamdad F, Ghanem A. Effect of seed and foliar application of nano-zinc oxide, zinc chelate, and zinc sulphate rates on yield and growth of pinto bean (*Phaseolus vulgaris*) cultivars. Journal of Plant Nutrition. 2018;**41**:2401-2412

[22] Deshpande P, Dapkekar A, Oak MD, Paknikar KM, Rajwade JM. Zinc complexed chitosan/TPP nanoparticles: A promising micronutrient nanocarrier suited for foliar application. Carbohydrate Polymers. 2017;**165**:394-401

[23] Mahajan P, Dhoke SK, Khanna AS. Effect of nano-ZnO particle suspension on growth of mung (*Vigna radiata*) and gram (*Cicer arietinum*) seedlings using plant agar method. Journal of Nanotechnology. 2011;**2011**:7

[24] Adhikari T, Kundu S, Biswas AK, Tarafdar JC, Subba Rao A. Characterization of zinc oxide nano particles and their effect on growth of maize (*Zea mays* L.) plant. Journal of Plant Nutrition. 2015;**38**:1505-1515

[25] Milani N, McLaughlin MJ, Stacey SP, Kirby JK, Hettiarachchi GM, Beak DG, et al. Dissolution kinetics of macronutrient fertilizers coated with manufactured zinc oxide nanoparticles. Journal of Agricultural and Food Chemistry. 2012;**60**:3991-3998

[26] Rameshaiah GN, Pallavi J, Shabnam S. Nano fertilizers and nano sensors—An attempt for developing smart agriculture. International Journal of Engineering Research Genetic Science. 2015;**3**:314-320

[27] Iavicoli I, Leso V, Beezhold DH, Shvedova AA. Nanotechnology in agriculture: Opportunities, toxicological implications, and occupational risks. Toxicology and Applied Pharmacology. 2017;**329**:96-111

[28] Dimkpa CO, Bindraban PS. Nanofertilizers: New products for the industry? Journal of Agricultural and Food Chemistry. 2017;**66**:6462-6473

[29] Prasad R, Bhattacharyya A, Nguyen QD. Nanotechnology in sustainable agriculture: Recent developments, challenges, and perspectives. Frontiers in Microbiology. 2017;**8**:1014

[30] Ramady E et al. Plant nano-nutrition: Perspectives and challenges. Nanotechnology, Food Security and Water Treatment. 2018:129-161

[31] Ma C, White JC, Zhao J, Zhao Q, Xing B. Uptake of engineered nanoparticles by food crops: Characterization, mechanisms, and implications. Annual Review of Food Science and Technology. 2018;**9**:129-153

[32] Cornelis G et al. Fate and bioavailability of engineered nanoparticles in soils: A review. Critical Reviews in Environmental Science and Technology. 2014;**44**:2720-2764

[33] S. Fan, Ending hunger and undernutrition by 2025: The role of

horticultural value chains. In: **XXIX** International Horticultural Congress on Horticulture: Sustaining Lives. Livelihoods and Landscapes. 2014. pp. 9-20

[34] Kah M, Kookana RS, Gogos A, Bucheli TD. A critical evaluation of nanopesticides and nanofertilizers against their conventional analogues. Nature Nanotechnology. 2018;**13**: 667-684

[35] López-Valdez F, Miranda-Arámbula M, Ríos-Cortés AM, Fernández-Luqueño F, de la Luz V. Nanofertilizers and their controlled delivery of nutrients. Agricultural Nanobiotechnology. 2018:35-48

[36] Srivastava AK, Malhotra SK. Nutrient use efficiency in perennial fruit crops: A review. Journal of Plant Nutrition. 2017;**40**:1928-1953

[37] Kyriacou MC, Rouphael Y. Towards a new definition of quality for fresh fruits and vegetables. Scientia Horticulturae. 2018;**234**:463-469

[38] Singh G, Rattanpal H. Use of nanotechnology in horticulture: A review. International Journal of Agricultural Science and Veterinary Medicine. 2014;**2**:34-42

[39] Pradhan S, Mailapalli DR. Interaction of engineered nanoparticles with the agrienvironment. Journal of Agricultural and Food Chemistry. 2017;**65**:8279-8294

[40] Yadav TP, Yadav RM, Singh DP. Mechanical milling: A top down approach for the synthesis of nanomaterials and nanocomposites. Nanoscience and Nanotechnology. 2012;**2**:22-48

[41] Ingale A, Chaudhari AN. Biogenic synthesis of nanoparticles and potential applications: An eco-friendly approach. Journal of Nanomedicine and Nanotechnology. 2013;**4**:2

[42] Khot LR, Sankaran S, Maja JM, Ehsani R, Schuster EW. Applications of nanomaterials in agricultural production and crop protection: A review. Crop Protection. 2012;**35**:64-70

Electrochemistry at the Interface

Chapter 17

Electrochemical Impedance Spectroscopy and Its Applications

Camila Pía Canales

Abstract

Electrochemistry has become an important and recognized field for the future since many of its approaches contemplate the establishment of stable energy supplies and the minimization of our impact on the environment. In this regard, electrochemistry can face both objectives by studying the electrode/solution interface. As a result, different electrochemical techniques can be used to study the interface to understand the electron transfer phenomena in different reactions. Considering this, one of the most useful techniques to understand the electrode/solution interface is electrochemical impedance spectroscopy. This technique allows us to describe the electrode behavior in the presence of a certain electrolyte in terms of electrical parameters such as resistances and capacitances, among others. With this information, we can infer the electrochemical behavior toward a specific reaction and the capacity of the electrode to carry on the electron transfer depending on its resistance (impedance) values. The aim of this chapter is to go from the theory, based on Ohm's Law and its derivations, to actual applications. This will lead us to characterize the solution, electrode, and the interface between these two phases based on their electrical components by using an equivalent electrical circuit, such as the Randles equivalent circuit.

Keywords: electrochemical impedance spectroscopy, Randles equivalent circuit, electron transfer, Ohm's law, interface, electrocatalysis

1. Introduction

Electrochemistry is a branch of chemistry, which is focused on the study of chemical processes that cause the movement of electrons. This phenomenon is called electricity and can be generated by the movements of electrons from one elements to another in a reaction known as an oxidation–reduction ("redox") reaction. This kind of reaction involves a change in the oxidation state of one or more elements or atoms. For instance, when an atom loses an electron, its oxidation state increases; thus, it is oxidized. On the other hand, when an atom gains an electron, its oxidation state decreases, and it is said to be reduced.

Many electrochemical techniques exist to study the movement of electrons in a redox reaction. Most of these reactions require the application of an external potential (E) since there is a gap in energy that the electrons need to move from one species to another, according to their Fermi levels. In this regard, most of the electrochemical techniques require the application of a certain potential to reach this energy gap and generate a current response (I). This external potential, which is "away" from the equilibrium between the involved species, is called "overpotential" [1]. With this information, we can correlate this potential with the amount of energy necessary for the redox reaction to occur. Nevertheless, there are other techniques in which a current is applied, and, for instance, the potential is measured. The selection of the technique will depend on the objective of the study.

Electrocatalysis is one of the most important fields within electrochemistry. This field aims to find materials (called electrodes) that can serve as electrocatalysts, by means that a certain redox reaction can occur as close as the equilibrium potential of a specific specie and its redox couple. For example, many researchers are focused on studying the capacity of different materials to improve the performance toward reactions of environmental and energetic interest. One of the main reactions that are studied is the hydrogen evolution reaction (HER), which implies the reduction of two protons by gaining two electrons and resulting in one mol of hydrogen gas. Another reaction that is highly studied in electrocatalysis is the oxygen reduction reaction (ORR), where one mol of the oxygen molecule is converted into hydrogen peroxide if it gains two electrons, or a water molecule if the material is good enough to go further and be able to give four electrons instead. In this latter case, hydrogen peroxide can be an intermediate of the whole reaction. Both HER and ORR are involved in proton exchange fuel cells, where hydrogen gas serves as fuel, and it is oxidized to protons in the anode. Then, these protons diffuse through a proton exchange membrane to the cathodic chamber and are part of the ORR in the cathode [2].

For these studies, electrochemical techniques such as voltammetries (linear, cyclic, square wave, and differential pulse) are highly used as starting studies, where a potential scan is applied in a certain range and the current response is measured. Then, other techniques such as chronoamperometry and chronopotentiometry are used as well, where a specific potential and a specific current are fixed over time, correspondingly. However, these techniques are focused on obtaining the desired product by either oxidizing or reducing the reactant. To be able to use these techniques, the following is required: (i) a working electrode (WE), which is where the target reaction occurs, (ii) a reference electrode (RE), which has a known potential, and every potential acquired in the WE are described as "versus" the RE, and (iii) a counter electrode (CE), where the opposite reaction is occurring. In this case, if a reduction process is occurring onto the WE, an oxidation process occurs in the CE. Then, the same if it was the opposite way. This setup of three electrodes is called the "electrochemical cell" (**Figure 1**) and it is connected to an external equipment called potentiostat in which the operator can control the parameters of the techniques and measure the results.

Then, the reactant is mostly found in an aqueous phase that contains a supporting electrolyte. A supporting electrolyte is commonly a salt dissolved in the aqueous phase and gives conductivity to the solution, but it is inert to react with the WE, by means that it does not interfere in the selectivity of the electrode toward the reactant. Then, if the electrode is polarized toward negative potentials, the WE is referred to as a cathode, while if it is polarized toward positive potentials, it is called an anode. In any case, the electron movement from the electrode to the reactant or *vice versa* occurs in the proximities of the electrode surface (**Figure 2**), or interface electrode/solution, and it is known as the "double layer."

To understand this, many models of the interface or double layer have been studied, where the closest layer corresponds to the Stern layer (region I), while other layers come afterward, such as the inner Helmholtz plane (IHP) and outer Helmholtz plane (OHP) that are located further (regions II and III, respectively)

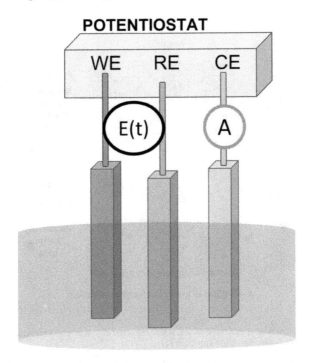

Figure 1.
An electrochemical system with a working electrode (WE), a counter electrode (CE), and a reference electrode (RE). The potential E(t) is applied between the working and reference electrode, and the resulting current is measured at location (A).

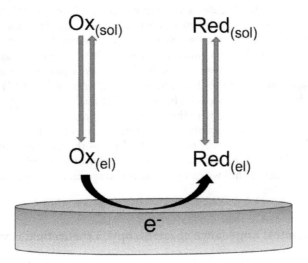

Figure 2.
Representation of a mass transfer controlled electrochemical reaction involving an oxidized species (ox) and a reduced species (red).

until we find the bulk solution (**Figure 3**). Once the WE is being polarized, the reactant diffuses toward the proximities of the electrode surface, forming the double layer; afterward, the electron transfer occurs.

Considering this, the electrocatalytic reactions should be highly focused on understanding what happens in the electrode/solution interface as an addition to

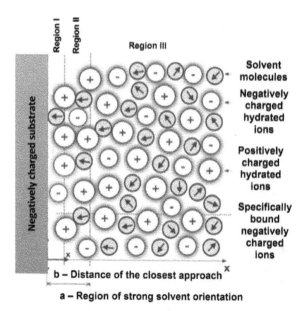

Figure 3.
Charge distribution in the generalized Stern model, where the interval 0 < x < a is the region of strong water orientation and b is the distance of closest approach [3].

the current or potential responses toward a particular redox reaction. The electro-chemical impedance spectroscopy (EIS) is a very powerful tool that allows us to study the double layer or interface in more detail and describes it as a function of an electrical circuit. Based on this, when the double layer is formed, we can refer to it as a capacitor (C_{dl}) and calculate its value in Farads (F). The double layer is the heart of electrochemistry: All electrochemical reactions occur in this region, and it determines one of the basic macroscopic relations of electrochemistry that between the electrode charge and the potential, or equivalently its interfacial capacitance [4]. Then, when the electrochemical reaction occurs, we can correlate this process in terms of the resistance of the charge transfer (R_{ct}) or impedance (Z), both in Ohms (Ω). Other parameters can also be obtained by analyzing two main graphs: (i) Nyquist plots, which correlate with the imaginary impedance (Z") *versus* the real impedance (Z'), and (ii) Bode plots, which show the correlation between the total impedance of the cell (Z) and the phase shift (°) *versus* the frequency of an applied potential. To understand this technique, we must follow the derivation of Ohm's law and its components until we can find the actual applications in different known processes.

2. Electrochemical impedance spectroscopy: a theoretical review

The electrochemical impedance spectroscopy (EIS) technique is an electro-chemical method used in many electrochemical studies, which is based on the use of an alternating current (AC) signal that is applied to the working electrode, WE, and determining the corresponding response. In the most common experimental pro-cedures, a potential signal (E) is applied to the WE and its current response (I) is determined at different frequencies. However, in other cases, it is possible to apply a certain current signal and determine the potential response of the system. Thus, the potentiostat used processes the measurements of potential *vs.* time and current *vs.* time, resulting in a series of impedance values corresponding to each frequency analyzed. This relationship of impedance and frequency values is called the "impedance spectrum." In studies using the EIS technique, the impedance spectra obtained are usually analyzed using electrical circuits, made up of components such as resistors (R), capacitances (C), inductances (L), combined in such a way as to reproduce the measured impedance spectra. These electrical circuits are called "equivalent electrical circuits."

Impedance is a term that describes electrical resistance (R), used in alternating current (AC) circuits. In a direct current (DC) circuit, the relationship is between current (I) and potential (E) is given by Ohm's law Equation. (1), where E is in volts (V), I is amperes (A), and R is Ohms (Ω):

$$I(A) = \frac{E\,(V)}{R\,(\Omega)}.$$ (1)

In the case of an alternate signal, the equivalent expression is as follows:

$$I(A) = \frac{E\,(V)}{Z\,(\Omega)}.$$ (2)

In Eq. (2), Z represents the impedance of the circuit, in units of Ohms. It is necessary to note that, unlike resistance, the impedance of an AC circuit depends on the frequency of the signal that is applied. The frequency (f) of an AC system is expressed in units of hertz (Hz) or the number of cycles per second (s^{-1}). In this way, it is possible to define the admittance (Y) of an AC circuit. Admittance is the reciprocal of impedance and is an important parameter in mathematical calculations involving the technique and on the other hand, the equipment used in EIS studies measures admittance.

The impedance of a system at each frequency is defined by the ratio between the amplitude of the alternating current signal and the amplitude of the alternating potential signal and the phase angle. A list of these parameters at different frequencies constitutes the "impedance spectrum" (**Figure 4**). The mathematical development of the theory underlying the EIS technique allows describing the impedance of a system in terms of a real component and an imaginary component (associated with the square root of −1).

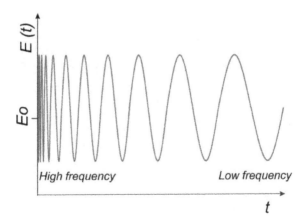

Figure 4.
Periodic perturbation signal with amplitude (ΔE) is applied between the WE and RE from high to low frequencies.

Since the EIS technique is based on the study of electrical networks, there is a great deal of information in the literature regarding electrical circuits. Thus, in understanding the theory that supports the EIS technique, it is convenient to describe current and voltage as rotating vectors or "phasor," which can be represented in a complex plane or "Argand Diagram." For instance, a sinusoidal voltage can be represented by the following expression.

$$E(V) = \Delta E(V) \; sen \; \omega t. \tag{3}$$

where E is the instantaneous value of the potential, ΔE is the maximum amplitude, and ω is the angular frequency, which is related to the frequency f according to

$$\omega = 2\pi f. \tag{4}$$

In this case, ΔE can be understood as the projection, on axis 0 of phasor E in a polar diagram, as shown in **Figure 5**.

In most cases, the current (I) associated with a sinusoidal potential signal is also sinusoidal, with the same frequency (ω) but with a different amplitude and phase than the potential. This can be represented according to Eq. (5).

$$I(A) = \Delta I(A) \; sen \; (\omega t + \varnothing). \tag{5}$$

This means that, in terms of phasors, the rotating vectors are separated in the polar diagram by an angle $\varnothing$ in degrees (°). This can be illustrated as shown in **Figure 6**.

The response to a potential E, of a simple circuit with a pure resistance R, can be described by Ohm's law Equation (1). This, in terms of phasors, corresponds to a situation where the phase angle is equal to zero.

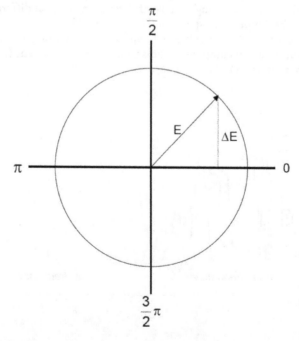

Figure 5.
Phasor diagram corresponding to the alternating potential of Eq. (3).

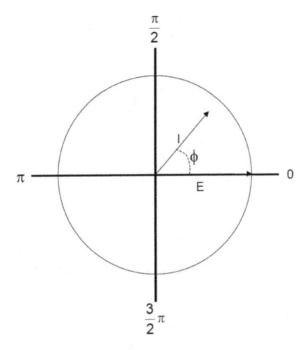

Figure 6.
Phasors of current (I) and potential (E) separated by a phase angle Ø.

When a capacitor is considered in the electrical circuit, different aspects must be known. The concept of "capacitance" (C) can be defined as the relationship between the potential applied (between the capacitor plates) and the total charge (q), according to Eq. (6):

$$q = CE. \tag{6}$$

Considering the current I(A) flowing through the capacitor, the current can be expressed as

$$I(A) = \frac{dq}{dt} = C\frac{dE}{dt}. \tag{7}$$

Thus, considering that current is obtained as a sinusoidal response, we can describe it in terms of the potential, capacitance, and angular frequency, as shown in Eq. (8):

$$I(A) = \omega \, C\Delta E(V) \, cos \, \omega t. \tag{8}$$

As Ohm's equation described in Eq. (1), it is possible to rewrite the equation in terms of Eqs. (3) and (5), giving the following expression:

$$Z(\Omega) = \frac{\Delta E(V) \, sen \, \omega t}{\Delta I(A) \, sen \, (\omega t + \text{Ø})}. \tag{9}$$

In mathematical terms, the real and imaginary components of phasor E and phasor I can be represented in an Argand diagram, with the abscissa corresponding to the real component and the ordinate axis referring to the imaginary component.

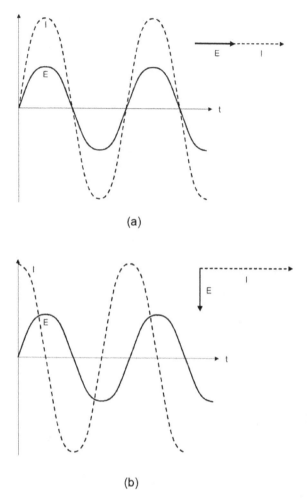

Figure 7.
Phasor representation of current (I) and potential (E) with time (t), for a relationship between current and potential in a circuit with a capacitive reactance of (a) phase angle (∅) = 0, and (b) phase angle (∅) = 90°.

Figure 7a shows the representation of phasor E and I for a purely resistive circuit and **Figure 7b** shows a circuit with a reactive capacitance.

In this regard, when an electrode is highly conductive, it is expected that the phase angle tends to zero, due to limited resistivity for the electron transfer to occur. However, in practice, the electrodes present a delay in the current response, which in this technique is represented by the phase angle (phase shift). This means that the system presents a certain resistivity toward the electron transfer, or in actual terms, impedance.

In **Figure 8**, we can observe the final representation of the terms explained in this section.

Therefore, to understand the practical applications, it is important to consider the following. Assume that we apply a sinusoidal potential excitation. The response to this potential is an AC signal. This current signal can be analyzed as a sum of sinusoidal functions (a Fourier series). Concisely, a conventional electrochemical impedance experimental setup involves an electrochemical cell, a potentiostat, and a frequency response analyzer (FRA). The FRA applies the sine wave and examines the response of the system to determine its impedance. The quality of the

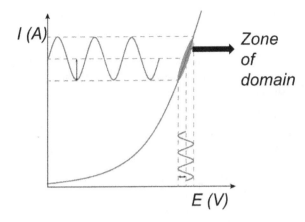

Figure 8.
Electrochemical response to the potential perturbation measured in the linear domain.

impedance measurement is controlled by a series of parameters, including the selection of potentiostatic or galvanostatic modulation, the perturbation amplitude, the frequency range, and the number of cycles used to measure each frequency. The parameters used are highly dependent on the experimental setup and are influenced by factors such as material type, electrolyte, and electrolyte concentration. According to Gamry Instruments [5], a simple explanation of these parameters is described below:

 i. Initial frequency: It defines the frequency of the first AC signal applied during data acquisition. The frequency is entered in Hertz. It is generally recommended to start at high frequency (e.g., 100,000 Hz) and sweep to low frequencies.

 ii. Final frequency: It is specified as the frequency of the last AC signal applied during data acquisition (usually 0.1 Hz).

 iii. Points per decade: It corresponds to the number of measurements that will take place within one decade of frequency. Each data point corresponds to a different frequency of the applied AC signal.

 Therefore, the initial frequency, final frequency, and points per decade parameters can help us to calculate the total number of data points acquired. It is recommended that the frequency range used is as wide as possible. Preferably, this implies a range of 6–7 decades (e.g., 10^{-2}–10^5 Hz), if mathematical tool such as Kramers–Kroning (KK) analysis is used. However, many systems do not allow analysis over a wide frequency range, without obtaining a significant amount of noise.

 iv. AC voltage: It determines the amplitude of the AC signal applied to the cell. The units are RMS (root mean square) millivolts. The resolution of the AC voltage and its range differs from system to system. They vary on both the frequency response analyzer (FRA) and the potentiostat. Generally, you can enter values from 1 mV to around 2 V, depending on the sensitivity of the equipment.

 v. DC voltage: This is specified as the constant potential offset that is applied to the cell throughout the data acquisition. This potential is chosen based on the study at hand. There are some studies where the DC voltage is set up at the open circuit potential (OCP), which is the potential where the total current in the system is zero. Other potentials can be chosen from the voltammetric profiles obtained from the initial studies, where the capacitive and faradaic zones can be recognized. The AC voltage is summed with the DC voltage.

vi. Estimated Z: It corresponds to a user-entered estimate of the cell's impedance at high frequencies. It is used to limit the number of trials required while the system optimizes the potentiostat settings for improvement, current range, offset, etc. Before taking the first data point, the system chooses the potentiostat settings that are ideal for the estimated Z value. If the estimate is precise, the first (or the following one) attempt to determine the impedance will succeed. Nevertheless, if the estimate lacks accuracy, the system will take a few trials while it optimizes the potentiostat settings.

2.1 Graphical representations of impedance measurements

As mentioned before, data obtained in electrochemical impedance spectroscopy tests are reported by the potentiostat equipment in the following two ways:

a. Real component of total impedance (Z real) and imaginary component of total impedance (Z imaginary), as shown in **Figure 9**.

Although the International Union of Pure and Applied Chemistry (IUPAC) conventions hold that the real part should be represented by Z' and the imaginary part denoted by Z'', the use of this notation can also be found as Z with a subscript notation of "r" for the real part and "j" for the imaginary part.

b. Impedance modulus ($|Z|$) and phase shift (angle) (°) versus the frequency (f), as shown in **Figure 10**.

These two methods of describing impedance measurements are the basis for two common ways of presenting data, called Nyquist and Bode plots, respectively.

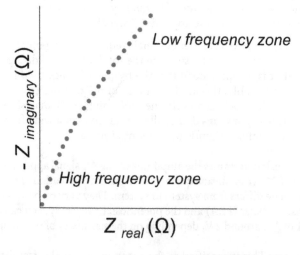

Figure 9.
Impedance data presented in $-Z_{imaginary}$ (Z") versus Z_{real} (Z') representation.

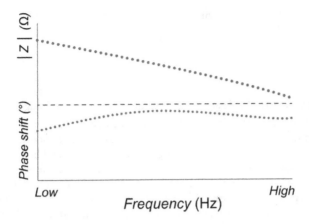

Figure 10.
Impedance data presented in phase shift and total impedance versus f representation.

Nyquist plots show the correlation between the real and imaginary parts of the impedance when the frequency varies. The difficulty of Nyquist plots is that frequency information cannot be directly shown. However, Bode plots show how the magnitude and phase angle of impedance change as a function of the frequency [6]. Because of this, it is important to have a complete understanding of both Nyquist and Bode plots, when analyzing impedance data.

The impedance modulus $|Z|$, the phase shift, $\emptyset$ (°), and the real and imaginary components of the total impedance are related to each other according to the following expressions.

$$|Z|^2 = Z'^2 + Z''^2. \tag{10}$$

$$\tan \emptyset = \frac{Z''}{Z'}. \tag{11}$$

$$Z' = |Z| \cos \emptyset. \tag{12}$$

$$Z'' = |Z| \mathrm{sen} \emptyset. \tag{13}$$

The previous analysis has not considered the fact that all electrodes show a capacitance, called "double layer capacitance" (C_{dl}), which is independent of Faradaic reactions. On the other hand, an electrical resistance, associated with the resistance of the electrolyte (R_{sol}), exists between the point at which the potential is measured (usually the tip of the Luggin capillary) and the working electrode. This resistance will also manifest itself in the total impedance of the system.

2.2 EIS data manipulation and fitting

All Kramers–Kronig-consistent impedance spectra may be fitted to an equivalent circuit, which is the point of using a measurement model. The effects of C_{dl} and R_{sol} can be considered in impedance analysis if their magnitudes are known. They can also be determined by measurements in the absence of the pair of electroactive species. However, determining the C_{dl} and R_{sol} values separately increases the complexity of experimentation and information analysis. An analysis method that avoids the need to separate measurements is derived from a process widely used in other areas, such as electrical engineering, and adapted for electrochemical applications. This method is called "complex plane impedance analysis."

Considering a simple series circuit of resistance and capacitance, the total impedance is equal to

$$Z = R + \frac{1}{j\omega C},\tag{14}$$

where the real part of Z is simply R and the corresponding imaginary part is $\frac{1}{j\omega C}$.

If the behavior described by Eq. (14) is represented in a diagram of $Z = Z' + Z''$ (Argand diagram), where Z' = real component of total impedance and Z'' = imaginary component of total impedance, the graph of **Figure 11** will be obtained. In this case, the corresponding graph is a series of points at different values of ω, where the value of the imaginary impedance component (Z'') tends to zero as the frequency tends to infinity. Here, the capacitance can be considered as short-circuited. Additionally, it is necessary to mention that, in electrochemical studies, the imaginary component of the total impedance (Z') is usually multiplied by -1. This is because, in strict mathematical rigor, in most of these systems, Z'' has negative a value (as shown in **Figure 9**).

The ohmic resistance-adjusted phase angle has an asymptotic value of $-90°$ at high frequency when an electrode is considered "ideally polarizable." If there is a constant-phase element (CPE), the asymptotic value at high frequency would be lower than 90°. Thus, plots of the ohmic resistance-adjusted phase angle present a direct demonstration of a capacitive behavior or frequency dispersion behavior.

The constant-phase element (CPE) is frequently used to improve the fit of models of impedance data. Then, a capacitance may be obtained from a distribution of time constants along the electrode surface. However, not all time-constant distributions lead to a CPE. The ohmic resistance-corrected phase angle gives a useful method to establish whether a time-constant distribution is represented by a CPE or a C_{dl}.

Every graphical result can be described in an equivalent circuit. For the results shown in **Figure 11**, two main elements appear: (i) R_{sol}, which corresponds to the resistivity of the electrolyte, and (ii) a capacitor in series to the R_{sol}, which assumes

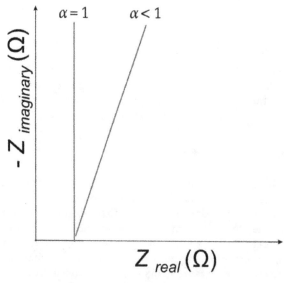

Figure 11.
Representation of Eq. (14).

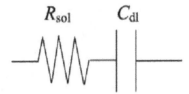

Figure 12.
Electrical equivalent circuit consisting of R_{sol} in series to a capacitor (C_{dl}).

the formation of the double layer or the deposition of charged species over the electrode surface. The equivalent circuit is shown in **Figure 12**.

In other cases, the real and imaginary components of the total impedance can behave as parallel combinations of resistors and capacitors. Thus, the response is characterized by the presence of a semicircle. At low frequencies, the impedance is purely resistive because the reactance of the capacitor is very large. The diagram in **Figure 13** corresponds to the simplest analogy of a Faradaic reaction on an electrode with an interfacial capacitance, C_{dl}. Graphically, in a Nyquist plot, the initial point of the semicircle corresponds to the R_{sol}, while the final point that reaches the x-axis minus the value of the R_{sol}, corresponds to an R_{ct}. Here, it is important to consider that, for a redox reaction to occur, the species must be close enough to allow the electron transfer. Then, a capacitor (C_{dl}) or CPE in parallel will always be used in this circuit model.

This representation allows obtaining a simile of an electrochemical reaction and increases the complexity of the analysis. Then, if the R_{sol} or R_e value is high, the semicircle will be shifted to higher values of the x-axis of the graph. The corresponding electrical circuit is shown in **Figure 14**.

Another representation that is common to see in practice is a diffusional component in series to the R_{ct}. This diffusional component is called "Warburg element" (denoted as W) and it is shown as a 45° linear response right after the semicircle

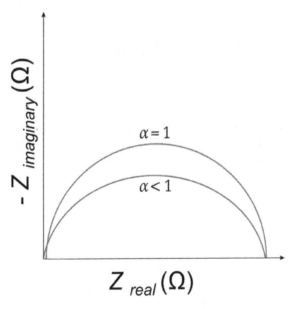

Figure 13.
Representation of the real and imaginary components of the total impedance of a parallel combination of a resistor and a capacitor. α, a parameter associated with the CPE.

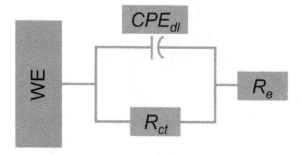

Figure 14.
Equivalent circuit in which a mechanistic interpretation of the system under study is used to extract a meaning for the faradaic impedance, R_{ct}; double-layer constant-phase element, CPE_{dl}; and ohmic resistance of the solution, R_e.

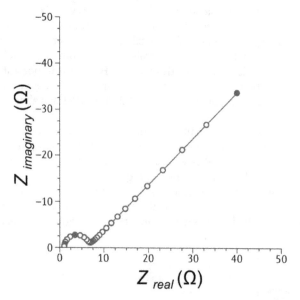

Figure 15.
Equivalent circuit corresponding to the R_{sol} in series with a Warburg element, and a capacitor in parallel.

closes at the x-axis (**Figure 15**). Warburg semi-infinite diffusion is an impedance element, which describes the diffusion behavior of the electrolyte in the absence of convection with a diffusion layer that can spread to infinity.

When this electrical component is shown, the equivalent circuit is called the "Randles equivalent circuit" (**Figure 16**).

The circuit representation of the process model shown in **Figure 16** has a corresponding mathematical expression given as

$$Z = R_e + \frac{R_{ct} + W_o}{1 + (j\omega)^{\alpha}Q(R_{ct} + W_o)}, \tag{15}$$

where R_e is the ohmic resistance, R_{ct} is the charge-transfer resistance, W_o is the diffusion impedance, and α and Q are parameters for a CPE, that is, $Z = ((j\omega)^{\alpha}Q)$ CPE^{-1}. The diffusion impedance is expressed in terms of a diffusion resistance, R_d, and a dimensionless diffusion impedance, $-1/\theta'(0)$, as:

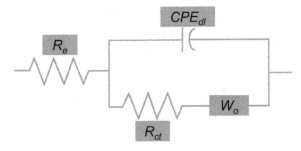

Figure 16.
Electrical circuit corresponding to a high-frequency constant-phase element (CPE) behavior, which is associated with the double layer. R_e (R_{sol}) is the ohmic resistance; R_{ct} is the charge-transfer resistance associated with electrode kinetics; W_o, diffusion impedance is associated with the transport of reactive species to the electrode surface.

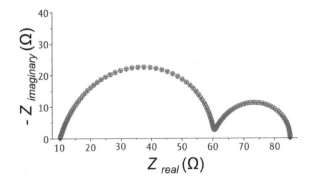

Figure 17.
Nyquist plot with more than one R_{ct} denoted by semi-circles.

$$W_o = R_d \left(\frac{-1}{(\theta)'(0)} \right), \tag{16}$$

where θ is the dimensionless concentration phasor scaled to its value at the electrode surface, and $\theta'(0)$ is the derivative with respect to the position evaluated at the electrode surface [7].

Once we obtain the measured data and the corresponding Nyquist and Bode plots, the data can be fitted with these corresponding electrical equivalent circuits. Most of the software that are in the potentiostats can fit the results by using common electrical circuits. Nevertheless, it is highly important to know what kind of materials we are using in practice to give the electrical circuit a physical meaning. The fitting software, such as ZView/ZPlot from Scribner Associates ®, can approach the fit of the results by graphing an electrical circuit and getting the "theoretical" graph out of the model, in addition to the precision of the electrical circuit. The quality of the fit can be defined using a graphical comparison of the impedance data or by the weighted χ^2 statistic. Both simplex regression and Levenberg–Marquardt regression strategies are used [7, 8].

Finally, when two or more semicircles are shown, it is possible to modify the electrical equivalent circuit shown in **Figures 14** and **16**, by adding more R_{cts} in the fitting. Usually, the R_{cts} that are away from the R_{sol} are the charge transfers occurring on the proximities of the surface of the electrode and are in the low-frequency zone (**Figure 17**). Here, many possibilities appear as options for fitting the data as shown in **Figure 18**.

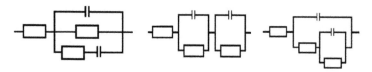

Figure 18.
Different examples of equivalent circuits when two charge transfers are involved plus the resistance of the electrolyte (in boxes).

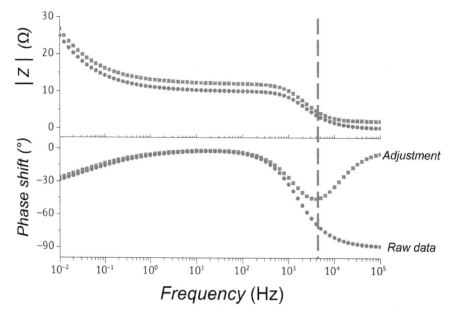

Figure 19.
Bode plot from circuit presented in Figure 14.

Regarding the Bode plots for these examples, **Figure 19** shows the expected diagram for the Randles equivalent circuit in **Figure 14**.

As it is observed in **Figure 19**, a typical Bode plot is shown for a simplified Randles circuit (without the diffusional component). In this case, the frequency values go from the lowest to the highest, which means that the first points correspond to the impedance of the electrolyte if we analyze the graph from the right to the left. When a charge transfer occurs, the $|Z|$ *vs. f* graph shows a slope (red dashed line), while the phase shift *vs. f* shows a valley at the same frequency value. The flattened areas correspond to the rearrangements of the double layer until we observe another slope and a semi-valley. At these frequencies, the electrical behavior may be attributed to diffusional effects. In an ideal system, where no resistance or impedance is contributing to the electrochemical system, the phase shift tends to zero, but, as mentioned earlier, all electrodes present a resistance toward the charge transfer, as well as the electrolyte, where the reaction is taking place. For instance, neither of these parameters can be ignored.

2.3 Applications

Electrochemical impedance spectroscopy can be used in many areas, and its application is highly important when describing the interface electrode/solution. As the results are represented in equivalent circuits, it must be considered that this electrical behavior occurs between the WE and theCE, where a differential in current is taking place.

2.3.1. Electrocatalysis

As mentioned earlier, electrocatalysis is one of the main fields within electro-chemistry. Many electrocatalysts are built with the aim to electrocatalytic reactions that are involved, for example, in batteries or fuel cells. In these cases, the studies are more complex since it is important to consider the performance of the anode, cathode, and membrane. All these components contribute to the final impedance value in the electrochemical system.

Figure 20 shows an example of a Nyquist plot resulting from a solid oxide fuel cell.

In this example, three semicircles are observed, meaning that three electron transfers are taking place over the range of measurement. An ideal panorama for the performance of the electrodes is to have the impedance values as low as possible, meaning that the electrodes are highly conductive and the application of the potential is low.

Another example, in fuel cells, is presented in **Figure 21**, where four zones can be identified. From left to right, we observe proton conduction in electrolyte mem-brane, charge transfer at the electrode-electrolyte interphase, gas diffusion through gas diffusion layer, and water transport across membrane or relaxation of adsorbed intermediates.

In this case, where a polymer electrolyte fuel cell is studied, we can observe a fourth zone that is attributed to the adsorption of species onto the surface of the electrode. When fitting the results of this diagram, an inductor (L) must be added to the equivalent circuit. An inductor considers both a capacitive and a resistive behavior due to the formation of another double layer. Thus, every time the semi-circles pass through the x-axis, it is considered that the system presents an induc-tance in its electrochemical performance. An inductive contribution is usually caused by the connecting wires in the high-frequency domain when low imped-ances are measured (for battery applications) or when a significant noise in the low-frequency range is found when measuring high impedance.

2.3.2. Corrosion

The number of equivalent circuits that can fulfill the behavior of a corrosion cell is practically infinite. However, there is an essential condition for the selection of an

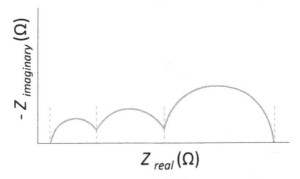

Figure 20.
Typical EIS spectra and the corresponding physical processes in a solid oxide fuel cell, where the first semi-circle corresponds to ion conduction within grains; the second to ion conduction across or along grain boundaries; and the third, to the charge transfer at the electrode-electrolyte interface.

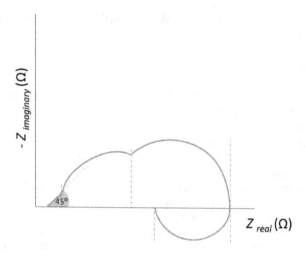

Figure 21.
Typical EIS spectra and the corresponding physical processes in a polymer electrolyte fuel cell.

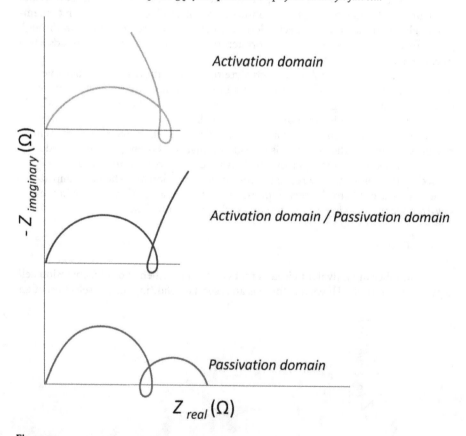

Figure 22.
Typical EIS of iron corrosion in sulfuric acid solution.

equivalent circuit: Both the components of the circuit, as well as the electrical circuit itself, must have a physical explanation. This is of particular importance since there can usually be several equivalent circuits that describe the experimental data with the same accuracy. Most of the EIS works related to corrosion processes are complementary tools for the Tafel Slopes obtained in the materials. In this regard, the circuits can become very complex, but they can help to understand the phenomena occurring during the oxidation processes such as crevice and pitting corrosion.

In general, galvanostatic impedance is more suitable for non-invasive probing of metal corrosion at the open-circuit potential and for measuring most high-energy electrochemical devices, where the impedance is low and current levels are elevated. Inductive loops are commonly seen in corrosion processes. For example, the inductive loops for the corrosion of iron in sulfuric acid are attributed to the coupling of electrochemical reactions by three intermediate species (**Figure 22**). However, many examples can be given in these processes, and they will highly depend on the material and the electrolyte.

3. Conclusions

Electrochemical impedance spectroscopy (EIS) is a highly used technique in electrochemistry and helps to understand the phenomena occurring at the interface electrode/solution. Nevertheless, in the case of an electrochemical system, the main complication is that the system must remain in a stationary state during the measurement. EIS uses a small-amplitude potential or current perturbation to excite the electrochemical system at different frequencies, as illustrated in the figures presented in the chapter. By measuring the response in the current of the system to this perturbation, a transfer function is calculated known as the electrochemical impedance of the system. The data acquisition helps us to understand the processes that the system is performing, by means of resistances, capacitances, inductances, etc. In this regard, Nyquist and Bode plots can be fitted in commercial software to obtain the actual values of every electrical component, based on the equivalent circuit chosen. This technique is highly used in electrocatalysis and corrosion studies, two important fields within electrochemistry, and serves as a complementary tool to other electrochemical techniques, such as voltammetry.

Acknowledgements

The author thanks Dr. J. Genescá, who provided some of the graphics shown in this chapter, and Dr. Orazem, for his knowledge and advice for writing this chapter.

Conflict of interest

The author declares no conflict of interest.

Notes/thanks/other declarations

In honor of my Ph.D. advisor, Professor Thomas E. Mallouk, and my beloved husband, Dr. Arnar Már Búason.

Acronyms and abbreviations

EIS	Electrochemical impedance spectroscopy
Z	impedance

R	resistance
C	capacitance
WE	working electrode
RE	reference electrode
CE	counter electrode
Rct	Charge transfer resistance
Re or Rsol	Electrolyte resistance
Cdl	Capacitance of the double layer
CPE	Constant Phase Element
L	Inductance
q	charge
t	time
w	radial frequency
f	frequency
Q	variable of the CPE constant
α	variable of the CPE constant
W	Warburg element

Author details

Camila Pía Canales
Faculty of Industrial Engineering and Science Institute, Mechanical Engineering
and Computer Science, University of Iceland, Reykjavík, Iceland

*Address all correspondence to: camila@hi.is

IntechOpen

References

[1] Brett C, Brett A. Electrochemistry: Principles, Methods, and Applications Vol. 427. Oxford: Oxford University Press; 1993

[2] Bard AJ, Faulkner LR, Leddy J, Zoski CG. Electrochemical Methods: Fundamentals and Applications. Vol. 2. New York: Wiley; 1980

[3] Gongadze E, van Rienen U, Iglič A. Generalized stern models of the electric double layer considering the spatial variation of permittvity and finite size of ions in saturation regime. Cellular & Molecular Biology Letters. 2011;**16**(4): 576-594

[4] Schmickler W. Double layer theory. Journal of Solid State Electrochemistry. 2020;**24**(9):2175-2176. Available from: https://link.springer.com/article/10.1007/s10008-020-04597-z

[5] Basics of EIS: Electrochemical Research-Impedance. [cited 2021 Sep 20]. Available from: https://www.gamry.com/application-notes/EIS/basics-of-electrochemical-impedance-spectroscopy/

[6] Huang J, Li Z, Liaw BY, Zhang J. Graphical analysis of electrochemical impedance spectroscopy data in Bode and Nyquist representations. Journal of Power Sources. 2016;**309**:82-98

[7] Wang S, Zhang J, Gharbi O, Vivier V, Gao M, Orazem ME. Electrochemical impedance spectroscopy. Nature Reviews Methods Primers covers. 2021;**1**(1):1-21. Available from: https://www.nature.com/articles/s43586-021-00039-w

[8] Orazem ME. Electrochemical impedance spectroscopy: The journey to physical understanding. Journal of Solid State Electrochemistry. 2020;**24**(9): 2151-2153

Printed in the USA
CPSIA information can be obtained
at www.ICGtesting.com
LVHW051331300923
759790LV00008B/186

9 781803 550848